D0082693

VIBRATION OF STRUCTURES

Applications in civil engineering design

VIBRATION OF STRUCTURES

Applications in civil engineering design

J. W. SMITH
B.Sc., Ph.D, M.I.Struct.E.

London New York
CHAPMAN AND HALL

First published in 1988 by
Chapman and Hall Ltd
11 New Fetter Lane, London EC4P 4EE
Published in the USA by
Chapman and Hall
29 West 35th Street, New York NY 10001

Printed in Great Britain by
J. W. Arrowsmith Ltd, Bristol

ISBN 0 412 28020 5

British Library Cataloguing in Publication Data

Smith, J. W. (John William)
Vibration of structures: applications
in civil engineering design.
1. Structural dynamics
I. Title
624.1'71 TA654

ISBN 0-412-28020-5

Library of Congress Cataloging in Publication Data

Smith, J. W. (John W.), 1941–
Vibration of structures: applications in civil engineering design
J. W. Smith.
p. cm.
Includes bibliographical references and index.
ISBN 0-412-28020-5
1. Structural dynamics. 2. Vibration. 3. Earthquake resistant design. I. Title.
TA654.S625 1988 87-31994
624.1'76—dc19 CIP

To Sheila my wife

Contents

Preface

This book is intended as a concise guide to vibration theory, sources of dynamic loading and methods of vibration control for civil and structural engineers. It has been written primarily for final year students specializing in advanced structural engineering, and for practising engineers needing some revision or updating in the area of vibrations. The book should also be useful for research engineers who are interested in practical applications of their work.

An elementary knowledge of differential equations and matrices is assumed but an effort has been made to keep the mathematical content to a minimum. The book endeavours to develop a sound physical understanding of the nature of vibration. This is particularly important in the case of finite element modelling which is readily available in the form of sophisticated commercial programs running on machines ranging from desk-top microcomputers to large mainframes. It is now possible to model complex structural problems realistically without having to understand the advanced mathematics which is embedded in the finite element program. However, the responsibility for correct modelling of the *physical* problem remains with the engineer.

Numerical examples are included in the text. In many cases these are modified versions of practical problems that the author has encountered as a vibrations consultant. They have been devised to illustrate the various principles outlined in the book.

A large number of references is listed at the back. This is partly to give credit where it is due. It also gives the reader the opportunity of following up certain areas of vibration knowledge more deeply. An attempt has been made to trace back to original papers as far as possible. This should benefit specialists and researchers since it is important to read the work of the originator of any concept. Inevitably, the works of some worthy authors have been omitted. No offence is intended. Liberal references are made to codes of practice and design regulations of various countries, in particular those of the United Kingdom, Europe and North America. Extracts from British Standards are reproduced by permission of BSI. Complete copies can be obtained from them at Linford Wood, Milton Keynes, MK14 6LE.

In writing the book, I am indebted to my wife and family who have supported me enthusiastically throughout. I would like to acknowledge the support of my colleagues at the University of Bristol, especially Professor Roy Severn, leader of the earthquake engineering research team. I am particularly grateful to Professor A. Heidebrecht and Dr Tom Lawson for their helpful reviews of certain chapters. I would also like to thank Professor Sir Alfred Pugsley who first suggested that I should write such a book.

Finally, I should emphasize that no creative work is possible without the presence of the living God. 'The Son is the radiance of God's glory and the exact representation of his being, sustaining all things by his powerful word' (Hebrews, 1:3). Praise Him!

Notation

a, a_i, a_f, a_c	half length of a crack, initial, final, critical
$\boldsymbol{A}$	dynamic matrix
C_L, C_D	lift, drag coefficients
C_S	seismic coefficient
c_R	Rayleigh wave velocity
$C(r, r'; n)$	covariance of wind velocity
C_z, C_ϕ, C_x, C_ψ	soil coefficients of uniform compression, non-uniform compression, uniform shear, non-uniform shear
c, $\boldsymbol{C}$	damping coefficient, matrix
D	flexural stiffness of a thin plate
EI	flexural stiffness of a beam
f	frequency
g_B, g_D	peak factors for non-resonant, resonant response
G	geometry factor in seismic risk analysis
G	Lamé's constant
h	focal depth of an earthquake
$H(n)$	mechanical admittance
i	$\sqrt{(-1)}$
I	Modified Mercalli intensity
I	$P(t)\mathrm{d}t$, impulse
I	importance factor (seismic design)
k, $\boldsymbol{K}$, K_m	stiffness – coefficient, matrix, generalized
k	frequency number in wave equation
K	configuration factor for footbridges
K_1, K_{1c}	stress intensity factor, fracture toughness
m, m_s, m_u, $\boldsymbol{M}$, M_m	mass or mass per unit length, sprung, unsprung, matrix, generalized
m	slope of S–N fatigue curve or index in Paris equation
$M(x)$	bending moment

M, M_0, $\bar{M}$	Richter magnitude for earthquakes, smallest of interest, average
n	frequency (wind engineering)
$\boldsymbol{N}$	isoparametric interpolation functions
$p(x,t)$, $P(t)$, $\boldsymbol{P}$	distributed load, loading function, vector
$q(x)$, $\boldsymbol{q}$, Q_m	distributed load, load vector, generalized force
$\boldsymbol{Q}$	matrix of subspace eigenvectors
r, R	radial distance, focal distance
R	factor to take account of ductility
R_e	Reynolds number
$R(\Delta r)$	cross-correlation coefficient
S	seismic response factor
$S(r,r';n)$	cross spectral density
S	Strouhal number
$S_x(n)$, S_{Q_m}, S_{Y_m}	spectral density of horizontal gustiness, generalized force, modal amplitude
S_a, S_d, S_v	spectral acceleration, displacement, velocity (strong ground motion)
t, T	time, period of vibration
u, $\dot{u}$, $\ddot{u}$	horizontal displacement, velocity, acceleration
u_g, u_r	ground displacement, relative displacement
$\boldsymbol{u}$	displacement vector
U	strain energy
v, $\dot{v}$, $\ddot{v}$	vertical displacement, velocity, acceleration
v_x, v_y, v_z	wind velocity components
$v(t)$, $\bar{V}$, v_*	wind speed, mean speed, friction velocity
$V(x)$, V	shear force, base shear
V	vehicle speed
$\boldsymbol{V}$	matrix of Lanczos vectors
V, V_x, V_y, V_z	peak particle velocity and components
w, $\dot{w}$, $\ddot{w}$	vertical displacement, velocity, acceleration
W_e, W_i	virtual work – external, internal
Y_m, $\dot{Y}_m$, $\ddot{Y}_m$, $\boldsymbol{Y}$	modal amplitude of displacement, velocity, acceleration, vector
z_0	roughness length (wind)
α	frequency constant
α	damping matrix coefficient
α	speed parameter $\pi V/L\omega$ (moving load)
α	stress intensity geometry factor
α	compression wave velocity
β	Newmark weighting factor

β	shear wave velocity
β	influence coefficient
β	bounce parameter (moving sprung mass)
γ_{xy}	shear strain
γ_n	mode participation factor
$\gamma^2(r, r'; n)$	coherence (wind)
δ	Newmark weighting factor
δ_c	midspan displacement
$\varepsilon_x, \varepsilon_y, \varepsilon_z, \boldsymbol{\varepsilon}$	strain components, vector
ζ	surface wave number
η	surface wave number
η	eddy viscosity
$\theta_x, \theta_y, \theta_z$	mass moments of inertia
λ	wavelength
λ	Lamé's constant
$\boldsymbol{\Lambda}$	matrix of subspace eigenvalues
μ	viscosity
μ	mass ratio
ν	Poisson's ratio
ξ, ξ_s, ξ_a	damping ratio, structural, aerodynamic
ρ	density
ρ	frequency ratio
$\sigma_x, \sigma_y, \sigma_z, \boldsymbol{\sigma}$	stress components, vector
$\sigma_x^2, \sigma_B^2(E), \sigma_D^2(E)$	variance of horizontal gustiness, non-resonant load effect, resonant load effect
τ	time
τ_{xy}	shear stress
ϕ	phase lag
$\phi(x), \boldsymbol{\phi}_n, \boldsymbol{\Phi}$	mode shape, vector, matrix of eigenvectors
$\chi_a(n)$	aerodynamic admittance
ψ	dynamic response factor (footbridges)
$\boldsymbol{\Psi}$	matrix of transformed eigenvectors
$\omega, \omega', \bar{\omega}$	natural circular frequency, damped, limiting frequency
Ω	forcing frequency

1

Introduction

1.1 EXAMPLES OF VIBRATION IN EVERYDAY EXPERIENCE

It is a common experience to observe a tall street lighting standard oscillating in a steady wind. It may be noticed that the phenomenon usually occurs in strong winds, and careful observation often reveals that the motion takes place at right angles to the direction of the wind. This is due to vortex shedding and is caused by small eddy currents as the wind flows around the pole. The natural frequency of vibration of the lighting standard happens to coincide with the rate at which the eddies are created. The Tacoma Narrows Bridge in the USA collapsed in 1940 partly because of vortex shedding. All long span bridges are now designed to resist the dynamic components of wind loading.

The throbbing of the deck of a passenger ship under one's feet has pleasant associations of holidays and travel. It is caused by the unbalanced forces of powerful engines being transmitted through the relatively flexible structure of the ship's hull. On the other hand, heavy industrial machinery can give rise to unpleasant vibrations in factories such that working conditions may be impaired, or even to the extent that damage may occur to the buildings themselves. Therefore the foundations of industrial machinery have to be designed to keep vibrations within acceptable limits.

Traffic rumble is often blamed for cracks in plaster and minor damage to buildings close to a busy road. The problem tends to be complicated by the accompanying noise together with emotional reactions stemming from ownership of the properties. However, where a road repair has been carried out, resulting in an abrupt discontinuity in the road pavement profile, heavy vehicles travelling at speed can generate a shock that is transmitted through the ground to the foundations of adjoining buildings. Similar problems can be caused by pile driving and blasting.

If not an everyday experience, in certain parts of the world earthquakes occur very frequently and affect the lives of the population. Countries such as the USA, Mexico, Chile, China and Japan have experienced many severe and devastating earthquakes during this century. In these countries the design of buildings is dominated by the need to provide earthquake resistance. Even in countries that are often thought to be aseismic, such as Great Britain, there is a small but finite risk of significant earthquake motion. This risk has to be considered in the design of nuclear plant or major industrial installations where the consequences of damage could be far reaching.

1.2 THE PHYSICAL NATURE OF VIBRATION

Vibration differs from static behaviour in two important respects. First, the applied forces vary with time. Fluctuating wind pressure and blast forces are obvious examples. Reciprocating machines produce pulsating loads. Moving vehicles and locomotives bounce on their suspension in addition to being time variant in terms of position on the structure. In the case of earthquakes we have ground motion imposed on buildings that are initially at rest.

Secondly, the motion of a structure gives rise to inertia forces. Motion of a fixed structure implies a to-and-fro oscillation. During the motion the various parts of the structure possess momentum which tends to cause the structure to overshoot its natural position of deflection, decelerate and come to rest momentarily before returning in an oscillatory manner. Inertia forces correspond to the changing momentum and are distributed along the structure in proportion to its mass. The applied loads, the inertia forces and the elastic resistance are in a continually changing state of dynamic equilibrium.

A good example is the classical problem of soldiers marching over a light footbridge. Pulsating forces are applied in time with the marching frequency causing motion of the structure. According to Newton's second law there must be inertia forces acting on each element of the bridge in proportion to its acceleration. These inertia forces will be distributed along the bridge and tend to make it overshoot as it approaches its point of maximum deflection. The elasticity of the structure acts as a restoring force that makes the bridge spring back, resulting in oscillation. This tendency to vibrate is aggravated if the marching frequency coincides with the natural frequency of the structure. Very large motions can quickly build up and hence the custom has arisen of breaking step.

1.3 SOURCES OF DYNAMIC EXCITATION

In practical engineering design the first requirement is to identify the sources of dynamic excitation and to assess their magnitude and significance compared with the static loads. Structural calculations for static loads are generally much easier than for dynamic excitation, and that is why structural engineers prefer to adopt equivalent static forces as far as possible for the analysis. However, most forms of loading have dynamic components and some forms of structure, especially if they are slender, are susceptible to the dynamic effects. Furthermore, the use of the structure, e.g. as a laboratory housing sensitive instruments, may require that vibration be considered. Thus there is an interrelation between the sources of dynamic excitation, the structural form and the purpose of the structure.

The different types of dynamic loading considered in this book are as follows: earthquakes; wind; industrial machinery; human forces; moving vehicles; and blasting and pile-driving. It is standard practice to use equivalent static horizontal forces when designing buildings for earthquake and wind resistance. This is the simplest way of obtaining the dimensions of structural members. Dynamic calculations may follow to check and perhaps modify the design. Human forces, in the form of crowd loading, are almost always treated as static distributed loads. But observations at a pop concert or football ground will demonstrate the highly dynamic nature of the loading, though the worst effects are restricted to a particular frequency range. Moving vehicles can be designed for by adding an allowance for impact to their static weight. This has proved to be satisfactory for the design of highway and railway bridges. But the procedure may not be justified for loads moving at ultra-high speeds. Vibrations caused by industrial machinery, blasting and pile-driving have to be assessed by methods of dynamical analysis or by experiment.

Some examples of dynamic loading are shown in Fig. 1.1. The first is a record of fluctuating wind velocity. Corresponding fluctuating pressures will be applied to the structure. The random nature of the loading is evident and it is clear that statistical methods are required for establishing an appropriate design loading. The next figure shows the regular pulsating force applied to the foundations of a reciprocating engine. In practice the fluctuation may not be perfectly sinusoidal but it will be at constant frequency and of known amplitude. The third figure shows the characteristic shape of the air pressure impulse caused by a sonic boom or blasting, in construction or open-cast mining. The shapes of air blast curves are usually quite similar, having an initial peak followed by an almost linear decay and often followed by some suction. The lengths of the pulses and their amplitudes depend on many factors, e.g. distance from blast, nature of rock being blasted or size, shape and altitude of aircraft.

Some forms of loading are quite well defined and may be quantified by observation or experiment. Many forms of loading are not at all well defined and require judgement on the part of the engineer. Data on certain types of dynamic loading, e.g. earthquakes and wind, are readily available in many design codes. Other types of loading are less well covered, though much data may be available in published research papers. One of the aims of this book is to discuss the nature of the most important types of dynamic loading and to direct the reader to relevant literature for further information.

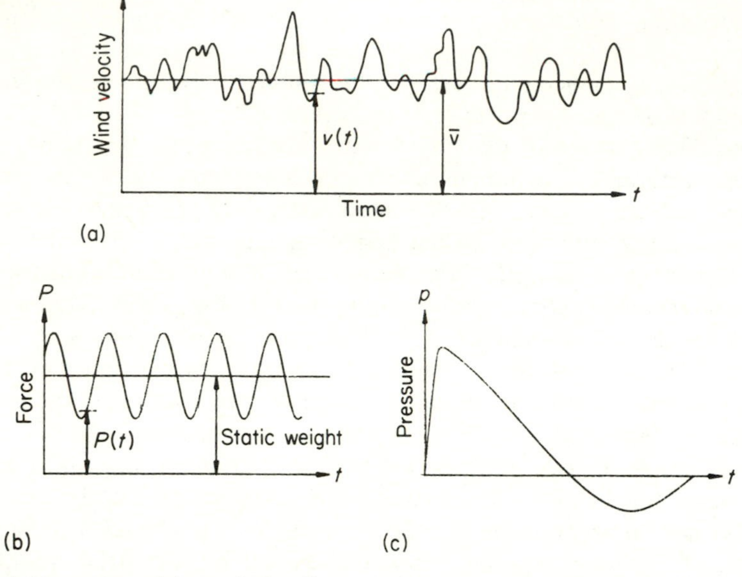

Fig. 1.1 Examples of dynamic loading. (a) Fluctuating wind velocity; (b) pulsating force due to reciprocating engine; (c) air pressure impulse due to blasting or sonic boom.

1.4 DYNAMIC ANALYSIS OF STRUCTURES

Dynamic analysis is similar to static analysis except that there is the extra dimension of time to take into account. Static analysis, using elastic theory, requires the solution of equilibrium equations while taking account of compatibility which relates the elastic deformations of different parts of the structure to each other. The formal theory usually reduces to a set of simultaneous stiffness equations. This is the basis of slope deflection, moment distribution and the stiffness matrix method. Dynamic analysis essentially involves solving for equilibrium in the same way but with variation in time. Consequently the stiffness equations have additional terms proportional to velocity and acceleration and form sets of second-order differential equations. Thus it is evident that detailed mathematical analysis of structural vibrations requires more calculation than is the case for static problems.

Fortunately, in many design problems it is not necessary to obtain detailed stress distributions at all instants of time. Furthermore, when structures vibrate they tend to adopt smoother deflected shapes than under static loads and the corresponding stresses are more evenly distributed. These factors make it possible to model structures in less fine detail than for static analysis and still obtain accurate results.

It is even possible to obtain useful practical information for many structures when they are reduced to *single-degree-of-freedom* systems. An example is shown in Fig. 1.2(a), where a single-storey dwelling is subjected to air blast loading. The structure may be idealized as shown in Fig. 1.2(b) where the pitched roof is reduced to a rigid beam and the wall framing is treated as two vertical cantilevers built in at the base. This system may be further reduced to the simple mass and spring of Fig. 1.2(c). The mass can only move horizontally, hence the single degree of freedom. The product of the pressure and the area over which it is effectively distributed resolves to a point load varying with time. A simplified model of this kind has, in fact, been used for calculating the vibrations of buildings under blast loading (Dowding, Fulthorpe and Langan, 1982).

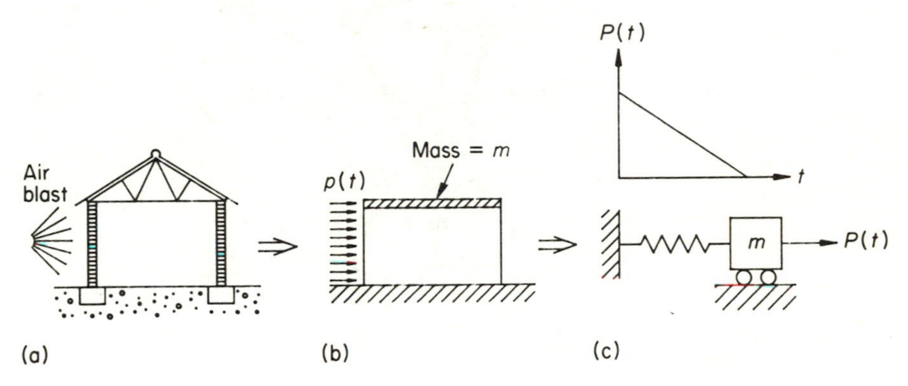

Fig. 1.2 Modelling of a structure for dynamic analysis.

A larger, more important structure may demand more detailed modelling. But even so a judicious engineer will start with a simple model to estimate the vibration approximately before proceeding to a more realistic model. An example of an offshore oil production structure is shown in Fig. 1.3. The varying-diameter tubular members are represented by thin rods connected together while the flexible soil foundation is represented by a system of springs. The stiffness properties of the members and the springs have to be assigned rather carefully, but the resulting simplified model is capable of being analysed on a computer of modest size.

With mention of computers, this is the right point to state that one of the features of this book will be its emphasis on computer analysis using the finite element method. Perhaps the most useful aspect of the finite element method is that it frees engineers from advanced mathematical analysis and enables them to concentrate on choosing appropriate and realistic models of structures using the excellent graphical facilities that are now available.

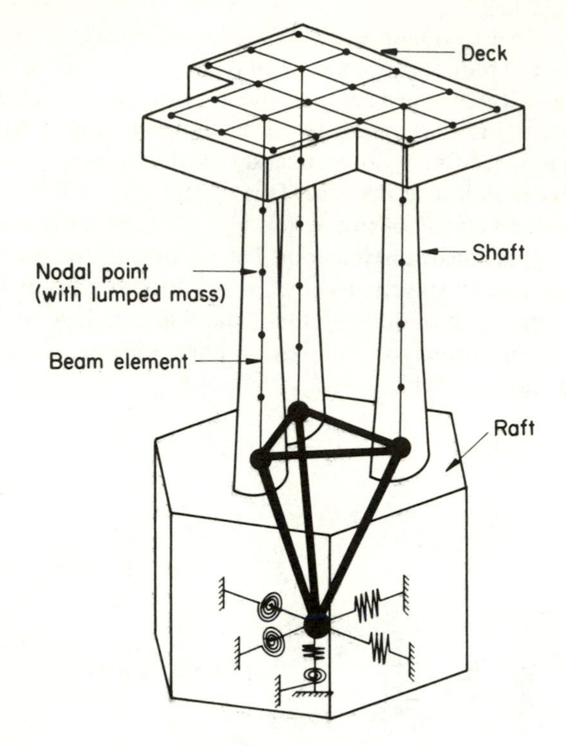

Fig. 1.3 Simplified modelling of a three-dimensional structure (with permission of *Offshore Engineer*).

1.5 CONSEQUENCES OF VIBRATION

Vibration of structures is undesirable for a number of reasons, as follows:

(a) overstressing and collapse of structures,
(b) cracking and other damage requiring repair,
(c) damage to safety-related equipment,
(d) impaired performance of equipment or delicate apparatus,
(e) adverse human response,
(f) fatigue fracture.

With modern forms of construction it is feasible to design buildings to resist the forces arising from major earthquakes. The essential requirement is to prevent total collapse and consequent loss of life. For economic reasons, however, it is the accepted practice to absorb the earthquake energy by ductile deformation, therefore accepting that repair might be required. Blasting, pile-driving and sonic booms can cause superficial damage in the form of cracked plaster and broken window panes. Minor structural damage is also possible from these sources and therefore a method of predicting such problems is desirable.

A most important consequence of vibration, mainly from earthquakes, is the risk of damage to safety-related equipment in nuclear plant. Seismic qualification occupies a large fraction of the working time of many vibrations consultants. Damage to equipment in industrial plant may not have the alarming secondary consequences of damage to nuclear equipment but may nevertheless incur major financial losses due to down-time of processes. Also, there are some exceedingly delicate manufacturing processes, such as the production of microelectronic chips, that impose stringent limits on their vibration environments.

It is not always realized that human beings are highly sensitive to vibration. In fact vibrations that are intolerable to occupants of buildings are usually accompanied by surprisingly small stresses. Thus tall buildings, in seismically quiet parts of the world, have to be designed so that wind-excited vibrations do not disturb the occupants to an unacceptable degree. Furthermore, the principal reason for checking the vibrations caused by industrial machinery is to ensure that human tolerance limits are not exceeded.

Fatigue is a phenomenon that affects structures subjected to millions of cycles of applied load. It most commonly occurs in welded steel structures where tiny cracks, invisible initially, grow in size under repetitions of stress until they are large enough to be seen or to cause rupture. Surprisingly small stresses, if repeated often enough, can cause fatigue failure. Tall steel lattice masts may experience many millions of cycles of significant stress under the action of turbulent wind. Fatigue fractures have occurred in the welded connections of steel offshore structures. Bridges are designed to resist fatigue under repeated axle loads, though this is not strictly related to vibration.

1.6 VIBRATION CONTROL IN THE DESIGN OF STRUCTURES

The previous sections have indicated that there are three steps necessary in the design of structures that are susceptible to vibration. These are:

(a) identifying the dynamic loads in terms of frequency and amplitude or measured variation with time,
(b) analysing the response of the structure to obtain dynamic deflections, stresses, frequencies and accelerations,
(c) checking the calculated or measured performance against specified criteria to ensure that there are no adverse consequences of vibration.

However, vibration analysis should not be left until the end. It is wise to think ahead during the early stage of conceptual design, making approximate calculations in the process, so that vibration susceptibility, or liveliness, can be minimized.

Sometimes vibration can be designed out altogether by eliminating dynamic loading. One example is the quite frequent problem of vibrations in a building frame caused by fork lift trucks. This is often due to a combination of hard wheels and uneven factory flooring. By spending a bit of money on providing a smooth riding surface for the floor, and careful detailing of expansion joints, the problem of vibration may be eliminated completely.

Some structures, such as dance floors, are affected over a confined frequency range. With foresight it may be economical to increase the structural depth to improve the overall stiffness and hence keep the frequency of the structure above the predominant dancing frequency. Braced structures tend to have higher natural frequencies than rigid frames and therefore may be beneficial for avoiding low critical frequencies. Structures in zones affected by frequent earthquakes should be designed to possess as much ductility as possible, to absorb the dynamic energy. Brittle forms of constructions, such as unreinforced masonry, have not performed well in major earthquakes and have often contributed to the scale of disasters.

Control over the natural frequency of buildings is possible, by increasing the stiffness and reducing the mass, but it is often difficult or uneconomic to achieve an optimum value. It may be more efficient to design a special vibration-absorbing device as part of the structure to reduce dynamic effects. Tuned vibration absorbers have been used in some major projects (Wargon, 1983). The principle of operation will be discussed in Chapter 3. Some forms of construction, e.g. welded steelwork, are more prone to vibration than others because of their lack of inherent damping capacity. It is sometimes possible to choose materials with high damping, or, failing that, to install artificial damping devices (Brown, 1977).

Finally, there are many applications where vibration isolation is required. Delicate equipment, such as is used in microelectronic manufacturing, must be mounted on sensitive suspensions. Buildings close to heavily trafficked railway lines have been mounted on rubber bearings so as to eliminate the transmission of rumble (Waller, 1966, 1969). Machines are sometimes mounted on springs to isolate them from their supports. Base isolation of buildings to minimize the effects of earthquakes is feasible and may be a cost-effective solution.

2

Vibration of systems with one degree of freedom

2.1 INTRODUCTION

When we think of an offshore drilling platform twisting and shuddering under the blows of a heavy sea, we realize that the lives of the hundred or so men on board depend on the competent dynamic calculations of a structural engineer. It is easy to be intimidated by the complexity of such a problem. The structure has hundreds of members, the ground conditions are uncertain, there are fluctuating wind and wave forces and there are many possible ways in which the structure may sway, pitch, twist and heave. However, experts in dynamic analysis will readily confirm that it helps to think of the problem as a much simpler system whose motions can be more easily understood. That is precisely what we are going to do in this chapter.

The most basic vibrating system consists of a single lumped mass and a spring. This is said to have *one degree of freedom*. That is to say, there is only one possible direction of movement for the lumped mass. This is in contrast with the offshore platform which may sway, twist, heave and so on, as we have already noted. It is, in reality, a system with *many degrees of freedom*. However, if we consider that horizontal sway is the predominant motion we can begin to see a way of simplifying the problem. In Fig. 2.1(a) the relatively heavy topside platform of the structure is replaced by a lumped mass while the braced steel jacket structure provides the necessary lateral stiffness against wind and wave loading and can be thought of as an elastic spring. Similarly, the bridge girder in Fig. 2.1(b) can be idealized as a leaf spring with its mass concentrated at midspan being dynamically loaded by a moving heavy lorry. In each case we are left with a lumped mass m, a spring of stiffness k, and an external loading that varies with time, $P(t)$. The possible motions are reduced to one, namely u.

It may be thought that such apparently crude approximations could yield

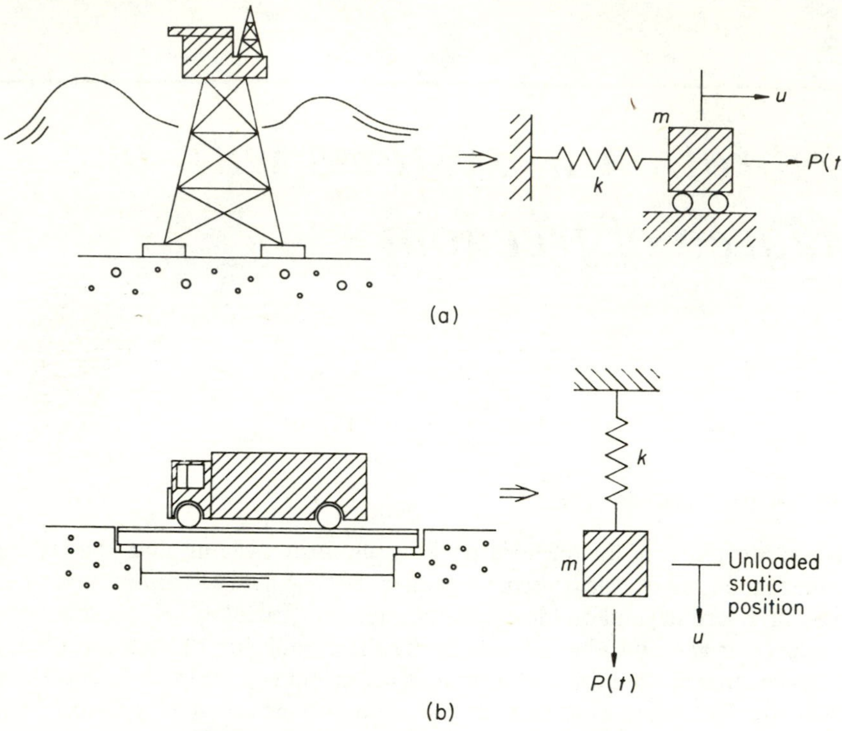

Fig. 2.1 Idealization of one degree of freedom.

little in the way of practical information. However, they are in fact very useful for two reasons. First, skilled engineers can make surprisingly accurate predictions of the behaviour of real structures by intelligent choice of the parameters of simple systems. Secondly, it is found that the more complicated motion of a real structure can be resolved into the simpler motions of a number of mass–spring systems and dealt with by superposition.

Finally, a sound physical understanding of the vibration of one-degree-of-freedom systems is an essential requirement for any further study. A useful aid to understanding vibrations is illustrated in Fig. 2.2. A few minutes of making the toy frog bob up and down will reveal some important characteristics of vibration behaviour.

2.2 EQUATION OF MOTION

Consider the system shown in Fig. 2.3(a). This system is similar to the one shown in Fig. 2.1(a) except that a viscous dashpot, with coefficient c, has been added to represent damping. Damping is a property of practical structures which causes vibrations to die away quite rapidly. There will be more discussion of this later.

Fig. 2.2 Understanding a mass–spring system.

The equation of motion may be obtained by considering the equilibrium of forces acting on the mass. If the force $P(t)$ causes the mass to accelerate to the right then according to Newton's second law there will be an inertia force in the opposite direction. There will also be a viscous damping force and a spring force resisting the motion.

These resisting forces will be proportional to the acceleration, velocity and displacement as follows:

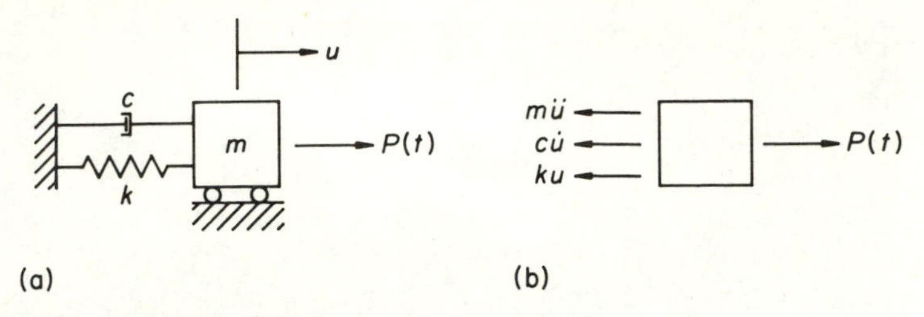

Fig. 2.3 System with one degree of freedom. (a) Light damping ($\xi < 1$); (b) critical damping ($\xi = 1$).

$$\text{inertia force} = \text{mass} \times \text{acceleration} \quad = m\,\mathrm{d}^2u/\mathrm{d}t^2 = m\ddot{u},$$

$$\text{damping force} = \text{coefficient} \times \text{velocity} \quad = c\,\mathrm{d}u/\mathrm{d}t = c\dot{u},$$

$$\text{spring force} = \text{stiffness} \times \text{displacement} = ku,$$

where the dots over the u denote first and second derivatives with respect to time. The forces acting on the mass are shown in Fig. 2.3(b), and hence for dynamic equilibrium of the system it follows that

$$m\ddot{u} + c\dot{u} + ku = P(t). \tag{2.1}$$

This is the equation of motion of the system.

2.3 FREE VIBRATION

Free vibration may occur in the absence of any external forcing function, $P(t)$. The equation of motion can thus be written

$$\ddot{u} + 2\xi\omega\dot{u} + \omega^2 u = 0 \tag{2.2}$$

where

$$2\xi\omega = c/m \quad \text{and} \quad \omega^2 = k/m.$$

This is a second-order differential equation of standard form. Methods of solution can be found in many appropriate textbooks. Suffice it to say that the solution for u should be a type of function that when differentiated retains the same form. The exponential function satisfies this condition. Therefore try

$$u = A\mathrm{e}^{nt} \tag{2.3}$$

and substitute (2.3) in the equation of motion. It is found that two solutions are possible since we obtain

$$n^2 + 2\xi\omega n + \omega^2 = 0 \tag{2.4}$$

and therefore

$$n = -\xi\omega \pm \omega\sqrt{(\xi^2 - 1)}. \tag{2.5}$$

The solution depends on the magnitude of damping. The most important case occurs when the damping in the structure is light so that $\xi < 1$. In this case there will be a negative sign under the square root, thus introducing complex numbers into the solution, which becomes

$$u = e^{-\xi\omega t}(A_1 e^{i\omega' t} + A_2 e^{-i\omega' t}) \tag{2.6}$$

where

$$\omega' = \omega\sqrt{(1-\xi^2)} = \text{natural circular frequency (radians/s)}. \tag{2.7}$$

If one notes that

$$e^{\pm i\omega t} = \cos \omega t \pm i \sin \omega t \tag{2.8}$$

Equation (2.6) may be written more conveniently as

$$u = e^{-\xi\omega t}(A \cos \omega' t + B \sin \omega' t). \tag{2.9}$$

The coefficients A and B depend on the initial conditions of vibration and may be determined if these are known.

Hence, if $u = u_0$ and $\dot{u} = \dot{u}_0$ at $t = 0$ it may be verified that

$$u = e^{-\xi\omega t}[u_0 \cos \omega' t + (\dot{u}_0 + \xi\omega u_0) \sin \omega' t/\omega']. \tag{2.10}$$

This is a decaying vibratory motion which is shown in Fig. 2.4(a). Between complete cycles there is a constant period $T = 2\pi/\omega'$.

Thus the frequency of vibration in cycles per second (Hertz) is

$$f = 1/T = \omega'/2\pi. \tag{2.11}$$

In most practical structures damping is very light and ω' is hardly distinguishable from the undamped natural frequency ω. Thus the usual

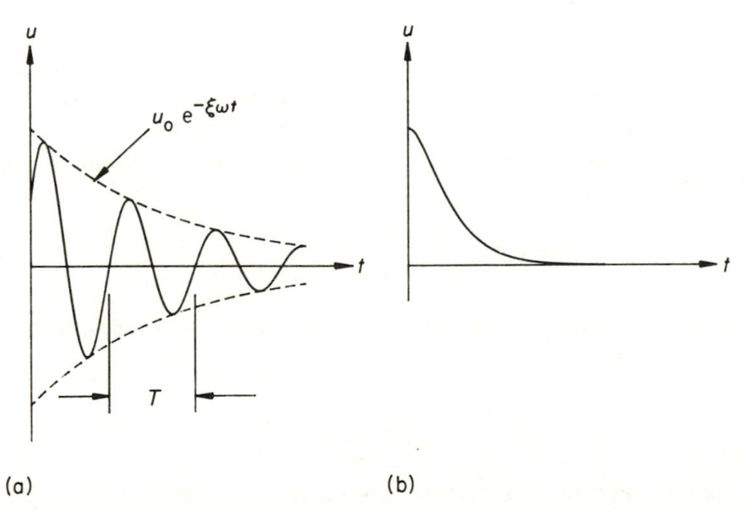

Fig. 2.4 Free vibration with damping. (a) Light damping ($\xi < 1$); (b) critical damping ($\xi = 1$).

formula for natural frequency is

$$f=(1/2\pi)\sqrt{(k/m)}. \tag{2.12}$$

In overdamped structures where $\xi>1$ there is no ensuing vibration and any initial displacement gradually creeps to zero. This is rare in structural engineering.

2.4 DAMPING

The intermediate case where $\xi=1$ is called critical damping. If this value for ξ is substituted in (2.10) it will be noted that $\omega'=0$ and the second term will be indeterminate, being equal to 0/0. However, if we consider the case very close to critical damping such that $\omega'=\delta\omega$ the second term in (2.10) will have the factor

$$\frac{\sin\delta\omega t}{\delta\omega}\rightarrow\frac{\delta\omega t}{\delta\omega}=t.$$

Therefore, with critical damping the displacement will be given by

$$u=\mathrm{e}^{-\omega t}[u_0+(\dot{u}_0+\omega u_0)t]. \tag{2.13}$$

This is shown in Fig. 2.4(b) and it can be seen that it is a limiting condition when vibratory motion is just prevented. Thus, from the notations used in (2.2) the critical damping coefficient is given by

$$c_{\mathrm{c}}=2\omega m=2\sqrt{(km)} \tag{2.14}$$

and

$$\xi=c/c_{\mathrm{c}}=\text{damping ratio}. \tag{2.15}$$

The damping ratio is a useful non-dimensional measure of damping and is often expressed as a percentage. It is worth noting that in most structures there is no more than a few per cent of critical damping.

Another well known measure of damping is the logarithmic decrement which is defined as

$$\delta=\log_{\mathrm{e}}\left(\frac{\text{amplitude at cycle } n}{\text{amplitude at cycle } n+1}\right). \tag{2.16}$$

Therefore, referring to Fig. 2.4(a), the ratio of the peak amplitudes of two successive cycles is given by

$$\frac{u_n}{u_{n+1}}=\frac{u_0\mathrm{e}^{-\xi\omega t}}{u_0\mathrm{e}^{-\xi\omega(t+T)}}=\mathrm{e}^{\xi\omega T}, \tag{2.17}$$

and consequently the logarithmic decrement and the damping ratio are related thus:

$$\delta=\xi\omega T=2\pi\xi\omega/\omega'\simeq2\pi\xi. \tag{2.18}$$

Damping in real structures is not strictly due to viscosity but is mostly caused by friction at interfaces such as in bolted connections, in joints of cladding and in the cracks of reinforced concrete. A simple model of frictional, or Coulomb, damping is shown in Fig. 2.5(a). In this case motion in the u direction is resisted by a constant frictional force proportional to the normal force. If one denotes this by F then Equation (2.1) is modified and becomes

$$m\ddot{u}+F+ku=P(t) \qquad \dot{u} \text{ positive}$$

or

$$m\ddot{u}-F+ku=P(t) \qquad \dot{u} \text{ negative}. \tag{2.19}$$

The solution of these equations (Biggs, 1964) in the case of free vibrations, $P(t)=0$, reveals that there is linear decay of the amplitude, as shown in Fig. 2.5(b).

Experiments show that the vibrations of real structures usually lie between the viscous and frictional response. However, the viscous assumption is convenient to use analytically and is sufficiently accurate for most purposes.

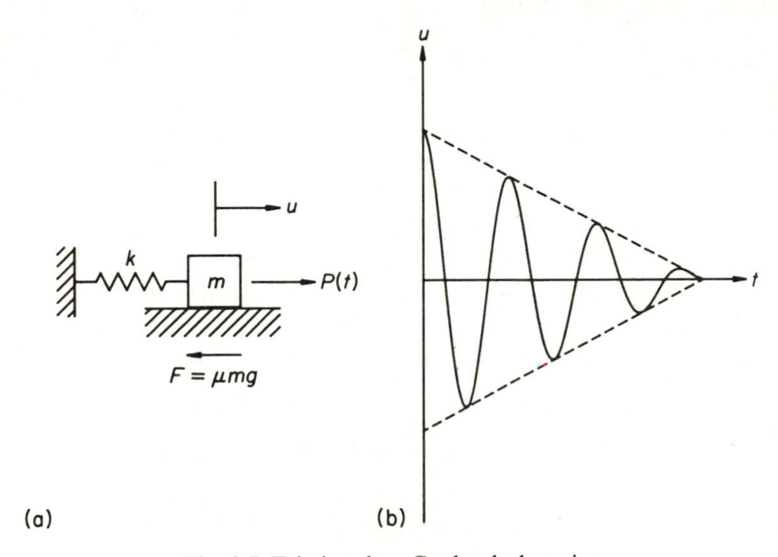

Fig. 2.5 Frictional or Coulomb damping.

EXAMPLE 2.1

The structural frame shown in Fig. 2.6(a) is rigid jointed and fixed to its supports. The mass of the structure of 5000 kg is concentrated on the beam which is assumed to be rigid. The columns are assumed weightless and each has flexural stiffness (EI) of 4.5×10^6 N m^2 in the plane of the structure. If the structure is assumed to have a viscous damping ratio of 4% calculate (a) the

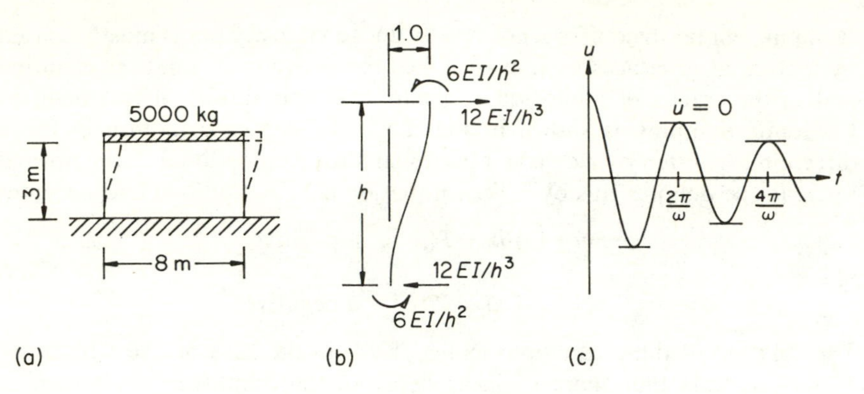

Fig. 2.6 Vibration of structural frame.

damped and undamped natural frequencies; (b) the peak displacements of the first five cycles of vibration given that the beam is displaced horizontally 25 mm and suddenly released.

Solution

The mass of the equivalent one-degree-of-freedom system is

$$m = 5000 \text{ kg}.$$

Note that when the frame is displaced, each column is deflected as shown in Fig. 2.6(b) with zero rotations at top and bottom. The forces corresponding to a unit lateral displacement can be calculated from elementary bending theory and are as shown. Hence the effective spring stiffness of the structure is twice the lateral stiffness of each column and is therefore

$$k = 2 \times 12EI/h^3 = 2 \times 12 \times 4.5 \times 10^6/3^3 = 4.0 \times 10^6 \text{ N/m}.$$

(a) Hence, the undamped natural frequency from (2.12) is

$$f = \frac{1}{2\pi}\sqrt{\left(\frac{4 \times 10^6}{5000}\right)} = 4.502 \text{ Hz}.$$

The damped natural frequency is calculated from (2.7) and is

$$f' = f\sqrt{(1-\xi^2)} = 4.502 \times \sqrt{(1-0.04^2)} = 4.498 \text{ Hz},$$

which is practically the same.

(b) Vibration peaks are calculated using (2.10). Ignoring the difference between the damped and undamped frequency and noting that initial velocity is zero we have

$$u = u_0 e^{-\xi\omega t}(\cos \omega t + \xi \sin \omega t)$$

Peaks occur when the velocity is zero. The velocity is given by

$$\dot{u}=u_0\omega e^{-\xi\omega t}(-\sin\omega t+\xi\cos\omega t-\xi\cos\omega t-\xi^2\sin\omega t)$$
$$=-u_0\omega(1-\xi^2)\sin\omega t e^{-\xi\omega t}.$$

Therefore, velocity is zero ($\dot{u}=0$) and displacement is maximum when

$$\sin\omega t=0,$$

i.e. at $t=n\pi/\omega$.

However, the peak displacement at each cycle must be calculated at $t=0,\ 2\pi/\omega,\ 4\pi/\omega,\ \ldots$ as follows:

$$u=25.0\exp(-0.04\times 2n\pi)=25.0\exp(-0.2513n);\quad (n=0,1,2,3,4,5).$$

Therefore the peak displacements are

$$u=25.0,\ 19.44,\ 15.12,\ 9.15,\ 7.12\ \text{mm}.$$

2.5 PERIODIC FORCING FUNCTION

The forcing function $P(t)$ in Equation (2.1) will now be considered. An important step in understanding vibration behaviour is achieved by considering a periodic forcing function,

$$m\ddot{u}+c\dot{u}+ku=P_0\sin\Omega t, \tag{2.20}$$

where Ω is the circular frequency of the applied force. The best practical example of this in civil engineering is in the design of foundation blocks for reciprocating machines (see Chapter 8).

It would be reasonable to expect the system to vibrate at the same frequency as the forcing function. Therefore, try a solution in the form

$$u=A_1\sin\Omega t+A_2\cos\Omega t. \tag{2.21}$$

Substitution in (2.20) yields

$$-mA_1\Omega^2\sin\Omega t-mA_2\Omega^2\cos\Omega t+cA_1\Omega\cos\Omega t-cA_2\Omega\sin\Omega t$$
$$+kA_1\sin\Omega t+kA_2\cos\Omega t=P_0\sin\Omega t. \tag{2.22}$$

This can be written as two separate equations, one factored by $\sin\Omega t$ and the other by $\cos\Omega t$. The factors may then be cancelled out, leaving

$$-mA_1\Omega^2-cA_2\Omega+kA_1=P_0 \tag{2.23a}$$

$$-mA_2\Omega^2+cA_1\Omega+kA_2=0, \tag{2.23b}$$

from which

$$A_1=\frac{(k-m\Omega^2)P_0}{(k-m\Omega^2)^2+c^2\Omega^2} \tag{2.24a}$$

$$A_2 = \frac{-c\Omega P_0}{(k - m\Omega^2)^2 + c^2\Omega^2}. \tag{2.24b}$$

It can then be verified quite easily that the solution (2.21) may be written in the form

$$u = \frac{P_0 \sin(\Omega t - \phi)}{\sqrt{[(k - m\Omega^2)^2 + c^2\Omega^2]}} \tag{2.25}$$

where

$$\tan\phi = c\Omega/(k - m\Omega^2). \tag{2.26}$$

It is now clear that vibration at the same frequency as the forcing function is a valid dynamic response. However, free vibrations may also occur and the full solution should include a term of the form given by Equation (2.9). Even if the system is initially at rest free vibrations will be generated at the onset of periodic forcing. These decay rapidly due to damping and may be neglected after the first few cycles, leaving only the steady-state response given by (2.25). A case where the first few cycles of transient response are important is the vibration of bridges under moving loads (see Chapter 9).

The behaviour of structures under steady-state forced vibration can best be described by evaluating the dynamic load factor (DLF) or magnification factor. Dividing (2.25) by the static deflection, P_0/k, and making $\sin(\Omega t - \phi) = 1$ for maximum response, we find that

$$\text{DLF} = \frac{\text{max. dynamic deflection}}{\text{max. static deflection}} = \frac{1}{\sqrt{[(1 - \rho^2)^2 + (2\xi\rho)^2]}} \tag{2.27}$$

where $\rho = \Omega/\omega$ = frequency ratio.

The DLF is plotted against frequency ratio for different values of damping in Fig. 2.7. Resonance is the phenomenon that occurs when the forcing frequency coincides with the natural frequency ($\rho = 1$). The amplitude reaches a maximum in this region. The position of the amplitude peak depends on the value of damping and is accurately given by $\rho^2 = 1 - 2\xi^2$ and is thus shifted slightly to the left when damping is large. It can also be seen that the resonance peak disappears altogether when $\xi = 1$.

However, since damping in structures is usually small the very large DLF at $\rho = 1$ is of much more practical importance. For example, if $\xi = 0.02$, which would be typical of many structures, then the ratio of dynamic to static deflection would reach 25 at resonance.

This is a fact that can be used in resonance testing. A known periodic forcing function can be applied to a structure by means of a rotating eccentric mass exciter (Severn, Jeary and Ellis, 1980). Then by measuring the dynamic motion of the structure at resonance the damping can be deduced from (2.27).

A second useful fact is derived from Equations (2.25) and (2.26). These indicate that the vibration response lags behind the forcing function, as can

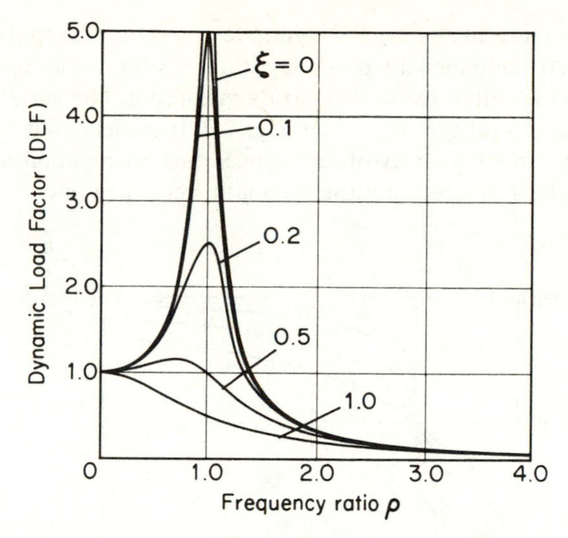

Fig. 2.7 Steady-state response to periodic forcing function.

be seen in Fig. 2.8. When the frequency of the forcing function is very low compared with the natural frequency of the system, the dynamic response follows the forcing function almost in phase and the displacement is not amplified. In fact, the system behaves pseudostatically. As the frequency of

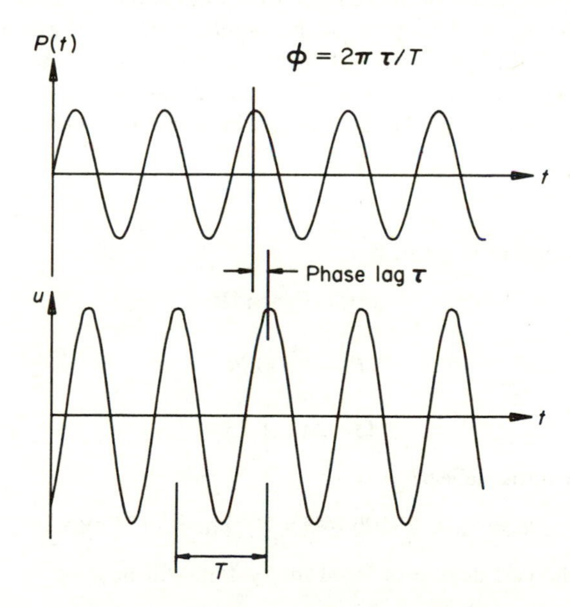

Fig. 2.8 Phase lag of steady-state response.

the forcing function increases, the dynamic response is amplified, and also begins to lag. At resonance the phase lag is exactly 90°, whatever the value of damping, and the amplitude is close to its maximum. The variation of phase lag with frequency ratio is shown in Fig. 2.9. It is clear that the phase lag changes rapidly in the vicinity of resonance, thus giving an accurate experimental method for determining the resonant frequency of a structure.

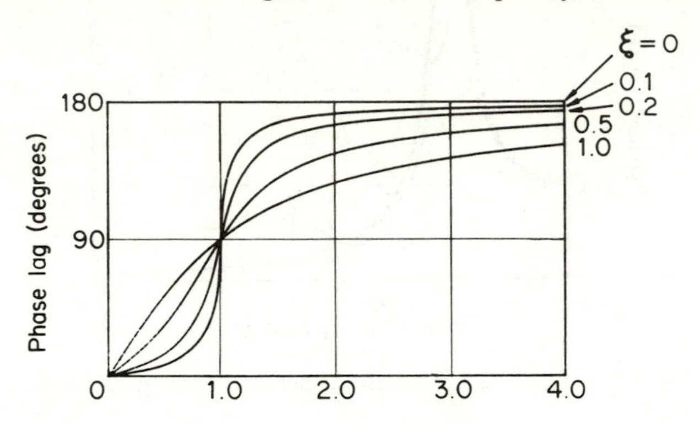

Fig. 2.9 Variation of phase lag with frequency ratio.

EXAMPLE 2.2

The structural frame shown in Fig. 2.6 supports a reciprocating machine which exerts a horizontal periodic force of 8.5 kN at a frequency of 1.75 Hz in the plane of the frame. The mass of the machine of 4000 kg is added to the existing mass of the structure. (a) What is the steady-state amplitude of vibration if the damping ratio is 4%? (b) What would the steady-state amplitude be if the forcing frequency was in resonance with the structure?

Solution

The loading function is given by

$$P(t) = P_0 \sin \Omega t$$

where

$$P_0 = 8500 \text{ N}$$

and

$$\Omega = 2\pi \times 1.75.$$

Therefore, the static deflection is

$$u_{st} = P_0/k = 8500/(4.0 \times 10^6) \text{ m} = 2.125 \text{ mm}.$$

The mass of the one-degree-of-freedom system will now be

$$m = 5000 + 4000 = 9000 \text{ kg},$$

giving a natural frequency of

$$f=\frac{1}{2\pi}\sqrt{\left(\frac{4\times 10^6}{9000}\right)}=3.355 \text{ Hz.}$$

Note the importance of using consistent units in the calculation of frequency. This is because of the presence of force and mass units in the same formula, stiffness k being force per unit displacement. Force is a combined unit. In the SI system one Newton is defined as the force required to give one kilogram an acceleration of one metre per second squared. Therefore, the expression for frequency would have the units

$$f\rightarrow\sqrt{\left(\frac{\text{stiffness}}{\text{mass}}\right)}\rightarrow\sqrt{\left(\frac{\text{force/deflection}}{\text{mass}}\right)}\rightarrow\sqrt{\left(\frac{\text{kg m/s}^2\text{)/m}}{\text{kg}}\right)}\rightarrow\frac{1}{\text{s}}.$$

Mistakes can be avoided by doing all the calculations in units of kg, m and s. Displacements may be converted to mm at the end of a calculation for convenience.

(a) The frequency ratio is

$$\rho=\Omega/\omega=1.75/3.355=0.522.$$

Therefore, from (2.27) the dynamic load factor is

$$\text{DLF}=1/\sqrt{[(1-0.522^2)^2+(2\times 0.04\times 0.522)^2]}=1.372,$$

giving a dynamic deflection of

$$u=u_{\text{st}}\times\text{DLF}=2.92 \text{ mm.}$$

(b) At resonance ($\rho=1$) the dynamic load factor is

$$\text{DLF}=1/(2\xi)=1/0.08=12.5,$$

giving a dynamic deflection of

$$u=2.125\times 12.5=26.56 \text{ mm.}$$

2.6 ARBITRARY FORCING FUNCTION

In most practical applications the dynamic loading $P(t)$ is irregular and non-periodic. Forces arising from the wind or earthquakes are good examples.

One way of analysing this problem is to assume that the irregular forcing function is made up of a sequence of very brief impulses as shown in Fig. 2.10. The vibrations caused by all the impulses are added together to obtain the total response.

First the vibration caused by a single impulse is required. Newton's second law states that the rate of change of momentum of a mass is equal to the

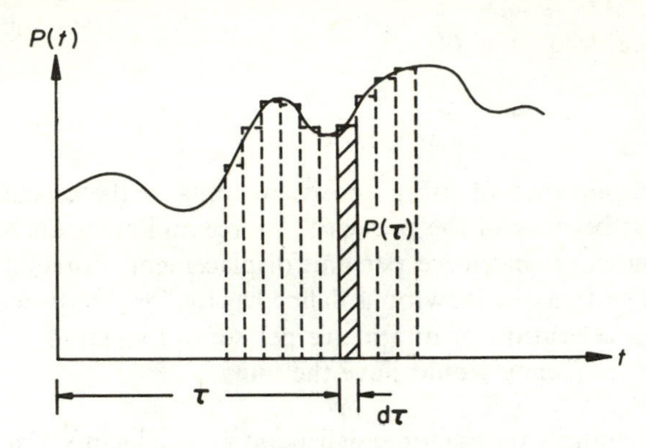

Fig. 2.10 Arbitrary forcing function.

applied force. In the case of the single-degree-of-freedom system this is represented by

$$\mathrm{d}(m\dot{u})/\mathrm{d}t = P(t). \tag{2.28}$$

Thus the change in momentum over a brief interval $\mathrm{d}\tau$, brought about by the instantaneous force $P(\tau)$, is given by

$$\mathrm{d}(m\dot{u}) = P(\tau)\mathrm{d}\tau. \tag{2.29}$$

It should be noted that the small changes in velocity and displacement occurring during the interval $\mathrm{d}\tau$ will make a negligible contribution to the change in momentum given by (2.29). Therefore the change in velocity during the interval is

$$\mathrm{d}\dot{u} = P(\tau)/m\ \mathrm{d}\tau. \tag{2.30}$$

This is equivalent to the velocity, $\dot{u}_0$, caused by an impulse that is applied for a duration of $\mathrm{d}\tau$ and then removed.

The vibration at a later time t can be calculated from Equation (2.10), which gives the motion of a single-degree-of-freedom system with initial velocity and displacement. In this equation the elapsed time t must be replaced by $(t-\tau)$, ω' can be assumed to be equal to ω for practical purposes, u_0 is zero and $\dot{u}_0$ is given by the expression for $\mathrm{d}\dot{u}$ in Equation (2.30).

Thus the displacement at t caused by the impulse at τ is given by

$$\delta u(t) = \mathrm{e}^{-\xi\omega(t-\tau)}\left[\frac{P(\tau)\mathrm{d}\tau}{m\omega}\sin\omega(t-\tau)\right]. \tag{2.31}$$

Each impulse in Fig. 2.10 will produce a vibration of this form. Because the vibration equations are linear, the effect of each impulse is independent of every other impulse and the total resulting motion can be obtained by the principle of superposition. Thus, in the limit, Equation (2.31) may be integrated from zero to t to give the vibration at time t. Hence

$$u(t)=\frac{1}{m\omega}\int_0^t P(\tau)\sin\omega(t-\tau)\mathrm{e}^{-\xi\omega(t-\tau)}\mathrm{d}\tau. \tag{2.32}$$

This is known as the convolution or Duhamel integral. Explicit solutions may be obtained for simple forms of forcing function such as rectangular or triangular impulses. These are useful for evaluating the response of structures to impact or shock loading (Biggs, 1964).

EXAMPLE 2.3

A simple example of some general interest is the case of a load applied suddenly at $t=0$ and remaining constant thereafter, as shown in Fig. 2.11(a). This is expressed by

$$P(t)=P_0;\quad t>0. \tag{2.33}$$

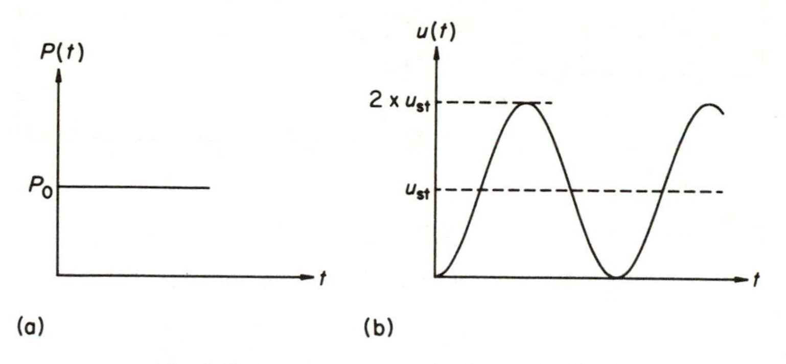

Fig. 2.11 Response to suddenly applied load.

If damping is neglected ($\xi=0$), substitution for $P(t)$ in (2.32) yields

$$\begin{aligned} u(t)&=(P_0/m\omega)\int_0^t \sin\omega(t-\tau)\mathrm{d}\tau \\ &=u_{\mathrm{st}}(1-\cos\omega t) \end{aligned} \tag{2.34}$$

since

$$P_0/m\omega=(P_0/k)\omega=u_{\mathrm{st}}\omega.$$

The displacement curve is shown in Fig. 2.11(b), where it is evident that the DLF is 2.0. This is a useful upper bound. With shock loading there is often a finite rise time before maximum load is reached, thus reducing the DLF to some extent.

EXAMPLE 2.4

Blast loading on a structure may be represented approximately by the triangular forcing function shown in Fig. 2.12(a), in which the force is applied suddenly and then decays linearly to zero. Compare it with Fig. 1.1(c). (a) Determine the expressions for displacement of a one-degree-of-freedom structure, neglecting damping. (b) Hence determine the maximum response of the structure in Fig. 2.6 when the peak load is 30 kN and the duration of the impulse is 0.16 s.

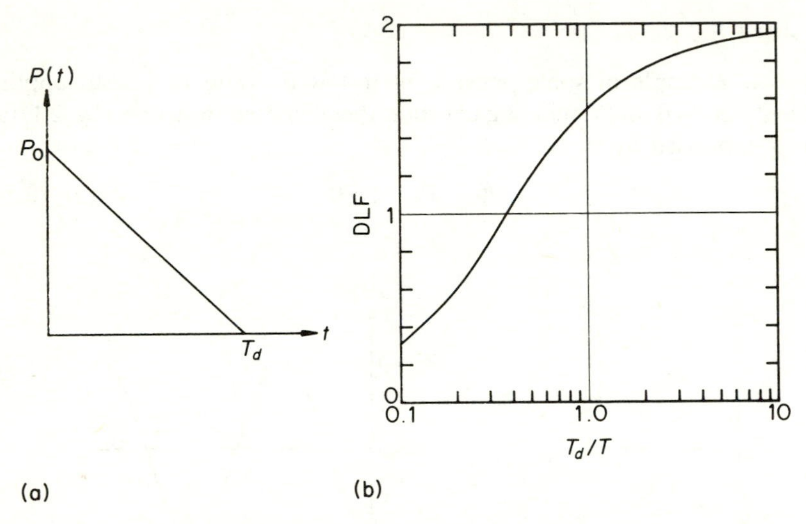

Fig. 2.12 Effect of triangular forcing function.

Solution

(a) The forcing function must be described in two parts, namely during the application of the force, and after it has decayed to zero. Therefore

$$P(t) = P_0(1 - t/T_d); \quad 0 \leqslant t \leqslant T_d \tag{2.35a}$$

$$= 0; \quad t > T_d. \tag{2.35b}$$

From (2.32) with neglect of damping, the displacement while the load is being applied is given by

$$u(t) = (P_0/m\omega) \int_0^t (1 - \tau/T_d) \sin \omega(t - \tau) d\tau$$

$$= u_{st}[1 - (t/T_d) + (1/\omega T_d) \sin \omega t - \cos \omega t]; \quad 0 \leqslant t \leqslant T_d. \tag{2.36}$$

The vibration after the loading has decayed to zero may be found by first obtaining the displacement and velocity at T_d. These can be determined from (2.36) and are

$$u(T_d) = u_{st}[(1/\omega T_d)\sin\omega T_d - \cos\omega T_d] \tag{2.37a}$$

$$\dot{u}(T_d) = u_{st}[\omega\sin\omega T_d + (1/T_d)(\cos\omega T_d - 1)]. \tag{2.37b}$$

Using (2.10) with the above values as initial conditions, neglecting damping and substituting $(t - T_d)$ for elapsed time we obtain

$$u = u_{st}\{[\sin\omega T_d + (1/\omega T_d)(\cos\omega T_d - 1)]\sin\omega(t - T_d) + [(1/\omega T_d)\sin\omega T_d - \cos\omega T_d]\cos\omega(t - T_d)\}; \quad t > T_d. \tag{2.38}$$

(b) The maximum value of u/u_{st}, i.e. DLF, Obtained from equations (2.36) and (2.38) is plotted against the ratio of blast duration to natural period of vibration in Fig. 2.12(b). When the blast duration is long compared with the period of vibration of the structure the loading is similar to a constant load suddenly applied, as outlined in Example 2.3, and the DLF tends towards 2.0. However, at the other extreme we have the case of a blast of very brief duration compared with the period of vibration, and the inertia of the structure prevents it from responding to the load quickly enough, so that the DLF is small.

The period of the given structure is

$$T = 1/f = 1/4.502 = 0.2221 \text{ s.}$$

The blast duration is

$$T_d = 0.16 \text{ s.}$$

Therefore the ratio is

$$T_d/T = 0.72.$$

Hence, from Fig. 2.12(b), the DLF is found to be

$$\text{DLF} = 1.40$$

and the maximum response is

$$u = \text{DLF} \times u_{st} = 1.4 \times (30 \times 10^3)/(4 \times 10^6)\text{ m} = 10.5 \text{ mm.}$$

2.7 NUMERICAL EVALUATION OF THE DUHAMEL INTEGRAL

It should now be clear that explicit evaluation of the Duhamel integral is only feasible for very simple load cases. Numerical integration is required for more general forcing functions (Clough and Penzien, 1975). Equation (2.32) should first be expressed in the form

$$u(t)=\frac{e^{-\xi\omega t}\sin\omega t}{m\omega}\int_0^t e^{\xi\omega\tau}P(\tau)\cos\omega\tau\,d\tau$$

$$-\frac{e^{-\xi\omega t}\cos\omega t}{m\omega}\int_0^t e^{\xi\omega t}P(t)\sin\omega\tau\,d\tau$$

$$=\frac{e^{-\xi\omega t}}{m\omega}\{A(t)\sin\omega t-B(t)\cos\omega t\} \tag{2.39}$$

where

$$A(t)=\int_0^t e^{\xi\omega t}P(\tau)\cos\omega\tau\,d\tau \tag{2.40a}$$

$$B(t)=\int_0^t e^{\xi\omega t}P(\tau)\sin\omega\tau\,d\tau. \tag{2.40b}$$

The integrals $A(t)$ and $B(t)$ have to be integrated numerically. A simple and sufficiently accurate procedure is to use the trapezoidal rule. Considering first the integral $A(t)$, we need to evaluate the function $y=e^{\xi\omega\tau}P(\tau)\cos\omega\tau$ at equally spaced intervals, as shown in Fig. 2.13. Then

$$A(t)=(\Delta\tau/2)(y_0+2y_1+2y_2+\ \ldots\ +2y_{N-1}+y_N). \tag{2.41}$$

This will yield the value of $A(t)$ at a time $t=N\Delta t$. However, usually the complete response history is required and therefore it is preferable to perform the summation incrementally using

$$A(t)=A(t-\Delta t)+(\Delta t/2)(y_{n-1}+y_n), \tag{2.42}$$

where $A(t-\Delta t)$ is the value of the integral at the previous time step. The integration of $B(t)$ is performed in exactly the same way except that the function contains $\sin\omega\tau$ instead of $\cos\omega\tau$. The displacement is evaluated using (2.39) at each time step.

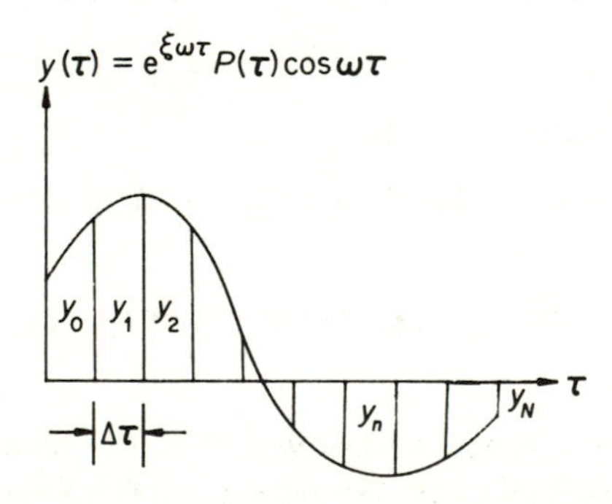

Fig. 2.13 Numerical evaluation of Duhamel integral.

EXAMPLE 2.5

The response of the structure subjected to blast loading as given in Example 2.4(b) will be obtained by evaluating the Duhamel integral numerically using a simple computer program written in BASIC. The listing of the program is as follows:

```
10 REM PROGRAM DUHAMEL
20 REM N = NO. OF STEPS
30 REM D = STEP LENGTH
40 REM W = CIRCULAR FREQU.
50 REM M = MASS
60 REM L = CRIT. DAMPING
70 REM P(I) = FORCE
80 REM U(I) = DISP.
90 DIM P(200)
100 DIM U(200)
110 READ N,D,W,M
120 DATA 32,0.01,28.2618,5000
130 FOR I = 1 TO N+1
140 READ P(I)
150 DATA 30,28.125,26.25,24.375,22.5,20.625,18.75,16.875
160 DATA 15,13.125,11.25,9.375,7.5,5.625,3.75,1.875,0,0
170 DATA 0,0,0,0,0,0,0,0,0,0,0,0,0,0,0
180 NEXT I
190 PRINT "Type in crit. damping"
200 INPUT L
210 IF L<0 GOTO 460
220 A = 0: B = 0
230 Y0 = P(1): Z0 = 0
240 U(I) = 0
250 FOR I = 2 TO N+1
260 T = W*(I-1)*D
270 E = EXP(L*T)
280 C = COS(T): S = SIN(T)
290 Y1 = E*P(I)*C
300 Z1 = E*P(I)*S
310 A = A + (Y0 + Y1)*D/2
320 B = B + (Z0 + Z1)*D/2
330 U(I) = (A*S - B*C)/(M*W*E)
340 Y0 = Y1: Z0 = Z1
350 NEXT I
360 PRINT "TIME      DISP."
370 J = 0
380 FOR I=1 TO N+1
390 J = J+1
400 IF J<20 GOTO 430
410 INPUT "--more--",J$
420 J = 0
430 PRINT USING "$$.$$$  "; (I-1)*D, U(I)*1000000!
440 NEXT I
450 GOTO 190
460 STOP
```

The program was run using different values of damping, namely zero damping and 5% of critical. Using a step length of 0.01 s the program was run for the first 0.32 s of motion. The results are shown in Fig. 2.14. The maximum displacement of the undamped case is close to the 'exact' value of 10.5 mm calculated in Example 2.4. Thus the step length, which is approximately 1/20 of the period of vibration, can be considered satisfactory for this analysis. The effect of damping should be noted, the amplitude of vibration being progressively depressed with time. After loading has ended the vibrations would gradually decay to zero.

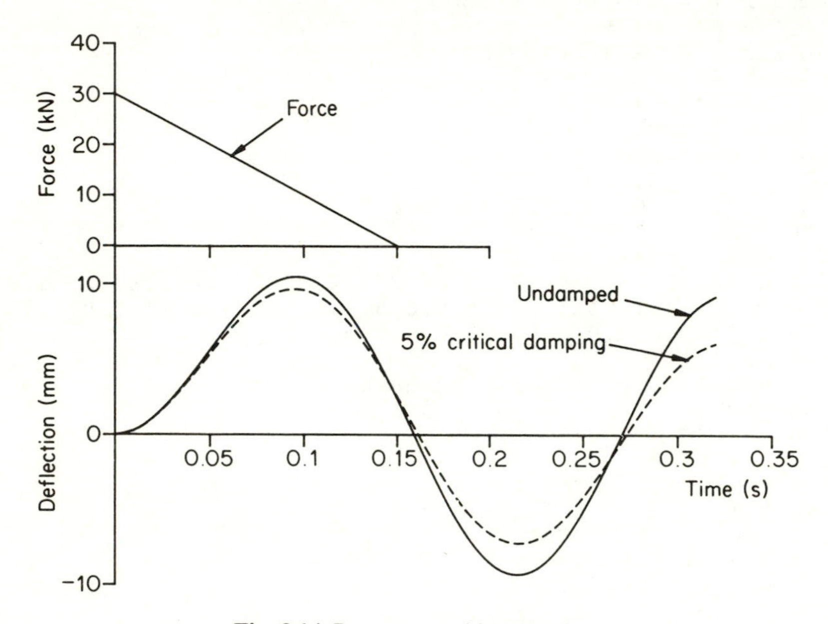

Fig. 2.14 Response to blast loading.

2.8 SUPPORT MOTION

Apart from external dynamic loads, vibration can be caused by ground motion or movement of the supports. This is of particular importance in relation to earthquakes. Other applications include the response of buildings to vibration transmitted by traffic or pile-driving, the design of vibration isolation systems, and the design of seismic detection devices. In fact the bouncing toy in Fig. 2.2 illustrates the effects of support motion.

Consider the simple system shown in Fig. 2.15 which is subjected to ground movement, denoted by $u_g(t)$. The equation of motion is

$$m\ddot{u} + c(\dot{u} - \dot{u}_g) + k(u - u_g) = 0, \tag{2.43}$$

in which it will be noted that the velocity of the dashpot and the extension of

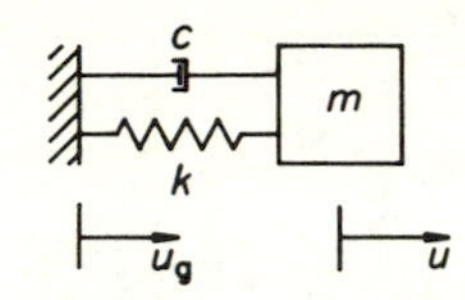

Fig. 2.15 Support motion.

the spring are both affected by the ground movement. The equation can be written

$$m\ddot{u}+c\ddot{u}+ku=ku_g(t)+c\dot{u}_g(t). \tag{2.44}$$

This is essentially the same as Equation (2.1) except that the external forcing function, $P(t)$, has now been replaced by the prescribed support motion. If the support motion is known then the ensuing vibration may be determined by the methods described previously in this chapter.

Take, for example, an imposed periodic motion of the support given by

$$u_g(t)=u_0\sin\Omega t. \tag{2.45}$$

Accordingly, the equation of motion becomes

$$\begin{aligned} m\ddot{u}+c\dot{u}+ku &= ku_0\sin\Omega t+c\Omega\, u_0\cos\Omega t \\ &= u_0\sqrt{(c^2\Omega^2+k^2)}\sin(\Omega t+\beta) \end{aligned} \tag{2.46}$$

where $\tan\beta=c\Omega/k$.

This is similar to Equation (2.20), which related to steady-state forced vibration. The only differences are in the amplitude of the forcing function and the existence of a phase difference β.

The latter can be ignored if we choose to measure time from a different origin. Thus the amplitude of vibration of the mass can be obtained using Equation (2.25), from which

$$(u)_{\max}=\frac{u_0\sqrt{(c^2\Omega^2+k^2)}}{\sqrt{[(k-m\Omega^2)^2+c^2\Omega^2]}}. \tag{2.47}$$

If we define the transmissibility or transmission ratio as

$$\mathrm{TR}=\frac{\text{amplitude of mass}}{\text{amplitude of support motion}}=\frac{(u)_{\max}}{u_0} \tag{2.48}$$

and substitute $k/m=\omega^2$, $c/m=2\xi\omega$, and $\Omega/\omega=\rho$, we obtain

$$\mathrm{TR}=\sqrt{\left(\frac{1+(2\xi\rho)^2}{(1-\rho^2)^2+(2\xi\rho)^2}\right)}. \tag{2.49}$$

The transmission ratio is plotted in Fig. 2.16 as a function of frequency ratio for various values of damping. The bouncing toy will illustrate the

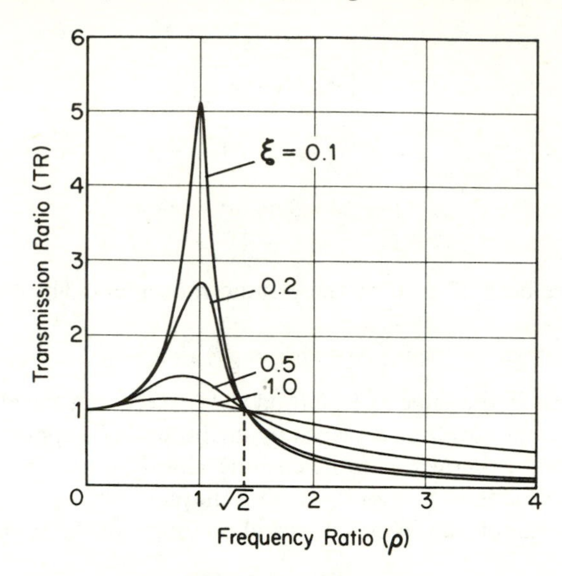

Fig. 2.16 Transmission ratio.

response most effectively. If you move your hand up and down very slowly then the toy frog will faithfully follow your motion. If you then speed up the motion of your hand a stage will be reached when you hit the resonant frequency of the frog on the spring and it will bounce up and down vigorously with little effort on your part. Finally, if you increase the frequency of vibration of your hand still further the oscillation of the toy frog will gradually diminish and eventually disappear altogether.

This behaviour illustrates the basic principle of vibration isolation. If an item of sensitive equipment needs to be protected from unwanted floor vibration, it can be mounted on a sprung suspension system. The stiffness of the springs should be chosen to ensure that the natural frequency of the instrument on its suspension is much lower than the frequency of the unwanted vibration. This will require a soft suspension which will permit high frequency motion at the support without transmission to the mass.

The principle of vibration isolation has been applied successfully to large buildings to protect them from vibration caused by nearby railway traffic (Waller, 1969). In order to achieve this it is necessary to support an entire building on springs at foundation level. Helical steel springs and laminated rubber bearings have been used effectively for this purpose. However, the problem is a bit more complicated than the simple theory given here because the ground itself is resilient and there is interaction between the vibration of the ground and the motion of a building. This known as *soil–structure interaction* and will be considered in Chapter 3.

A similar remedy can be adopted in the case of a machine which applies oscillating forces to a floor owing to the presence of unbalanced rotating masses. In this case, if we return to Equations (2.25) and (2.26) and work out the force exerted on the fixed support, taking due account of the force transmitted by both the spring and the dashpot, we find that the ratio of the maximum transmitted force to the applied force is given by exactly the same formula as Equation (2.49).

Both cases are shown in Fig. 2.17 and it can be seen that the transmission ratio is given by

$$\mathrm{TR}=(u)_{\max}/u_0=(P)_{\max}/P_0. \tag{2.50}$$

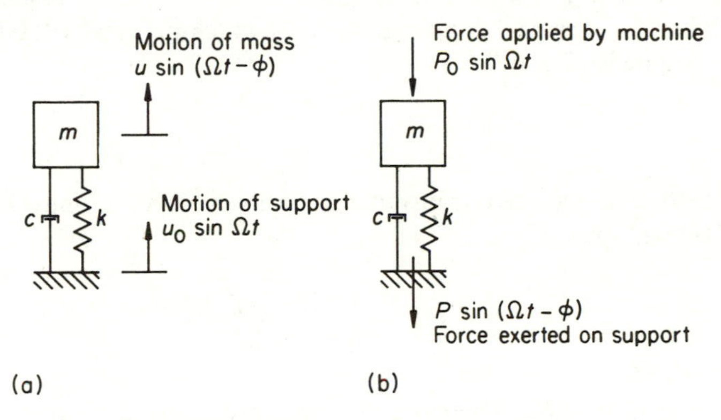

Fig. 2.17 Examples of vibration isolation.

A slightly different formulation of Equation (2.43) is required for earthquake engineering purposes. This is because earthquake ground motions are usually obtained in the form of records of ground acceleration measured by seismographs. Furthermore, we are not generally interested in the absolute displacement of the upper parts of a building but rather the stresses in the members which result from relative displacement between the various floors and the foundation movement.

Therefore we will introduce the following notation:

$$u_{\mathrm{r}}=u-u_{\mathrm{g}}, \quad \dot{u}_{\mathrm{r}}=\dot{u}-\dot{u}_{\mathrm{g}}, \quad \ddot{u}_{\mathrm{r}}=\ddot{u}-\ddot{u}_{\mathrm{g}}. \tag{2.51}$$

Substituting into (2.43), we find that

$$m\ddot{u}_{\mathrm{r}}+c\dot{u}_{\mathrm{r}}+ku_{\mathrm{r}}=-m\ddot{u}_{\mathrm{g}}. \tag{2.52}$$

This is now similar to (2.1) except that absolute displacement of the mass has been replaced by relative displacement of the spring, and the time-dependent external force $P(t)$ has been replaced by a ground acceleration. In the case of earthquake engineering the ground acceleration is obtained as a seismographic record of the transient ground motion. This may be treated as an arbitrary forcing function and the problem analysed by numerical evaluation of Duhamel's integral using the program given in Example 2.5.

EXAMPLE 2.6

A machine for manufacturing microelectronic circuits is to be mounted on a factory floor using a vibration-isolating suspension. Vibration of the floor during normal usage has a predominant frequency of 10 Hz with maximum amplitude of 20×10^{-4} mm. The greatest amplitude that can be tolerated by the machine for reliable operation is 1.0×10^{-4} mm. (a) Determine the required natural frequency of the machine on its suspension assuming that there will be 2% of critical damping. (b) If the machine weighs 1500 kg what will its static deflection be?

Solution

(a) Evidently the required transmission ratio is 1/20. Using (2.49) with $\xi = 0.02$ we obtain

$$\frac{1}{20} = \sqrt{\left(\frac{1 + (0.04\rho)^2}{(1-\rho^2)^2 + (0.04\rho)^2}\right)}$$

from which $\rho = 4.62$. Therefore the required natural frequency of the machine on its suspension is 2.16 Hz.

(b) From Equation (2.12) the required stiffness of the suspension must be

$$k = m(2\pi f)^2 = 276.3 \text{ kN/m}.$$

Hence, the deflection of the machine under its own weight will be

$$u_{st} = mg/k = 53.2 \text{ mm}.$$

Clearly a very soft suspension is required, and in practice compressed-air systems are often employed.

2.9 DIRECT NUMERICAL INTEGRATION OF EQUATION OF MOTION

We have already suggested that explicit solutions are not always available for practical problems. A case in point arose with irregular forcing functions. Numerical integration of Duhamel's integral is required in most cases. An alternative, and often preferable, approach is direct integration of the equation of motion by a numerical procedure. This is the only feasible method, in fact, if the structural behaviour is non-linear due to inelastic spring stiffness, plasticity or frictional damping.

There are several convenient methods of direct numerical integration of the equation of motion. They all require that the time-varying force and the response are specified at discrete times 0, Δt, $2\Delta t$, ..., $t-\Delta t$, t, $t+\Delta t$, ... with a regular interval of Δt. Assumptions are made about the nature of the acceleration over each time interval. We shall now consider two well known methods.

2.9.1 Acceleration impulse procedure

The simplest procedure is to assume that displacement varies in the form of a series of linear segments, as shown in Fig. 2.18(b). It is evident that the velocity must be constant over the time intervals, as indicated in Fig. 2.18(c). The velocities on either side of the instant t are therefore given by

$$\dot{u}_t^+ = \frac{u_{t+\Delta t} - u_t}{\Delta t}$$

and (2.53)

$$\dot{u}_t^- = \frac{u_t - u_{t-\Delta t}}{\Delta t}.$$

The sudden changes in velocity occurring at each time step must have been produced by a series of impulse accelerations as shown in Fig. 2.18(d). The acceleration impulses would be very brief but of sufficiently great magnitude to produce the sudden change in velocity. However, the summed areas of all the impulses must equal the area under the acceleration curve. Thus it is possible to deduce the change in velocity at t by considering an impulse of finite width Δt shown shaded. The magnitude of the impulse can be calculated from the change in velocity at that instant, thus

$$\ddot{u}_t = \frac{\dot{u}_t^+ - \dot{u}_t^-}{\Delta t}. \tag{2.54}$$

Substituting for $\dot{u}_t^+$ and $\dot{u}_t^-$ from (2.53) we obtain

$$\ddot{u}_t = \frac{u_{t+\Delta t} - 2u_t + u_{t-\Delta t}}{\Delta t^2}. \tag{2.55}$$

We can rewrite this to obtain the recurrence formula

$$u_{t+\Delta t} = 2u_t - u_{t-\Delta t} + \Delta t^2 \ddot{u}_t. \tag{2.56}$$

This expression requires the acceleration at the previous time step, $\ddot{u}_t$, which may be obtained from the equation of motion (2.1), written as follows:

$$m\ddot{u}_t = P(t) - c\dot{u}_t - ku_t. \tag{2.57}$$

If damping is present the velocity at the previous time step, $\dot{u}_t$, will also be

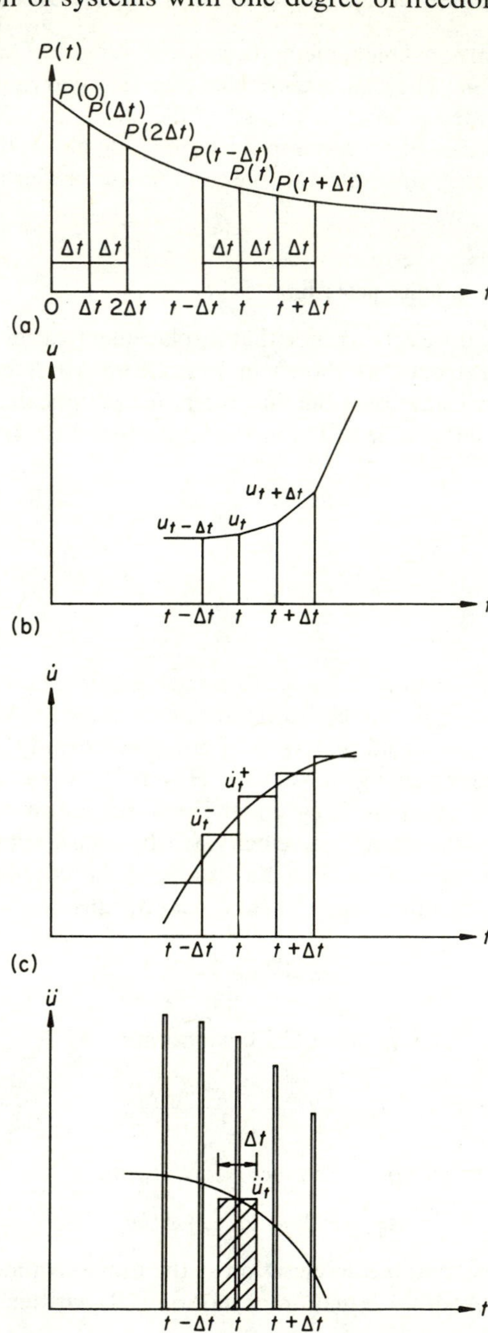

Fig. 2.18 Acceleration impulse procedure. (a) Force; (b) displacement; (c) velocity; (d) acceleration.

required. This is approximately given by

$$\dot{u}_t = \frac{u_t - u_{t-\Delta t}}{\Delta t} + \frac{\Delta t}{2}\ddot{u}_t \tag{2.58}$$

where the first term is the average velocity during the previous interval and the second term is the additional velocity given by the impulse during the latter half of that interval. Equation (2.58) may be combined with (2.57) to obtain

$$\ddot{u}_t = \frac{P(t) - ku_t - c(u_t - u_{t-\Delta t})/\Delta t}{m + c\Delta t/2}. \tag{2.59}$$

The dynamic response may then be computed in a stepwise manner as follows:

(a) Use Equation (2.59) to calculate the acceleration at time step t. A knowledge of the instantaneous displacement u_t and dynamic force $P(t)$ is required, together with the displacement at the previous time step, $u_{t-\Delta t}$.
(b) Use (2.58) to calculate the current velocity $\dot{u}_t$. This step is optional but is often included in computer codes because some design criteria are related to the dynamic velocity of a structure.
(c) Compute the displacement at next time step using the recurrence relation (2.56).
(d) Repeat the sequence of calculations for time step $t + \Delta t$ and so on.

It will be observed in Fig. 2.18(c) that the approximation in this procedure involves assuming constant velocity over successive time intervals. The method is therefore frequently referred to as the *constant-velocity procedure*. It is simple to apply and is even practicable for hand calculation. If a pocket calculator with programmable keystrokes is employed then it is possible to compute quite lengthy transient responses with very little effort. Clearly the accuracy depends on the interval Δt. Stability and accuracy of direct integration procedures will be discussed in Chapter 4. Meanwhile it is recommended that the step length Δt should not be greater than 1/20 of the period of vibration.

2.9.2 Starting procedures

One disadvantage of the acceleration impulse procedure is that the recurrence relation of (2.56) cannot be used at time step 0 to predict displacement at time step Δt. This is because no value exists for the displacement $u_{-\Delta t}$. A different method of estimating the first step must therefore be adopted.

If the structure is initially at rest then the acceleration at $t = 0$ is found from (2.57) to be

$$\ddot{u}_0 = P(0)/m. \tag{2.60}$$

The simplest starting procedure is achieved by assuming that the acceleration is constant during the first time step and equal to this initial value. This assumption yields

$$u_{\Delta t} = \tfrac{1}{2}\ddot{u}_0 \Delta t^2 = P(0)\Delta t^2/2m. \tag{2.61}$$

It is now possible to carry out steps (a) to (c) of the numerical integration in the normal way.

In the case of a dynamic load that is initially zero but increases with time, as in Fig. 1.1(c), the above procedure results in no change of acceleration or displacement over the first interval. The error incurred in this case is small, especially if the time steps are chosen to allow a reasonably faithful representation of the forcing function.

2.9.3 Linear acceleration method

The preceding discussion suggests that a better approximation might be achieved by assuming that the acceleration varies linearly over each time interval, as shown in Fig. 2.19. If τ denotes increase in time during an interval the acceleration at an instant $t+\tau$ is given by

$$\ddot{u}_{t+\tau} = \ddot{u}_t + (\tau/\Delta t)(\ddot{u}_{t+\Delta t} - \ddot{u}_t). \tag{2.62}$$

Integrating twice we obtain expressions for velocity and displacement at $t+\tau$ as follows:

$$\dot{u}_{t+\tau} = \dot{u}_t + \ddot{u}_t\tau + (\tau^2/2\Delta t)(\ddot{u}_{t+\Delta t} - \ddot{u}_t) \tag{2.63a}$$

$$u_{t+\tau} = u_t + \dot{u}_t\tau + \tfrac{1}{2}\ddot{u}_t\tau^2 + (\tau^3/6\Delta t)(\ddot{u}_{t+\Delta t} - \ddot{u}_t). \tag{2.63b}$$

At the end of the interval ($\tau = \Delta t$) these become

$$\dot{u}_{t+\Delta t} = \dot{u}_t + (\Delta t/2)(\ddot{u}_{t+\Delta t} + \ddot{u}_t) \tag{2.64a}$$

$$u_{t+\Delta t} = u_t + \Delta t\dot{u}_t + (\Delta t^2/6)(\ddot{u}_{t+\Delta t} + 2\ddot{u}_t). \tag{2.64b}$$

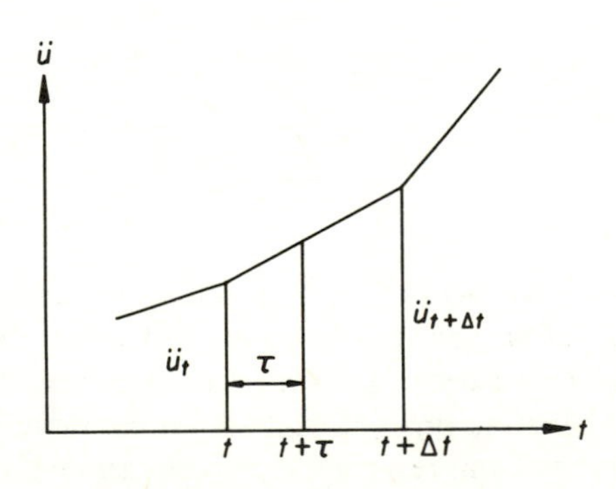

Fig. 2.19 Linear acceleration method.

It is convenient to use displacement as the basic variable during the step-by-step solution and therefore solving (2.64a) and (2.64b) for $\ddot{u}_{t+\Delta t}$ and $\dot{u}_{t+\Delta t}$ we obtain

$$\ddot{u}_{t+\Delta t}=(6/\Delta t^2)(u_{t+\Delta t}-u_t)-(6/\Delta t)\dot{u}_t-2\ddot{u}_t \tag{2.65a}$$

$$\dot{u}_{t+\Delta t}=(3/\Delta t)(u_{t+\Delta t}-u_t)-2\dot{u}_t-(\Delta t/2)\ddot{u}_t. \tag{2.65b}$$

The equation of motion (2.1) at $t+\Delta t$ is

$$m\ddot{u}_{t+\Delta t}+c\dot{u}_{t+\Delta t}+ku_{t+\Delta t}=P(t+\Delta t). \tag{2.66}$$

By substituting Equations (2.65) into (2.66) the equation of motion may be written in the form

$$\left(\frac{6}{\Delta t^2}m+\frac{3}{\Delta t}c+k\right)u_{t+\Delta t}$$
$$=P(t+\Delta t)+\left(\frac{6}{\Delta t^2}m+\frac{3}{\Delta t}c\right)u_t+\left(\frac{6}{\Delta t}m+2c\right)\dot{u}_t+\left(2m+\frac{\Delta t}{2}c\right)\ddot{u}_t. \tag{2.67}$$

In effect this is an equilibrium equation at $t+\Delta t$ in which the dynamic stiffness has been modified by the mass and damping, and the applied load depends on the motion at the previous time step. Therefore, if the loading is known at any time step then Equation (2.67) can be solved for the displacement. Equations (2.65) must then be used to determine the acceleration and velocity for input to the next time step. It should be noted that no special starting procedure is needed since Equation (2.67) does not require a knowledge of motion prior to the current time step.

Errors tend to accumulate during the calculation. This problem is discussed in Chapter 4, but meanwhile it is recommended that the step length should not be greater than 1/20 of the period of vibration.

EXAMPLE 2.7

The structural frame shown in Fig. 2.20(a) is subjected to an impulsive horizontal ground motion as given in Fig. 2.20(b). The properties of the equivalent single-degree-of-freedom system are as follows:

$$m=4000 \text{ kg}$$

$$c=20\,000 \text{ N s/m}$$

$$k=4\times 10^6 \text{ N/m}.$$

(a) Calculate the relative displacement of the columns during the first 0.24 s of motion using the acceleration impulse method. (b) Write a program in BASIC to integrate the equation of motion using the linear acceleration method and compare the results with (a).

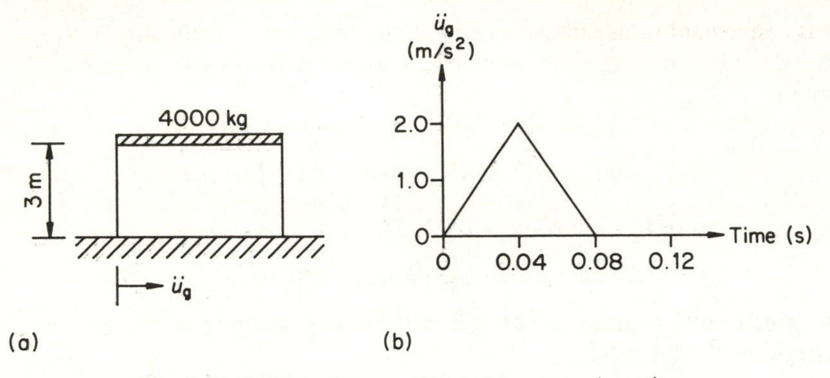

Fig. 2.20 Building frame subjected to ground motion.

Solution

First calculate the undamped natural frequency and period of vibration of the structure from (2.11) and (2.12):

$$f=\frac{1}{2\pi}\sqrt{\left(\frac{4\times 10^6}{4000}\right)}=5.03 \text{ Hz}, \quad T=\frac{1}{f}=0.199 \text{ s}.$$

An integration step length of 0.01 s is approximately 1/20 of the period and will therefore be adopted.

Out of interest, the damping value expressed as a percentage of critical will be given by (2.15) and is

$$\xi=\frac{20\,000}{2\sqrt{(4\times 10^6\times 4000)}}=0.079=7.9\%.$$

This is a rather high value but makes the example calculations easier.

Since this is a problem of support motions, Equation (2.52) must be used and relative motions will be obtained. The forcing function will be given by

$$P(t)=-m\ddot{u}_g$$

which will reach a peak of -8000 N after 0.04 s and then reduce to zero at 0.08 s.

(a) Equation (2.59) will be written in the form

$$(m+c\Delta t/2)\ddot{u}_t=P(t)-ku_t-c\,\Delta u_t/\Delta t$$
$$=\text{RHS}$$

where

$$\Delta u_t=u_t-u_{t-\Delta t}.$$

Values of u, Δu and RHS have been calculated and are given in Table 2.1, from which $\ddot{u}$ may be obtained. Forward integration at each step is

Table 2.1 Calculations using direct numerical integration (Example 2.7)

			Acceleration impulse procedure				*Linear acceleration procedure*	
t sec	$\ddot{u}_g$ m/s^2	$\frac{-m\ddot{u}_g}{10^3}$	$\frac{u}{10^3}$	$\frac{\Delta u}{10^3}$	RHS	$\ddot{u}$ m/s^2	$\frac{u}{10^3}$	$\ddot{u}$ m/s^2
0	0	0	0	–	0	0	0	0
0.01	0.5	−2	0	0	−2	−0.488	−0.001	−0.480
0.02	1	−4	−0.049	−0.049	−3.707	−0.904	−0.063	−0.891
0.03	1.5	−6	−0.188	−0.139	−4.970	−1.212	−0.205	−1.196
0.04	2	−8	−0.448	−0.260	−5.688	−1.387	−0.465	−1.373
0.05	1.5	−6	−0.847	−0.399	−1.814	−0.442	−0.843	−0.448
0.06	1	−4	−1.290	−0.443	2.046	0.499	−1.267	0.474
0.07	0.5	−2	−1.683	−0.393	5.518	1.346	−1.644	1.307
0.08	0	0	−1.941	−0.258	8.280	2.020	−1.894	1.975
0.09	0	0	−1.997	−0.057	8.102	1.976	−1.958	1.941
0.10	0	0	−1.856	0.141	7.142	1.742	−1.830	1.722
0.11	0	0	−1.541	0.315	5.534	1.350	−1.534	1.348
0.12	0	0	−1.091	0.450	3.464	0.845	−1.104	0.863
0.13	0	0	−0.556	0.534	1.156	0.282	−0.589	0.318
0.14	0	0	0.006	0.562	−1.148	−0.280	−0.042	−0.230
0.15	0	0	0.540	0.534	−3.228	−0.787	0.483	−0.731
0.16	0	0	0.995	0.455	−4.890	−1.193	0.936	−1.138
0.17	0	0	1.331	0.336	−5.996	−1.462	1.278	−1.415
0.18	0	0	1.521	0.190	−6.464	−1.576	1.480	−1.544
0.19	0	0	1.553	0.032	−6.276	−1.531	1.531	−1.518
0.20	0	0	1.432	−0.121	−5.486	−1.338	1.432	−1.348
0.21	0	0	1.177	−0.255	−4.198	−1.024	1.200	−1.056
0.22	0	0	0.820	−0.357	−2.566	−0.626	0.865	−0.677
0.23	0	0	0.400	−0.420	−0.760	−0.185	0.462	−0.251
0.24	0	0	−0.038	−0.438	1.028	0.251	0.035	0.178

achieved using (2.56) written as

$$u_{t+\Delta t} = u_t + \Delta u_t + \Delta t^2 \ddot{u}_t.$$

Note that a starting value using (2.61) needs to be calculated but in this instance it will be zero since the input acceleration is initially zero.

(b) A program was written in BASIC to compute the motion using the linear acceleration procedure. Equation (2.67) was written in the form

$$k1 \times u_{t+\Delta t} = P(t+\Delta t) + p1 \times u_t + p2 \times \dot{u}_t + p3 \times \ddot{u}_t,$$

where the coefficients $k1$, $p1$, $p2$, $p3$ were evaluated prior to the step-by-step solution. The results are compared with the acceleration impulse method in Table 1.1. The BASIC program is as follows:

```
10 REM PROGRAM: LINEAR ACCEL.
20 REM N = NO. OF STEPS
30 REM M = MASS
40 REM C = DAMPING
50 REM K = STIFFNESS
60 REM D = STEP LENGTH
70 REM P(I) = FORCE
80 REM U0 = DISP. AT START OF D
90 REM U1 = DISP. AT END OF D
100 REM V0,V1,A0,A1 = SIMILAR FOR VEL. AND ACCEL.
110 DIM P(100)
120 READ N,M,C,K,D
130 DATA 24,4000,20E3,4E6,.01
140 FOR I = 1 TO N+1
150 READ P(I)
160 DATA 0,-2E3,-4E3,-6E3,-8E3,-6E3,-4E3,-2E3
170 DATA 0,0,0,0,0,0,0,0,0,0,0,0,0,0,0,0,0
180 NEXT I
190 U0 = 0: V0 = 0: A0 = 0
200 D1=6/D: D2=D1/D: D3=D1/2
210 P1 = D2*M + D3*C
220 P2 = D1*M + 2*C
230 P3 = 2*M + D*C/2
240 K1 = K+P1
250 FOR I=2 TO N+1
260 U1 = (P(I) + P1*U0 + P2*V0 + P3*A0)/K1
270 V1 = D3*(U1-U0) - 2*V0 - D*A0/2
280 A1 = D2*(U1-U0) - D1*V0 - 2*A0
290 PRINT USING"$$$$$$.$$$$ ";P(I),U1*1000,V1,A1
300 U0=U1: V0=V1: A0=A1
310 NEXT I
320 STOP
```

3

Structures with many degrees of freedom

3.1 INTRODUCTION

In the previous chapter the theory of vibration and dynamic response of a single-degree-of-freedom system was developed. It was said that a lumped mass and spring approximation of a real structure can often give a useful estimate of its vibration sensitivity. While this approach is ideal for preliminary design it may not give results that are sufficiently accurate for final design. Since most practical structures are made of many members having distributed mass and stiffness, the resulting vibratory motions are more difficult to describe. Instead of there being one predominant direction of deformation, as in the case of a mass–spring system, there are additional directions of motion and there are also relative deformations of different parts of the structure. Thus we say that a structure has many degrees of freedom.

In this chapter the theory will first be extended to the vibration of lumped mass systems with two degrees of freedom for which there are some important practical implications. The vibration of beams and plates will be introduced next, as examples of simple structures with distributed material properties, such as bridges and floor slabs. It will be shown that there are characteristic natural frequencies and modes of vibration. Then more complex motion under arbitrary dynamic loading will be considered, leading naturally into a discussion of the theory of modal analysis.

Structures with many members or with irregular geometry can seldom be idealized as simple elements for which standard solutions exist. Therefore, their geometry has to be approximated by systems of lumped masses and discrete springs. The mathematical analysis is therefore developed using matrix notation which lends itself to numerical solution using a digital computer. The natural progression is towards the finite element method by which a structure of arbitrary geometry may be subdivided into a large number of small elements of simple shape. The finite element method will be discussed in more detail in Chapter 4.

In civil engineering the behaviour of structures is often intimately related to the flexibility of the ground on which they are supported. This is particularly true in the case of vibrations, since the distributed stiffness and mass of a body of soil or rock of infinite extent results in wave motion of the medium. The theory of wave motion of elastic solids will be discussed briefly with some useful formulae for calculating ground vibrations due to impacts. Wave motion is also important in relation to the propagation of earthquakes. The chapter will conclude with a discussion on the phenomenon of ground–structure interaction which is important when evaluating the vibration of buildings and other structures on flexible foundations.

3.2 SYSTEMS WITH TWO DEGREES OF FREEDOM

Some important examples of systems with two degrees of freedom are tuned vibration absorbers, impact forge hammers, buildings on springs, and foundation blocks for industrial machinery. The first three of these are illustrated in Fig. 3.1. A tuned vibration absorber is illustrated shown attached to a bridge. The natural frequency of the auxiliary mass on its spring is chosen to be close to the natural frequency of the bridge. The result is that when dynamic loads are applied to the bridge it is the auxiliary mass that does most of the vibrating instead of the bridge. This device will be discussed in detail in § 3.2.3. Next, an impact forge hammer is illustrated in Fig. 3.1(b). The hammer strikes the workpiece on the anvil and the masses m_0 and m_2 effectively adhere together. The anvil is separated from the foundation block by a

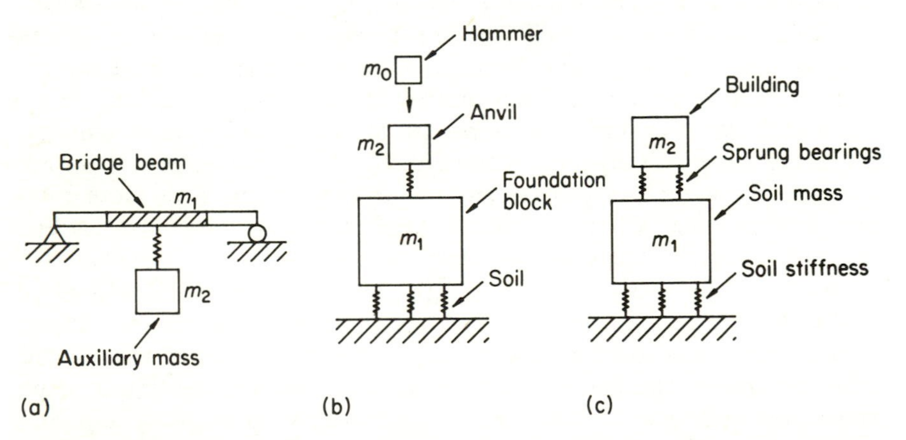

Fig. 3.1 Examples of systems with two degrees of freedom. (a) Vibration absorber; (b) forge hammer; (c) building on springs.

resilient layer, usually hardwood. The ground supporting the block can be treated as having an effective spring stiffness. Thirdly, there have been some examples of mounting large buildings on sprung bearings at foundation level. It has been found that the vibration of the building on its springs interacts with the effective mass and flexibility of the ground. The problem can be analysed by considering a notional two-degrees-of-freedom lumped mass system as shown in Fig. 3.1(c) (see Waller, 1969). The fourth example, that of a foundation block for industrial machinery, is not illustrated here because the motion is somewhat more subtle and will be discussed in detail in Chapter 8.

3.2.1 Free vibration with no damping

Consider the lumped mass system shown in Fig. 3.2(a). Each mass has a possible motion in a single direction, with the displacements being measured from the unloaded equilibrium position. The dynamic equilibrium of the masses is illustrated in Fig. 3.2(b). Note that the force in the spring k_2 depends on both displacements. There will therefore be an equation of motion for each mass, as follows:

$$m_1\ddot{u}_1 + k_1u_1 + k_2(u_1 - u_2) = 0 \tag{3.1a}$$

$$m_2\ddot{u}_2 + k_2(u_2 - u_1) = 0. \tag{3.1b}$$

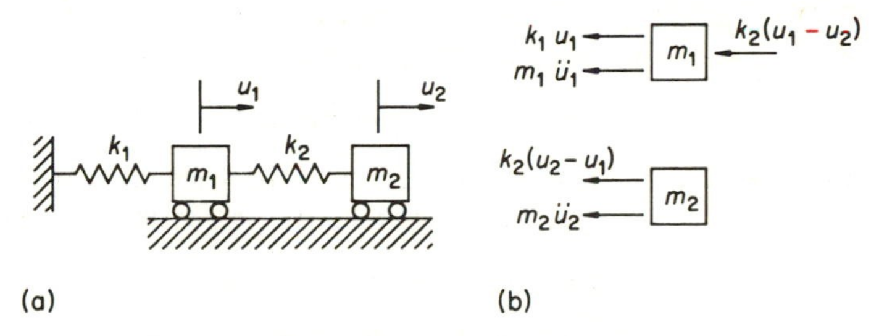

Fig. 3.2 Lumped mass system with two degrees of freedom.

In order to solve this problem it will be assumed that the masses vibrate in phase with each other. That is, they follow an identical variation with time but with their own characteristic amplitudes. Their motions may therefore be given by

$$u_1 = a_1 Y(t), \quad u_2 = a_2 Y(t), \tag{3.2}$$

where $Y(t)$ is the time variation, as yet unknown, and a_1 and a_2 are the amplitudes to be determined.

Substituting in the equations of motion (3.1) we obtain

$$m_1a_1\ddot{Y}(t) + (k_1 + k_2)a_1Y(t) - k_2a_2Y(t) = 0 \tag{3.3a}$$

$$m_2a_2\ddot{Y}(t) - k_2a_1Y(t) + k_2a_2Y(t) = 0 \tag{3.3b}$$

where $\ddot{Y}(t)$ denotes the forcing function differentiated twice with respect to time. These equations may be written alternatively thus:

$$\frac{(k_1+k_2)a_1-k_2a_2}{m_1a_1}=-\frac{1}{Y(t)}\ddot{Y}(t)=\text{constant}=\omega^2 \tag{3.4a}$$

$$\frac{-k_2a_1+k_2a_2}{m_2a_2}=-\frac{1}{Y(t)}\ddot{Y}(t)=\text{constant}=\omega^2. \tag{3.4b}$$

This is justified by the fact that the same time function appears in each equation. Therefore the left-hand sides must be equal and constant because of the lack of any time-dependent parameter. The time-varying part of the equations is thus represented by the equation

$$\ddot{Y}(t)+\omega^2 Y(t)=0. \tag{3.5}$$

By comparing (3.5) with (2.2) we can see that this is the same as the equation of free vibration of an undamped single-degree-of-freedom system. Hence, by analogy with (2.9), it may be seen that the solution is given by

$$Y(t)=A\cos\omega t+B\sin\omega t. \tag{3.6}$$

This time variation, therefore, satisfies the equations of motion (3.1) provided that the left-hand sides of equations (3.4) are equal. The equality depends on the value of ω^2 so that

$$(k_1+k_2)a_1-k_2a_2=\omega^2m_1a_1 \tag{3.7a}$$

$$-k_2a_1+k_2a_2=\omega^2m_2a_2. \tag{3.7b}$$

These equations relate the physical properties of the system, i.e. masses and stiffnesses, to the natural frequency. They may be written in the form

$$(k_1+k_2-\omega^2m_1)a_1-k_2a_2=0 \tag{3.8a}$$

$$-k_2a_1+(k_2-\omega^2m_2)a_2=0. \tag{3.8b}$$

It ought to be noted that these equations are in the standard form of an eigenvalue or characteristic value problem. A non-trivial solution will require a certain combination of values for a_1, a_2 and ω to satisfy the equations. The trivial solution is when a_1 and a_2 are zero, implying no motion. Finite values for a_1 and a_2 exist when the determinant of their coefficients is equal to zero. Therefore

$$\begin{vmatrix} k_1+k_2-\omega^2m & -k_2 \\ -k_2 & k_2-\omega^2m_2 \end{vmatrix}=0, \tag{3.9}$$

which can be expanded to

$$(k_1+k_2-\omega^2m_1)(k_2-\omega^2m_2)-k_2^2=0$$

or

$$\omega^4-\omega^2\left[\frac{k_2}{m_2}+\frac{(k_1+k_2)}{m_1}\right]+\frac{k_1k_2}{m_1m_2}=0. \tag{3.10}$$

This can be expressed more conveniently in terms of the mass ratio, $\mu=m_2/m_1$, and the limiting frequencies

$$\bar{\omega}_1^2=\frac{k_1}{(m_1+m_2)}, \quad \bar{\omega}_2^2=\frac{k_2}{m_2}. \tag{3.11}$$

The frequency equation then becomes

$$\omega^4-\omega^2[(\bar{\omega}_1^2+\bar{\omega}_2^2)(1+\mu)]+\bar{\omega}_1^2\bar{\omega}_2^2(1+\mu)=0. \tag{3.12}$$

This is a quadratic equation in ω^2 for which there are two roots,

$$\omega^2=\tfrac{1}{2}(\bar{\omega}_1^2+\bar{\omega}_2^2)(1+\mu)\pm\tfrac{1}{2}\sqrt{\{[(\bar{\omega}_1^2+\bar{\omega}_2^2)(1+\mu)]^2-4\bar{\omega}_1^2\bar{\omega}_2^2(1+\mu)\}}. \tag{3.13}$$

These two roots will be designated ω_1^2 and ω_2^2. They represent two possible natural frequencies of the system.

It now turns out that there are two values of ω^2 that will satisfy Equations (3.8). For each value of ω^2 substituted back into (3.8), there will be a corresponding ratio of a_1 to a_2. Note that absolute values of a_1 and a_2 can only be determined if the initial conditions are known. This was also true of a single-degree-of-freedom system. Consequently each natural frequency of the system has its own characteristic displaced shape, or *mode* of vibration. These mode shapes have a property of orthogonality, which will be discussed later in the chapter, and are referred to as *normal modes* of vibration. Since both modes are solutions to Equation (3.1), a general solution can be envisaged in which both modes are combined. This would certainly be the case under an arbitrary external forcing function.

EXAMPLE 3.1

Determine the natural frequencies and modes of vibration of the system shown in Fig. 3.2(a) given that both masses are equal and both stiffnesses are equal.

Solution

Given that $m_1=m_2=m$ and $k_1=k_2=k$, the limiting frequencies must be

$$\bar{\omega}_1^2=k/2m, \quad \bar{\omega}_2^2=k/m=2\bar{\omega}_1^2.$$

Hence the frequency equation (3.12) becomes

$$\omega^4-6\bar{\omega}_1^2\omega^2+4\bar{\omega}_1^2=0$$

from which

$$\omega^2 = 3\bar{\omega}_1^2 \pm \tfrac{1}{2}\sqrt{(36\bar{\omega}_1^4 - 16\bar{\omega}_1^4)}$$

yielding

$$\omega_1^2 = 0.764\,\bar{\omega}_1^2 \quad \text{and} \quad \omega_2^2 = 5.236\,\bar{\omega}_1^2.$$

The normal modes are obtained by substituting ω_1^2 and ω_2^2 back into either of the frequency equations (3.8). Thus it will be found that with $\omega_1^2 = 0.764\,\bar{\omega}_1^2$, the first mode is given by $a_1 = 0.62\ a_2$. By substituting $\omega_2^2 = 5.236\,\bar{\omega}_1^2$, the second mode is given by $a_1 = -1.62\,a_2$. The vibrations of the two masses in these modes are indicated in Fig. 3.3 where the displaced positions are shown dashed. In the first, or fundamental, mode both masses move in the same direction at a low frequency, whereas in the second mode they vibrate in opposite directions at a higher frequency.

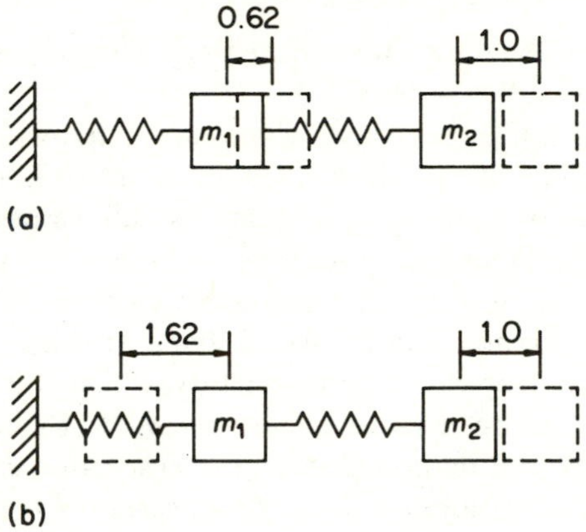

Fig. 3.3 Modes of vibration. (a) First or fundamental mode; (b) second mode.

3.2.2 Forced vibration with damping

We shall now consider the forced vibration of a system with two lumped masses as shown in Fig. 3.4. A damping element is placed between the masses and a periodic forcing function is applied to the first mass. This configuration has an important application in the design of tuned vibration absorbers.

The equations of motion of the system will be similar to (3.1) but with the addition of the applied force acting towards the right on mass m_1 and damping forces due to the relative velocities of the two masses. It is easy to see that in this case the equations of motion will be

$$m_1\ddot{u}_1 + c(\dot{u}_1 - \dot{u}_2) + k_1 u_1 + k_2(u_1 - u_2) = P_0 \sin \Omega t \qquad (3.14a)$$

$$m_2\ddot{u}_2 + c(\dot{u}_2 - \dot{u}_1) + k_2(u_2 - u_1) = 0. \qquad (3.14b)$$

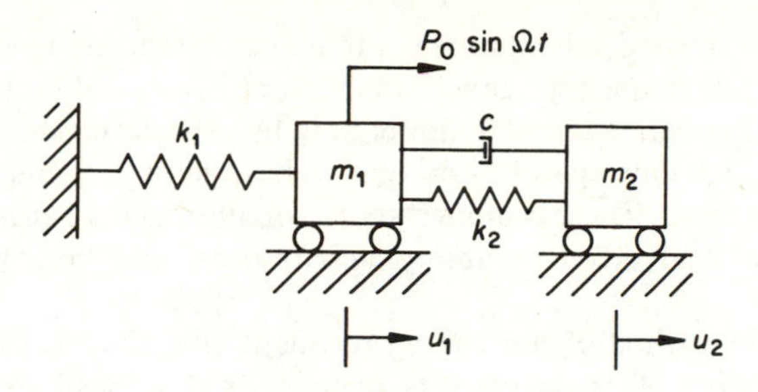

Fig. 3.4 Forced vibration of a two-degree-of-freedom system.

As in the case of the single degree of freedom (see § 2.5), it would be reasonable to assume that vibration would occur at the same frequency as the forcing function. Thus trial solutions for the displacements of the two masses are given by

$$u_1 = A_1 \sin \Omega t + A_2 \cos \Omega t \tag{3.15a}$$

$$u_2 = A_3 \sin \Omega t + A_4 \cos \Omega t. \tag{3.15b}$$

Substitution of these expressions into Equations (3.14) yields four equations from which the common factors $\sin \Omega t$ and $\cos \Omega t$ may be cancelled. This procedure is very similar to the one followed for forced vibration of a single-degree-of-freedom system in Chapter 2.

The solution of these equations for the constants A_1, A_2, A_3 and A_4 involves some tedious algebraic manipulations which have been omitted. What are finally required are the amplitudes of forced vibration of the two masses. These are given by

$$(u_1)_{\max} = \sqrt{(A_1^2 + A_2^2)}$$
$$= P_0 \left\{ \frac{c^2\Omega^2 + (k_2 - m_2\Omega^2)^2}{c^2\Omega^2[k_1 - m_1\Omega^2 - m_2\Omega^2]^2 + [k_2 m_2 \Omega^2 - (k_1 - m_1\Omega^2)(k_2 - m_2\Omega^2)]^2} \right\}^{1/2} \tag{3.16a}$$

and

$$(u_2)_{\max} = \sqrt{(A_3^2 + A_4^2)}$$
$$= P_0 \left\{ \frac{c^2\Omega^2 + k_2^2}{c^2\Omega^2[k_1 - m_1\Omega^2 - m_2\Omega^2]^2 + [k_2 m_2 \Omega^2 - (k_1 - m_1\Omega^2)(k_2 - m_2\Omega^2)]^2} \right\}^{1/2} \tag{3.16b}$$

3.2.3 The tuned vibration absorber

Occasionally it is difficult to avoid vibration of structures at resonant frequencies. In the case of wind loading of tall buildings, for example, the random nature of wind velocity means that there are likely to be significant loads over a wide frequency range. Stiffening of a very tall building to keep wind vibration within acceptable limits may be expensive. Another example is the tendency for long-span footbridges to have natural frequencies similar to walking frequency. There is limited scope for altering the natural frequency of long bridges and therefore annoying vibrations can be a problem (see Chapter 9).

An economic method of controlling resonant vibrations is to use a tuned vibration absorber. This device was principally developed by mechanical engineers but it has also been used with great success in civil engineering.

The principle of the vibration absorber must be attributed to Frahm (1911) who conceived of the frequency splitter. He found that the natural frequency of a structure, or mechanical body, could be split into a lower and a higher frequency by attaching a small sprung mass tuned to the same frequency as the structure. This had the effect of controlling vibration when periodic loading was applied at the natural frequency. Frahm applied the principle to the problem of stabilizing ships by the use of a system of interconnected tanks, filled with water, so that the resonant frequency of water in the tanks coincided with the rolling frequency of a ship.

Theoretical analysis of the vibration absorber was first achieved by Ormondroyd and Den Hartog (1928), who also showed that if the vibration absorber included a damping element, such as that shown in Fig. 3.4, good control of vibration could be obtained over a wide frequency range.

Following Den Hartog's method (Den Hartog, 1947), we shall focus our attention on the motion of the main system, i.e. Equation (3.16a). A non-dimensional version of this equation is useful. In this case the main dynamic system is represented by m_1 and k_1 while the vibration absorber is represented by the secondary system m_2, k_2 and c. The following notations will be introduced:

$(u_1)_{st} = P_0/k_1$ = static deflection of main system,
$\mu = m_2/m_1$ = ratio of mass of absorber to that of the main system,
$\rho_1 = \Omega/\sqrt{(k_1/m_1)}$ = ratio of forcing frequency to natural frequency of main system,
$\rho_2 = \Omega/\sqrt{(k_2/m_2)}$ = ratio of forcing frequency to natural frequency of secondary system,
$\xi = c/2\sqrt{(k_2 m_2)}$ = critical damping ratio of secondary system.

With these notations the non-dimensional displacement of mass m_1 becomes

$$\frac{(u_1)_{max}}{(u_1)_{st}} = \left\{ \frac{4\xi^2\rho_2^2 + (1-\rho_2^2)^2}{4\xi^2\rho_2^2[1-\rho_1^2-\mu\rho_1^2]^2 + [\mu\rho_1^2-(1-\rho_1^2)(1-\rho_2^2)]^2} \right\}^{1/2}. \quad (3.17)$$

In the design of a vibration absorber good results may be obtained when the natural frequency of the absorber is equal to the natural frequency of the main system, i.e. $\rho_2 = \rho_1 = \rho$. Also, it is usually necessary to keep the mass of the absorber small relative to the main mass. Taking for example $\mu = 0.05$, the performance of a vibration absorber may be illustrated by plotting a graph of Equation (3.17) for different values of ρ, as shown in Fig. 3.5. Curves are drawn for three different values of damping. If there is zero damping then resonance occurs at the two undamped resonant frequencies of the system. Using Equation (3.13) these may be found to occur at $\rho = 0.894$ and 1.118 respectively. The other extreme case occurs when there is infinite damping, which has the effect of locking the spring k_2. In this case the system has one degree of freedom with a stiffness of k_1 and a mass of $m_1 + m_2$. Resonance occurs when $\rho = 1/\sqrt{(1+\mu)} = 0.976$. Using an intermediate value of damping, somewhere between these extremes, it is possible to control the vibration of the main system over a wide frequency range, as can be seen in the figure.

It will be noted that all the curves intersect at S_1 and S_2 regardless of the value of damping. These points may be located by substituting the extreme cases of $\xi = 0$ and $\xi = \infty$ into (3.17) and equating. Hence

$$\rho^4 - 2\rho^2\left(\frac{1+\beta^2+\mu\beta^2}{2+\mu}\right) + \frac{2\beta^2}{2+\mu} = 0 \tag{3.18}$$

where $\beta = \rho_1/\rho_2$.

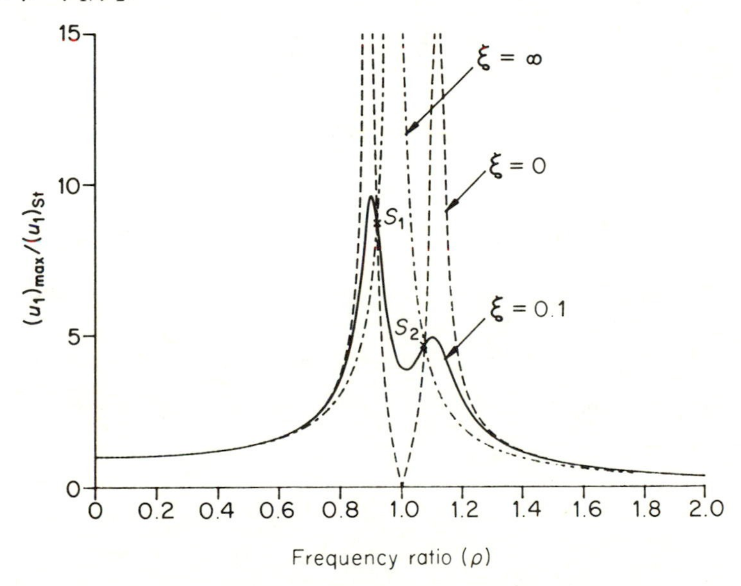

Fig. 3.5 Performance of a vibration absorber.

The two roots of Equation (3.18) represent the abscissae of S_1 and S_2. By substituting these values into (3.17) the corresponding amplitudes may be found. Hahnkamm (1933) suggested that the most efficient vibration absorber is achieved when the amplitudes at S_1 and S_2 are equal. He showed that this condition requires that

$$\frac{\text{Natural frequency of absorber}}{\text{Natural frequency of main system}} = \frac{\sqrt{(k_2/m_2)}}{\sqrt{(k_1/m_1)}} = \frac{1}{1+\mu}. \tag{3.19}$$

A device designed to this principle would be correctly termed a 'tuned' vibration absorber. Since the efficiency of the device depends on an accurate value of the ratio given by (3.19), it is important that its frequency is capable of adjustment after installation.

It was further noted by Brock (1946) that for optimum operation of a tuned vibration absorber the damping should be carefully chosen so that the peaks in the frequency response curve are as flat as possible. This may be achieved by making the curve horizontal at either S_1 or S_2 as shown in Fig. 3.6. By differentiating (3.17) and setting to zero at S_1 it may be shown that

$$\xi^2 = \frac{\mu\{3 - \sqrt{[\mu/(2+\mu)]}\}}{8(1+\mu)^3} \tag{3.20a}$$

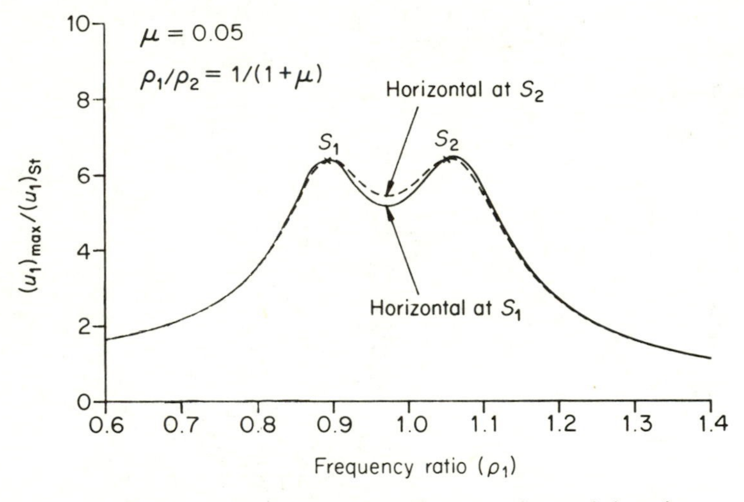

Fig. 3.6 Vibration absorber with optimum tuning and damping.

and by differentiating and setting to zero at S_2 it follows that

$$\xi^2 = \frac{\mu\{3 + \sqrt{[\mu/(2+\mu)]}\}}{8(1+\mu)^3}. \tag{3.20b}$$

A convenient average value for use in design would therefore be

$$\xi^2 = \frac{3\mu}{8(1+\mu)^3}. \tag{3.21}$$

Under these conditions the dynamic load factor becomes

$$\frac{(u_1)_{max}}{(u_1)_{st}} = \sqrt{(1+2/\mu)}. \tag{3.22}$$

It should be noted that the amplitude of the secondary system is always many times greater than that of the main system. This has two important implications in the design. First, allowance should be made to accommodate the quite large excursions of the secondary mass. Secondly, the stresses in the spring of the secondary system should be considered from the point of view of fatigue life. Coil springs have been found to be suitable for this purpose. The relative motion of the absorber mass to the mass of the main system was given by Den Hartog (1947) as

$$\left[\frac{u_{rel}}{(u_1)_{st}}\right]^2 = \frac{(u_1)_{max}}{(u_1)_{st}} \frac{1}{2\mu\rho_1\xi}. \tag{3.23}$$

This expression is difficult to derive, however.

Vibration absorbers have been installed in a number of footbridges in the United Kingdom. Footbridges are often very light, slender structures with minimal damping and there have been cases of excessive vibration under pedestrian loading. Jones, Pretlove and Eyre (1981) designed prototype vibration absorbers for two bridges, tuned in accordance with Equations (3.19) and (3.21). They adopted a mass ratio of about 1/100 in order to minimize the bulk of the devices. The performance of each bridge, with absorber installed, was most satisfactory. The peak amplitude, measured when a person walked over the bridge in step with its natural frequency, was reduced by a factor of up to four after the absorber had been installed.

The principle has also been used for tall buildings subjected to transient wind loading. One example is the Sydney Tower, Australia (Wargon, 1983), in which a massive water tank in the elevated turret was suspended by 10 m long cables giving it a pendulum frequency of 0.16 Hz, which was the calculated frequency of the fundamental mode of vibration of the tower. The required damping was provided by commercial Koni shock absorbers. A second vibration absorber was added to the structure to reduce the across-wind response in the second mode, due to vortex shedding (see Chapter 7). The absorbers in this case were not 'tuned' in the sense implied by Equations (3.19) and (3.21). However, it was reported by Kwok (1984) that the damping increase achieved by the installation of these devices was nearly double in the fundamental mode and nearly four times in the second mode.

EXAMPLE 3.2

A steel box girder footbridge may be assumed to be represented by a single-degree-of-freedom system with a mass of 17 500 kg and a stiffness of

3.0×10^6 N/m. The worst case of dynamic loading is considered to be when two people walk across the bridge in step in time with its natural frequency. This is almost equivalent to a sinusoidal loading with a constant amplitude of 0.48 kN. Design a suitable tuned vibration absorber and predict its performance.

Solution

Let the bridge be the main system with

$$m_1 = 17\,500 \quad \text{and} \quad k_1 = 3.0 \times 10^6 \text{ N/m.}$$

Therefore its natural frequency is

$$\omega_1 = \sqrt{\left(\frac{3 \times 10^6}{17\,500}\right)} = 13.09 \text{ rad/s} \quad (f_1 = 2.08 \text{ Hz}).$$

Choose a mass ratio (μ) of 1/100 giving

$$m_2 = 175 \text{ kg.}$$

This is a reasonable size of auxiliary mass to contain within the box girder.
 Therefore, using (3.19) optimum tuning requires that

$$\omega_2 = \omega_1/1.01 = 12.96 \text{ rad/s.}$$

Hence the stiffness of the absorber spring must be given by

$$k_2 = \omega_2^2 m_2 = 29\,393 \text{ N/m.}$$

The optimum damping in the vibration adsorber is obtained from (3.21), yielding

$$\xi^2 = \frac{3 \times 0.01}{8(1.01)^3}$$

or

$$\xi = 0.06, \text{ i.e. } 6\%.$$

The required damping coefficient is therefore given by

$$c = 2\xi\sqrt{(k_2 m_2)} = 272.2 \text{ N s/m.}$$

Under these conditions the dynamic load factor is given by (3.22), which yields

$$(u_1)_{\max}/(u_1)_{\text{st}} = \sqrt{(1 + 2/\mu)} = 14.2.$$

This is not, in fact, large compared with the DLF that would occur in a lightly damped structure subjected to periodic loading at resonance, as calculated by Equation (2.27). The static deflection under a load of 0.5 kN would be

$$(u_1)_{\text{st}} = (480/3 \times 10^6) \text{ m} = 0.16 \text{ mm,}$$

giving a maximum dynamic deflection of

$$(u_1)_{max} = 0.16 \times 14.2 = 2.27 \text{ mm}.$$

The maximum relative deflection of the absorber mass and the bridge may be calculated from (3.23), giving

$$\left[\frac{u_{rel}}{(u_1)_{st}}\right]^2 = \frac{14.2}{2 \times 0.01 \times 1 \times 0.06} = 11\,833.$$

Therefore

$$u_{rel} = 0.16 \times 108.8 = 17.4 \text{ mm}.$$

This is a motion that can easily be accommodated within the box girder using coil springs.

3.3 SYSTEMS WITH DISTRIBUTED MASS AND STIFFNESS

Most civil engineering structures are continuous and have distributed material properties. A bridge can be considered as a beam with uniform mass per unit length and uniform flexural stiffness. A slab or plate has two-dimensional flexural stiffness properties with mass distributed over its area. A curved shell is similar to a plate in its distribution of mass and stiffness but its geometry is three-dimensional, and this gives rise to membrane deformation in addition to flexure. Solid objects are fully three-dimensional with stiffness and mass being distributed throughout the volume of the material.

The theory will be developed first for beams and plates. Shells will not be considered here since they are much more complicated mathematically and therefore would require more extensive treatment, which is available elsewhere (Seide, 1975). Furthermore, shell structures can be analysed more conveniently by the finite element method which will be introduced in Chapter 4. Infinite or semi-infinite solids will be discussed later in this chapter since their behaviour is relevant to ground–structure interaction and to the propagation of earthquakes.

3.3.1 Equation of motion for a uniform beam

Consider a uniform beam with distributed mass m per unit length and flexural stiffness EI, as shown in Fig. 3.7(a). The moment–curvature relationship in the coordinate system shown is given by

$$M(x) = EI\, \mathrm{d}^2v/\mathrm{d}x^2, \qquad (3.24)$$

where v is the displacement in the y direction and $M(x)$ is the bending moment at x from the origin.

This equation can be written in terms of the applied distributed load by considering the differential element of the beam shown in Fig. 3.7(b), where the intensity of the distributed load is denoted by q and the shear force by V. Equilibrium of this element in the vertical direction requires that

$$q\mathrm{d}x - \mathrm{d}V = 0. \tag{3.25}$$

Equilibrium of moments about the midpoint of the element yields

$$V\mathrm{d}x - \mathrm{d}M = 0, \tag{3.26}$$

ignoring the negligible term $\mathrm{d}V\mathrm{d}x/2$.

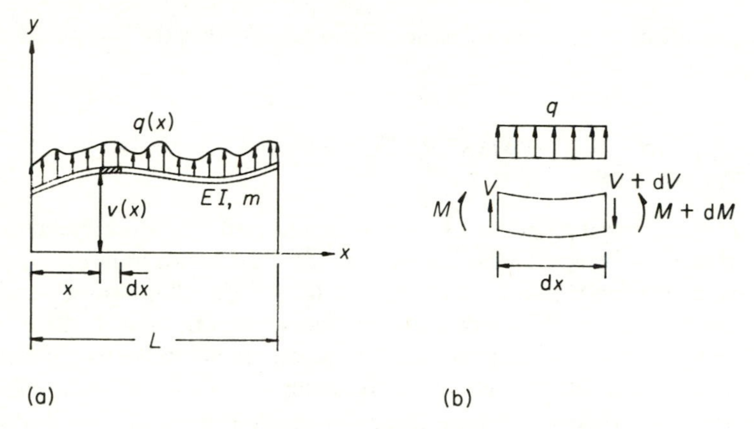

Fig. 3.7 Beam flexure with dynamic loading.

From (3.25) and (3.26) it therefore follows that

$$q = \mathrm{d}^2M/\mathrm{d}x^2. \tag{3.27}$$

Hence, differentiating (3.24) twice we find that

$$q(x) = EI\,\mathrm{d}^4v/\mathrm{d}x^4. \tag{3.28}$$

In a dynamic problem the distributed load consists of the externally applied dynamic force together with the inertia force which is proportional to the acceleration of the beam in the y direction. Both will vary with time and distance along the beam. Therefore, the former will be denoted by $p(x, t)$ and will be positive when acting in the y direction. The latter will be the product of acceleration and mass per unit length, and will be opposed to the positive direction of acceleration. Hence the forces are related by

$$q(x) = p(x, t) - m\,\mathrm{d}^2v/\mathrm{d}t^2. \tag{3.29}$$

This may be substituted into (3.28) to obtain the differential equation of motion of a beam subjected to dynamic loading. Noting that the existence of two variables, x and t, requires the introduction of partial derivatives, the equation may be written finally in the form

$$m\frac{\partial^2 v}{\partial t^2}+EI\frac{\partial^4 v}{\partial x^4}=p(x, t). \tag{3.30}$$

This equation should be compared carefully with the equation of motion of a lumped mass system with one degree of freedom as given by (2.1). Inertia, stiffness and loading terms occur in both equations. Damping has been neglected in the case of beam flexure. However, viscous damping may be included in the equation of motion by assuming that there will be a force counteracting motion proportional to velocity of the beam in the y direction. On this basis Equation (3.29) would be rewritten

$$q(x)=p(x, t)-c\,\mathrm{d}v/\mathrm{d}t-m\,\mathrm{d}^2v/\mathrm{d}t^2 \tag{3.31}$$

and the equation of motion would then become

$$m\frac{\partial^2 v}{\partial t^2}+c\frac{\partial v}{\partial t}+EI\frac{\partial^4 v}{\partial x^4}=p(x, t). \tag{3.32}$$

This form of the equation corresponds closely to the single-degree-of-freedom version given by (2.1). The above assumption for damping implies that notional viscous dashpots are distributed uniformly along the beam connecting it to the ground. While this assumption lacks physical realism it provides a mechanism for dissipating energy and, in the case of light damping, gives results that are sufficiently accurate for engineering purposes. Clough and Penzien (1975) suggested an alternative form in which viscous damping is proportional to rate of change of strain within the beam, thus modelling material damping more correctly.

3.3.2 Free vibration of beams

We have already seen that free vibration of an undamped two-degrees-of-freedom lumped mass system resolved into *modes* of vibration at certain *natural* frequencies. These are important when analysing the forced vibration of a structure, as we shall see later. Therefore, we shall begin by considering the free vibration of an undamped beam.

Referring to Equation (3.32) we can see that the equation of free vibration, in the absence of damping, reduces to

$$m\frac{\partial^2 v}{\partial t^2}+EI\frac{\partial^4 v}{\partial x^4}=0. \tag{3.33}$$

One method of solving this equation is by separation of variables. It would be reasonable to expect the displacement to vary with x and t as distinctly separate functions. We shall therefore assume that

$$v(x, t) = \phi(x)\, Y(t) \tag{3.34}$$

where ϕ is a function of distance along the beam defining its deflected shape when it vibrates, and $Y(t)$ defines the amplitude of vibration with time. Substituting (3.34) in (3.33) we obtain

$$m\phi(x)\frac{\partial^2 Y(t)}{\partial t^2} + EI\, Y(t)\frac{\partial^4 \phi(x)}{\partial x^4} = 0. \tag{3.35}$$

We may now rewrite (3.35) so that the x and t variables are collected together into separate terms as follows:

$$\frac{EI}{m}\frac{1}{\phi(x)}\frac{\partial^4 \phi(x)}{\partial x^4} = -\frac{1}{Y(t)}\frac{\partial^2 Y(t)}{\partial t^2} = \text{constant} = \omega^2. \tag{3.36}$$

It should be noted that, since the expression on the left-hand side is a function of x only and the expression on the right is a function of t only, it follows that both must be constant terms. Denoting the constant by ω^2 we may now write down two ordinary differential equations that need to be satisfied, these being

$$EI\frac{\mathrm{d}^4 \phi(x)}{\mathrm{d}x^4} = \omega^2 m\phi(x) \tag{3.37}$$

and

$$\ddot{Y}(t) + \omega^2\, Y(t) = 0, \tag{3.38}$$

where we return to the dot notation to indicate differentiation with respect to time.

Equation (3.38) should be compared with (2.2) and (3.5), which are the equations of free vibration of one- and two-degree-of-freedom systems respectively. It is therefore clear that the time-dependent function must be the same and given by

$$Y(t) = A\cos\omega t + B\sin\omega t. \tag{3.39}$$

In order to evaluate ω we now turn to Equation (3.37) which relates ω to the stiffness and mass of the beam. We require a function for $\phi(x)$ that retains its form after being differentiated four times. The exponential functions sin, cos, sinh and cosh all possess this property. Therefore, using the notation

$$\alpha^4 = \omega^2 m/EI \tag{3.40}$$

it can be verified that the general solution is given by

$$\phi(x) = C_1\sin\alpha x + C_2\cos\alpha x + C_3\sinh\alpha x + C_4\cosh\alpha x \tag{3.41}$$

where C_1, C_2 etc. are coefficients which depend on the boundary conditions of the beam.

The complete solution for a particular structure requires expressions for the displacement, slope, moment and shear at the supports which must be substituted into (3.41). This procedure will yield three of the coefficients in terms of the fourth and will also yield a frequency equation from which ω may be evaluated. The final coefficient, expressing amplitude of vibration, would require a knowledge of the initial conditions of motion.

EXAMPLE 3.3 Simply supported beam

The boundary conditions of a simply supported beam consist of zero deflection and zero bending moment at each end. These may be written in the form

$$v(0, t)=0 \quad \text{and} \quad EI\frac{d^2v}{dx^2}(0, t)=0, \tag{3.42a}$$

$$v(L, t)=0 \quad \text{and} \quad EI\frac{d^2v}{dx^2}(L, t)=0, \tag{3.42b}$$

where L is the length of the beam.

Substituting the conditions (3.42a) into (3.41) we obtain the two equations

$$0=0+C_2+0+C_4 \tag{3.43a}$$

$$0=0-\alpha^2C_2+0+\alpha^2C_4 \tag{3.43b}$$

from which $C_2=C_4=0$.

Substituting (3.42b) into (3.41) we obtain

$$0=C_1 \sin \alpha L+C_3 \sinh \alpha L \tag{3.44a}$$

$$0=-\alpha^2C_1 \sin \alpha L+\alpha^2C_3 \sinh \alpha L. \tag{3.44b}$$

Cancelling α^2 from the second equation and adding to the first we find that

$$0=2C_3 \sinh \alpha L. \tag{3.45}$$

But since $\sinh \alpha L$ cannot be zero it follows that $C_3=0$. Hence we are left with the relation

$$0=C_1 \sin \alpha L. \tag{3.46}$$

A non-trivial solution ($C_1 \neq 0$) only exists if

$$\sin \alpha L=0. \tag{3.47}$$

This is the frequency equation which will be satisfied only when $\alpha L=n\pi$.

Thus the natural frequencies are obtained from (3.40), giving

$$\omega_n=\frac{n^2\pi^2}{L^2}\sqrt{\left(\frac{EI}{m}\right)} \tag{3.48}$$

and the corresponding modes of vibration are therefore

$$\phi_n(x) = C_1 \sin(n\pi x/L). \tag{3.49}$$

C_1 is arbitrary and is usually taken equal to 1.

There is an infinite number of mode shapes and natural frequencies, each corresponding to a different value of n.

The frequencies and mode shapes of the first three modes of vibration are shown in Fig. 3.8(a). It is found experimentally that these are the frequencies and deformed shapes that the beam naturally adopts at resonance. In the case of the simply supported beam it can be seen that there are $n+1$ nodes, or points of zero deflection, in each mode of vibration. Furthermore, the frequencies increase with greater contortion of the structure.

EXAMPLE 3.4 Cantilever beam

The vibration of a cantilever is governed by the following boundary conditions:

$$v(0, t) = 0 \quad \text{and} \quad \frac{dv}{dx}(0, t) = 0 \tag{3.50a}$$

$$EI\frac{d^2v}{dx^2}(L, t) = 0 \quad \text{and} \quad EI\frac{d^3v}{dx^3}(L, t) = 0. \tag{3.50b}$$

These represent zero displacement and slope at the fixed end ($x = 0$), together with zero bending moment and shear at the free end ($x = L$). Substitution of the conditions (3.50a) in the general solution (3.41) gives

$$0 = 0 + C_2 + 0 + C_4 \tag{3.51a}$$

$$0 = \alpha C_1 + 0 + \alpha C_3 + 0, \tag{3.51b}$$

whereas substitution of (3.50b) in (3.41) gives

$$0 = -\alpha^2 C_1 \sin\alpha L - \alpha^2 C_2 \cos\alpha L + \alpha^2 C_3 \sinh\alpha L + \alpha^2 C_4 \cosh\alpha L \tag{3.52a}$$

$$0 = -\alpha^3 C_1 \cos\alpha L + \alpha^3 C_2 \sin\alpha L + \alpha^3 C_3 \cosh\alpha L + \alpha^3 C_4 \sinh\alpha L. \tag{3.52b}$$

Equations (3.51) state that $C_4 = -C_2$ and $C_3 = -C_1$. Substituting these equalities in (3.52) we find that

$$C_1(\sin\alpha L + \sinh\alpha L) + C_2(\cos\alpha L + \cosh\alpha L) = 0 \tag{3.53a}$$

$$C_1(\cos\alpha L + \cosh\alpha L) + C_2(-\sin\alpha L + \sinh\alpha L) = 0. \tag{3.53b}$$

Solving for C_1 it is found that

$$C_1[(\cos\alpha L + \cosh\alpha L)^2 - (\sin\alpha L + \sinh\alpha L)(-\sin\alpha L + \sinh\alpha L)] = 0. \tag{3.54}$$

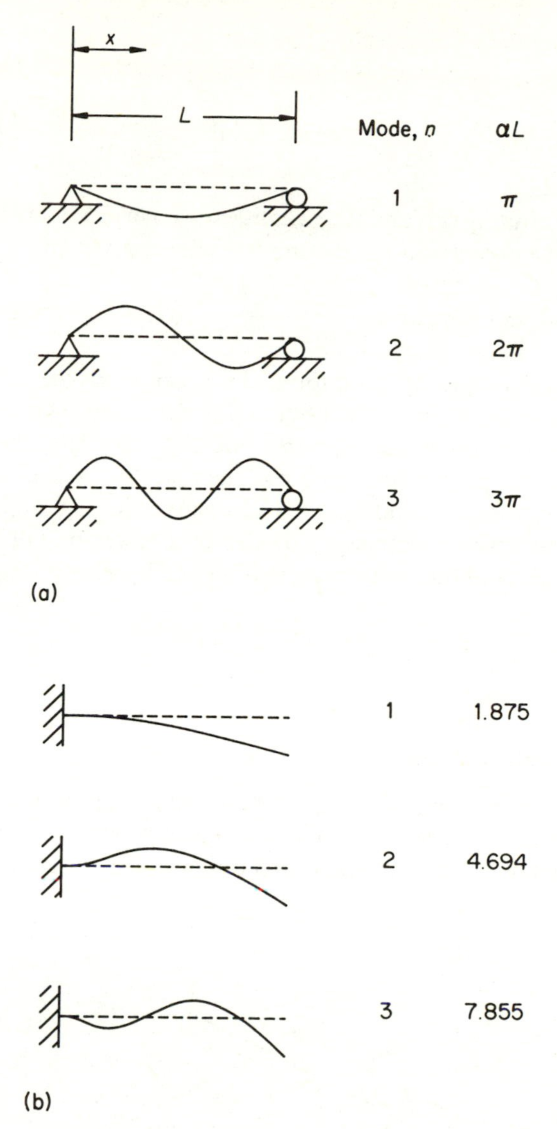

Fig. 3.8 Modes of vibration of simple beams. (a) Simply supported beam; (b) cantilever.

Exactly the same expression is obtained if we solve for C_2 except that C_2 replaces C_1. In order that neither C_1 nor C_2 are zero, the expression within the brackets must be zero. This reduces to

$$\cos \alpha L \cosh \alpha L + 1 = 0, \tag{3.55}$$

which is the frequency equation.

The mode shape is obtained by expressing C_2 in terms of C_1 by using (3.53a) which yields

$$C_2 = -\frac{\sin \alpha L + \sinh \alpha L}{\cos \alpha L + \cosh \alpha L} C_1. \tag{3.56}$$

Hence, by substituting this and the equalities given in (3.51) into (3.41) it can be verified that the expression for the mode shape is given by

$$\phi(x) = C_1 \left[\sin \alpha x - \sinh \alpha x + \frac{\sin \alpha L + \sinh \alpha L}{\cos \alpha L + \cosh \alpha L} (\cosh \alpha x - \cos \alpha x) \right]. \tag{3.57}$$

Returning to the frequency equation (3.55) it can easily be seen that in this case the solution for αL must be obtained numerically. The simplest procedure is to plot a graph of the left-hand side and hence find the values of αL where the function passes through zero. The first three values of αL together with the corresponding mode shapes are shown in Fig. 3.8(b). This latter example is descriptive of the sway modes of a tower or tall building. The natural frequencies of the cantilever are obtained by substituting values of αL into (3.40), giving

$$\omega_n = \frac{(\alpha L)^2}{L^2} \sqrt{\left(\frac{EI}{m}\right)}. \tag{3.58}$$

3.3.3 Free vibration of plates

The equation of bending of a thin plate is considerably more complicated than that of a uniform beam. This may be explained with reference to Fig. 3.9. Stresses caused by bending in the x direction give rise to strains in both the x

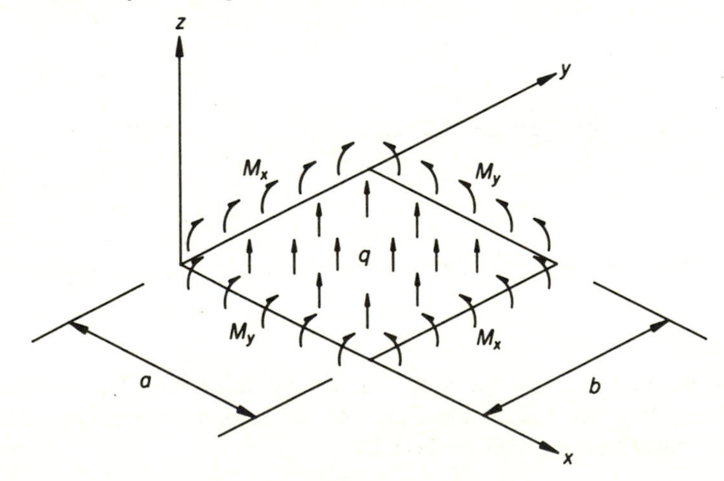

Fig. 3.9 Bending of a thin plate.

and y directions because of Poisson's ratio v which is the ratio of transverse strain to strain in the direction of stress. The same applies to bending in the y direction, and therefore the stresses and strains along orthogonal axes in a plate interact with each other.

The reader is referred to specialist texts for a full derivation of the theory of flexure of thin plates (Timoshenko and Woinowsky-Krieger, 1959). Using the coordinate axes shown in Fig. 3.9 the equilibrium equation is given by

$$D\left(\frac{\partial^4 w}{\partial x^4}+2\frac{\partial^4 w}{\partial x^2 \partial y^2}+\frac{\partial^4 w}{\partial y^4}\right)=q(x, y), \tag{3.59}$$

where w is the displacement in the z direction and $q(x, y)$ is the intensity of loading normal to the surface of the plate. The flexural stiffness is given by

$$D=Et^3/12(1-v^2) \tag{3.60}$$

where t is the thickness of the plate.

In the case of free vibration, and ignoring damping, the only loading will be due to inertia forces which will be opposed to acceleration of the plate in the z direction. Therefore

$$q(x, y, t)=-m\,\partial^2 w/\partial t^2 \tag{3.61}$$

where m is the mass per unit area of the plate.

Thus the equation of free vibration becomes

$$m\frac{\partial^2 w}{\partial t^2}+D\left(\frac{\partial^4 w}{\partial x^4}+2\frac{\partial^4 w}{\partial x^2\,\partial y^2}+\frac{\partial^4 w}{\partial y^4}\right)=0. \tag{3.62}$$

This is analogous to the equation of free vibration of a beam given by (3.33).

The solution is achieved, once more, by the method of separation of variables. Assume that the transverse displacement may be expressed as a product of separate functions of space and time thus:

$$w(x, y, t)=\phi(x, y)\,Y(t). \tag{3.63}$$

Then, following the same procedure outlined in §3.3.2, this is substituted into (3.62) from which the following two equations may be obtained:

$$\ddot{Y}(t)+\omega^2\,Y(t)=0 \tag{3.64}$$

$$\frac{\partial^4 \phi(x, y)}{\partial x^4}+2\frac{\partial^4 \phi(x, y)}{\partial x^2 \partial y^2}+\frac{\partial^4 \phi(x, y)}{\partial y^4}=\alpha^4 \phi(x, y) \tag{3.65}$$

where

$$\alpha^4=m\omega^2/D. \tag{3.66}$$

The solution again consists of an undamped periodic vibration, from (3.64), whose frequency is determined by solution of (3.65) together with the boundary conditions of the plate.

There is no convenient general solution to (3.65) for all plates. Explicit formulae are possible in the cases of circular and rectangular plates (Timoshenko, 1955).

In the case of the rectangular plate shown in Fig. 3.9, the displacement function

$$\phi(x, y) = \sin(r\pi x/a) \sin(s\pi y/b), \tag{3.67}$$

where r and s are integers, satisfies the boundary conditions for simple supports at all four edges. By substituting (3.67) into (3.65) it is found that

$$\alpha^2 = \pi^2(r^2/a^2 + s^2/b^2) \tag{3.68}$$

and, using (3.66), the natural frequency is given by

$$\omega = \pi^2\left(\frac{r^2}{a^2} + \frac{s^2}{b^2}\right)\sqrt{\left(\frac{D}{m}\right)}. \tag{3.69}$$

The lowest natural frequency will occur when $r = s = 1$. The corresponding mode shape will be given by (3.67). This will be a half sine curve in both x and y directions and therefore the deflected shape will resemble a mound at one extreme of oscillation and a dish at the other. The next two natural frequencies will occur when either r or s is increased to 2, the other remaining unchanged. This will have the effect of making (3.67) a half sine curve in one direction and a full sine curve in the other, thus dividing the plate in two with a line of zero displacement at midspan. The plate will move in opposite directions on each side of the nodal line. Other combinations of the integers r and s will produce more complicated patterns. Some examples, using finite element graphical output, are shown in Fig. 3.10.

3.3.4 Free vibration of other continuous structures

The preceding examples have indicated that the mathematical difficulty of deriving frequency equations and mode shapes increases rapidly as boundary conditions become more general. However, this should not deter the practical engineer because standard solutions exist for a wide range of useful structural configurations.

Blevins (1979) has compiled a comprehensive reference of formulae for natural frequencies and mode shapes. This reference includes information on the following systems: spring and pendulum systems; cables and cable trusses; beams – straight and curved; frames; membranes; plates; shells; fluid systems, e.g. floating objects; and vibrations of immersed structures.

Where it is not possible to find a standard configuration similar enough to the problem under investigation, recourse to the finite element method is required (see Chapter 4).

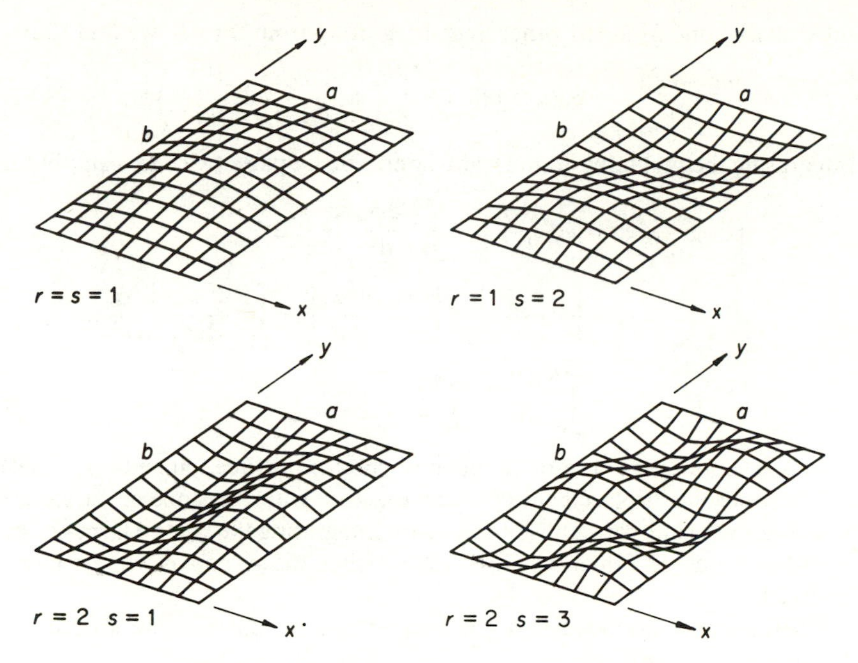

Fig. 3.10 Mode shapes for a rectangular plate.

3.3.5 The orthogonality of modes of vibration

All natural modes of vibration have the property of orthogonality. That is, if they are interpreted as vectors, any two are found to be normal to one another. For this reason they are often referred to as *normal modes* of vibration. This turns out to be a very useful property in the analysis of dynamic response of structures to various forcing functions. We shall now demonstrate the existence of this property in the case of beams in flexure.

Consider the equation of undamped free vibration of a uniform beam as given by (3.33):

$$m\frac{\partial^2 v}{\partial t^2}+EI\frac{\partial^4 v}{\partial x^4}=0. \tag{3.70}$$

If we substitute trial solutions for v, in terms of two different normal modes $\phi_n(x)$ and $\phi_m(x)$, into the above equation we obtain two equations of the form (3.37) as follows:

$$\frac{\mathrm{d}^4\phi_n(x)}{\mathrm{d}x^4}=\frac{\omega_n^2 m}{EI}\phi_n(x) \tag{3.71a}$$

$$\frac{\mathrm{d}^4\phi_m(x)}{\mathrm{d}x^4}=\frac{\omega_m^2 m}{EI}\phi_m(x). \tag{3.71b}$$

Multiplying the first of these equations by $\phi_m(x)$ and the second by $\phi_n(x)$, substracting one from the other and integrating from 0 to L, we find that

$$\frac{(\omega_n^2-\omega_m^2)m}{EI}\int_0^L \phi_m(x)\phi_n(x)\mathrm{d}x=\int_0^L\left(\phi_m\frac{\mathrm{d}^4\phi_n}{\mathrm{d}x^4}-\phi_n\frac{\mathrm{d}^4\phi_m}{\mathrm{d}x^4}\right)\mathrm{d}x. \tag{3.72}$$

Integrating the first term on the right-hand side by parts twice we can obtain

$$\begin{aligned}\int_0^L \phi_m\frac{\mathrm{d}^4\phi_n}{\mathrm{d}x^4}\mathrm{d}x&=\left|\phi_m\frac{\mathrm{d}^3\phi_n}{\mathrm{d}x^3}\right|_0^L-\int_0^L\frac{\mathrm{d}\phi_m}{\mathrm{d}x}\cdot\frac{\mathrm{d}^3\phi_n}{\mathrm{d}x^3}\mathrm{d}x\\&=\left|\phi_m\frac{\mathrm{d}^3\phi_n}{\mathrm{d}x^3}\right|_0^L-\left|\frac{\mathrm{d}\phi_m}{\mathrm{d}x}\cdot\frac{\mathrm{d}^2\phi_n}{\mathrm{d}x^2}\right|_0^L+\int_0^L\frac{\mathrm{d}^2\phi_m}{\mathrm{d}x^2}\cdot\frac{\mathrm{d}^2\phi_n}{\mathrm{d}x^2}\mathrm{d}x\\&=\int_0^L\frac{\mathrm{d}^2\phi_m}{\mathrm{d}x^2}\cdot\frac{\mathrm{d}^2\phi_n}{\mathrm{d}x^2}\mathrm{d}x.\end{aligned} \tag{3.73}$$

This is true for every case of simple end conditions (free, pinned and fixed) where either the displacement or shear together with the slope or curvature will be zero. A similar result is obtained on integrating the second term on the right-hand side of Equation (3.73) except that the m and n will be interchanged.

Consequently, provided that the frequencies are different we obtain the orthogonality condition

$$\int_0^L \phi_m(x)\phi_n(x)\mathrm{d}x=0;\quad \omega_n\neq\omega_m. \tag{3.74}$$

Further orthogonality conditions can be demonstrated in the case of a beam with non-uniform mass or stiffness. These are

$$\int_0^L \phi_m(x)\phi_n(x)m(x)\mathrm{d}x=0;\quad \omega_n\neq\omega_m \tag{3.75}$$

$$\int_0^L \frac{\mathrm{d}^2\phi_m}{\mathrm{d}x^2}\cdot\frac{\mathrm{d}^2\phi_n}{\mathrm{d}x^2}EI(x)\mathrm{d}x=0;\quad \omega_n\neq\omega_m. \tag{3.76}$$

Orthogonality conditions are not available for structures with two modes having coincident frequencies. This situation is possible but unusual in structural engineering. For a deeper treatment of orthogonality the reader is referred to Warburton (1976) or Clough and Penzien (1975).

3.3.6 Forced vibration of beams

We shall now return to the equation of forced vibration derived earlier and given by Equation (3.30) as follows:

$$m\frac{\partial^2 v}{\partial t^2}+EI\frac{\partial^4 v}{\partial x^4}=p(x, t). \tag{3.30}$$

The forcing function $p(x, t)$ tends to excite several natural frequencies and modes of the structure simultaneously. In fact, every mode is a possible solution to the differential equation and therefore the dynamic deflection under arbitrary loading will be a linear combination of all possible modes. Thus

$$v(x, t) = \sum_{n=1}^{\infty} \phi_n(x)\, Y_n(t) \tag{3.77}$$

where $Y_n(t)$ are time-varying modal amplitudes.

Substituting (3.77) into (3.30) we obtain

$$\sum_{n=1}^{\infty} \phi_n(x) \frac{\partial^2 Y_n(t)}{\partial t^2} + \frac{EI}{m} \sum_{n=1}^{\infty} \frac{\partial^4 \phi_n(x)}{\partial x^4} \cdot Y_n(t) = \frac{p(x, t)}{m}. \tag{3.78}$$

However, by separating the variables, we found previously that

$$\frac{EI}{m\phi_n(x)} \cdot \frac{\partial^4 \phi_n(x)}{\partial x^4} = \omega_n^2. \tag{3.79}$$

Thus (3.78) reduces to

$$\sum_{n=1}^{\infty} \phi_n(x) \frac{\mathrm{d}^2 Y_n(t)}{\mathrm{d}t^2} + \sum_{n=1}^{\infty} \phi_n(x) \omega_n^2 Y_n(t) = \frac{p(x, t)}{m}. \tag{3.80}$$

Now this equation can be decomposed by premultiplying by $\phi_m(x)$ and integrating from 0 to L, yielding

$$\sum_{n=1}^{\infty} \int_0^L \phi_m(x)\, \phi_n(x)\, \mathrm{d}x\, \ddot{Y}_n(t) + \sum_{n=1}^{\infty} \int_0^L \phi_m(x) \phi_n(x) \mathrm{d}x\, \omega_n^2\, Y_n(t) = (1/m) \int_0^L \phi_m(x) p(x,t) \mathrm{d}x. \tag{3.81}$$

Recalling the orthogonality condition

$$\int_0^L \phi_m(x)\, \phi_n(x) \mathrm{d}x = 0; \quad m \neq n,$$

it is clear that every term in the summations above must be zero except for the case where $m = n$. Therefore, Equation (3.81) is reduced to the simple form given by

$$\ddot{Y}_n(t) + \omega_n^2\, Y_n(t) = \frac{Q_n(t)}{M_n}; \quad n = 1, 2, \ldots, \infty \tag{3.82}$$

where

$$Q_n(t) = \int_0^L \phi_n(x)\, p(x, t) \mathrm{d}x \tag{3.83}$$

and

$$M_n = m \int_0^L \phi_n^2(x)\,dx. \tag{3.84}$$

The form of Equation (3.82) should be inspected carefully. First, since an arbitrary mode $\phi(x)$ is being considered, it is clear that there will be one such equation for every mode of vibration. In theory there is an infinite number of modes for distributed parameter structures, though usually only the first few are of any importance. Secondly, the equation is similar to the equation of forced vibration of a single-degree-of-freedom system as given by (2.1), but neglecting damping. The displacement coordinate $u(t)$ has been replaced by mode shape amplitudes $Y_n(t)$, the point force has been replaced by *generalized* forces $Q_n(t)$, and the mass has been replaced by *generalized* masses M_n.

It is evident that the equation of forced motion of a beam, Equation (3.30), has been reduced to a set of single-degree-of-freedom equations (3.82). These equations are uncoupled and can be solved independently for any kind of forcing function using the methods of Chapter 2. Finally, the total dynamic response of the structure is obtained by summation of the modal responses using Equation (3.77).

The above procedure is known as *modal analysis* and is a most powerful method of dynamic analysis of structures. It relies upon the property of orthogonality of vibration modes which enables the equation of motion of the beam to be uncoupled into a set of independent single-degree-of-freedom equations. Each of these equations is in terms of a modal amplitude function $Y_n(t)$. What has been done, in effect, is to transform the equation of motion from spatial coordinates, $v(x,t)$, to a new set of coordinates $Y_n(t)$. These are often referred to as *generalized* coordinates and each is associated with a *generalized* force $Q_n(t)$ which excites that particular mode, together with a corresponding *generalized* mass M_n. Modal analysis relies on the principle of superposition and therefore is only strictly applicable to linear elastic vibrations.

EXAMPLE 3.5 Beam subjected to impulsive load

An important problem in the design of long span lightweight floors is the tendency for annoying vibrations to occur under the effect of foot impacts. This will be discussed in Chapter 9. Meanwhile, it will be assumed that a foot impact can be represented by an impulsive load of large magnitude $P(t)$ and very brief duration dt such that the product $P(t)dt$ is an impulse of a finite amount. We shall calculate the vibrations of a simply supported beam of 8 m span when subjected to an impulse of 70 N s at midspan. The mass of the beam is 750 kg/m and its EI is 30×10^6 N m^2.

Solution

From (3.48) the natural frequencies are given by

$$\omega_n = \frac{n^2 \pi^2}{L^2} \sqrt{\left(\frac{EI}{m}\right)} = n^2 \omega_1 = n^2 \times 30.84 \text{ rad/s.}$$

The mode shapes are given by (3.49) with $C_1 = 1$ and are

$$\phi_n(x) = \sin(n\pi x/L).$$

The generalized mass may be calculated using (3.84), which yields

$$M_n = m \int_0^L \sin^2(n\pi x/L)\mathrm{d}x = mL/2 = 3000 \text{ kg.}$$

Note that the generalized mass is the same in every mode of vibration and is exactly one half the mass of the beam. Different end conditions would yield different values of generalized mass.

The generalized force may be evaluated from (3.83), which yields

$$Q_n(t) = \int_0^L \phi_n(x)\, p(x,t)\mathrm{d}x = \sin(n\pi/2)P(t)$$

$$= \begin{cases} P(t); & n = 1, 5, 9, \ldots \\ 0 \quad ; & n \text{ even} \\ -P(t); & n = 3, 7, 11, \ldots. \end{cases}$$

Note that the load at midspan will not excite any mode which has a node at its point of application.

We may now consider the problem as independent single-degree-of-freedom lumped mass systems and obtain their solutions one by one.

In §2.6, an arbitrary forcing function was treated as consisting of a series of brief impulses of different magnitude. The initial velocity imparted by an impulse was given by (2.30) as

$$\dot{u}_0 = I/m$$

where $I = P(t)\,\mathrm{d}t$. The ensuing vibrations may be calculated from (2.10), giving

$$u(t) = (I/m\omega)\sin\omega t.$$

Therefore, using the generalized parameters calculated above, the modal amplitudes in (3.82) may be calculated separately and will be

$$Y_n(t) = (I_n/M_n \times n^2\omega_1)\sin n^2\omega_1 t.$$

Combining the modes using (3.77), the midspan deflection is given by

$$v(L/2,t)=\sum_{n=1}^{\infty}\frac{I_n\sin n\pi/2}{M_n n^2\omega_1}\sin n^2\omega_1 t$$

$$=\frac{70}{3000\times 30.84}\left(\sin\omega_1 t+\frac{1}{9}\sin 9\omega_1 t+\frac{1}{25}\sin 25\omega_1 t+\ \dots\right).$$

The first three terms of the deflection response are shown in Fig. 3.11. It is evident that higher modes make less contribution to the dynamic response. Furthermore, damping would normally be present, with the result that the higher-mode contributions would decay to negligible proportions after their first few cycles.

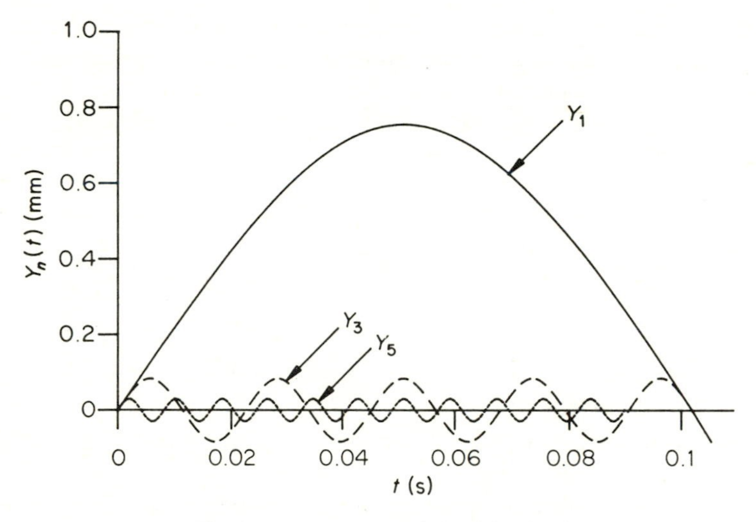

Fig. 3.11 Response to impulsive load.

3.3.7 Equivalent single-degree-of-freedom systems

In Chapter 2 single-degree-of-freedom systems were always considered to be made up of lumped masses and discrete springs. However, the principles just established provide us with a means of representing systems with distributed mass and stiffness as equivalent lumped mass systems. All we need to do is to assume a mode shape for the vibration under consideration and then determine the lumped mass and discrete spring that have the same energy terms.

The method will be illustrated by considering the system shown in Fig. 3.12. This consists of a floor beam AB of span L, distributed mass m, and stiffness EI. It supports a rigid beam CD, which is part of a balcony structure,

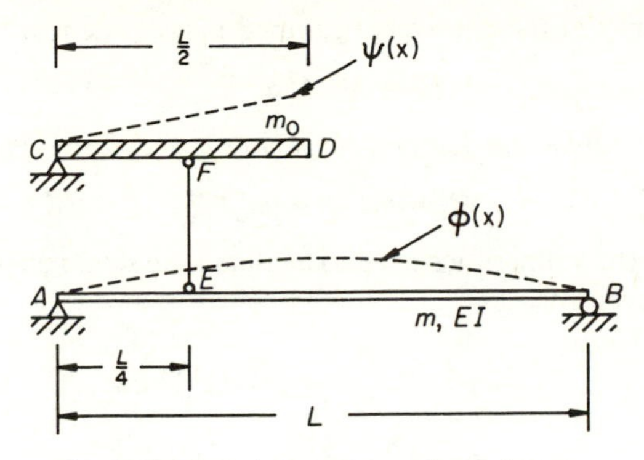

Fig. 3.12 Floor beam supporting a balcony.

with distributed mass m_0. The supporting column EF is assumed to be weightless.

First, it is assumed that the system vibrates in the characteristic mode indicated, being $\phi(x)$ for the floor and $\psi(x)$ for the balcony. The displacement of the floor is then given by

$$v(x,t)=\phi(x)\ Y(t) \tag{3.85}$$

where $Y(t)$ is the generalized coordinate and is, in effect, the amplitude of motion at a convenient reference point, which in this case will be taken as midspan.

The kinetic energy of the floor will be given by

$$T_{\mathrm{f}}=\tfrac{1}{2}\int_0^L m[\dot{v}(x,t)]^2\,\mathrm{d}x. \tag{3.86}$$

Assuming that the mode shape is given by $\phi(x)=\sin(\pi x/L)$ and the midspan motion is given by $Y(t)=A\sin\omega t$, then the kinetic energy at the instant of maximum velocity will be

$$T_{\mathrm{f}}=\frac{mA^2\omega^2}{2}\int_0^L \sin^2\frac{\pi x}{L}\,\mathrm{d}x=\frac{A^2\omega^2}{2}\frac{mL}{2}. \tag{3.87}$$

The amplitude of vibration at E will be $A/\sqrt{2}$ and therefore the motion of the balcony will be given by

$$v(x,t)=\frac{A}{\sqrt{2}}\frac{x}{L/4}\sin\omega t. \tag{3.88}$$

Hence, the kinetic energy of the balcony will be

$$T_{\mathrm{b}}=\frac{m_0A^2\omega^2}{2}\frac{8}{L^2}\int_0^{L/2} x^2\,\mathrm{d}x=\frac{A^2\omega^2}{2}\frac{m_0L}{3}. \tag{3.89}$$

The kinetic energy of the equivalent lumped mass system will be

$$T_e = M_e A^2 \omega^2/2 \tag{3.90}$$

where M_e is the equivalent lumped mass and is therefore given by

$$M_e = (mL/2) + (m_0 L/3). \tag{3.91}$$

Turning now to the stiffness, we need to evaluate the strain energy in the floor beam. This is given by

$$U_f = \tfrac{1}{2}\int_0^L EI[\mathrm{d}^2 v(x,t)/\mathrm{d}x^2]^2 \,\mathrm{d}x. \tag{3.92}$$

At the point of maximum displacement this will be

$$U_f = \frac{EI\,A^2}{2}\frac{\pi^4}{L^4}\int_0^L \sin^2\frac{\pi x}{L}\,\mathrm{d}x = \frac{EI\,A^2}{2}\frac{\pi^4}{2L^3}. \tag{3.93}$$

The strain energy in a spring of stiffness k_e and deflection A is $k_e A^2/2$, and therefore the equivalent stiffness will be

$$k_e = \pi^4 EI/2L^3. \tag{3.94}$$

3.4 LUMPED PARAMETER MULTI-DEGREE-OF-FREEDOM SYSTEMS

So far we have only considered the analysis of simple forms of structure, these being one- and two-degree-of-freedom lumped parameter systems, and some simple types of distributed parameter systems. However, many practical structures do not conform conveniently to simple descriptions because of their irregular geometry, many members or multiple boundary conditions. We have already seen that the analysis of even quite simple structural forms can require a considerable amount of mathematics. What is required, therefore, is a general procedure for modelling a structure as a system of lumped masses and discrete springs such that its dynamic behaviour can be evaluated approximately.

The finite element method is the most versatile and well known method that meets this requirement. An irregular structure is subdivided into a large number of small regular elements, the force–displacement relationships for which are relatively simple. The interactions between elements are analysed by a systematic and completely general mathematical procedure, employing matrix algebra, which is implemented on a digital computer. The finite element method is presented in more detail in Chapter 4. Meanwhile, the vibration behaviour of lumped parameter systems will now be discussed in general terms.

Consider the system of masses, springs and dashpots shown in Fig. 3.13. By considering the dynamic equilibrium of each mass in turn, the following equations of motion may be written down:

$$\left.\begin{aligned}
&m_1\ddot{u}_1+c_1\dot{u}_1+k_1u_1+c_2(\dot{u}_1-\dot{u}_2)+k_2(u_1-u_2)=P_1(t)\\
&m_2\ddot{u}_2+c_2(\dot{u}_2-\dot{u}_1)+k_2(u_2-u_1)+c_3(\dot{u}_2-\dot{u}_3)+k_3(u_2-u_3)=P_2(t)\\
&\dots\\
&m_N\ddot{u}_N+c_N(\dot{u}_N-\dot{u}_{N-1})+k_N(u_N-u_{N-1})=P_N(t).
\end{aligned}\right\} \tag{3.95}$$

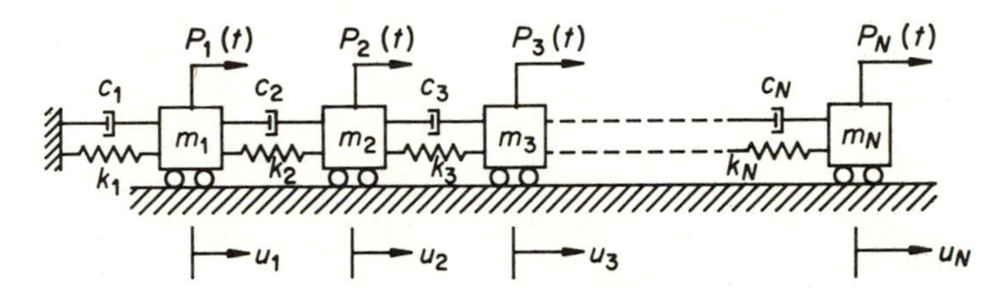

Fig. 3.13 Multi-degree-of-freedom system.

These equations may be arranged in the matrix form

$$\boldsymbol{M}\ddot{\boldsymbol{u}}+\boldsymbol{C}\dot{\boldsymbol{u}}+\boldsymbol{K}\boldsymbol{u}=\boldsymbol{P} \tag{3.96}$$

where

$$\boldsymbol{M}=\begin{bmatrix} m_1 & & & & \\ & m_2 & & & \\ & & \cdot & & \\ & & & \cdot & \\ & & & & \cdot \\ & & & & & m_N \end{bmatrix} \tag{3.97a}$$

$$\boldsymbol{C}=\begin{bmatrix} c_1+c_2 & -c_2 & & & & \\ -c_2 & c_2+c_3 & -c_3 & & & \\ & & \cdot & & & \\ & & & \cdot & & \\ & & & & \cdot & \\ & & & & -c_N & c_N \end{bmatrix} \tag{3.97b}$$

$$K = \begin{bmatrix} k_1+k_2 & -k_2 & & & \\ -k_2 & k_2+k_3 & -k_3 & & \\ & \cdot & & & \\ & & \cdot & & \\ & & & \cdot & \\ & & & -k_N & k_N \end{bmatrix} \tag{3.97c}$$

$$u = \begin{Bmatrix} u_1 \\ u_2 \\ \vdots \\ u_N \end{Bmatrix} \tag{3.97d}$$

$$P = \begin{Bmatrix} P_1(t) \\ P_2(t) \\ \vdots \\ P_N(t) \end{Bmatrix}. \tag{3.97e}$$

It should be noted that $\boldsymbol{M}$ is a diagonal matrix, with all off-diagonal terms equal to zero, whereas $\boldsymbol{C}$ and $\boldsymbol{K}$ are in banded form. Different arrangements of springs and dashpots in Fig. 3.13 are possible, with corresponding differences in the forms of $\boldsymbol{C}$ and $\boldsymbol{K}$. In the finite element method, structures with distributed mass can be represented more consistently, leading to a banded form of $\boldsymbol{M}$ also. $\boldsymbol{u}$ is the displacement vector and $\boldsymbol{P}$ is the load vector. Equation (3.96) is, in fact, the multi-degree-of-freedom equivalent of Equation (2.1). The mass and stiffness matrices can, in general, be readily evaluated, and the load vector $\boldsymbol{P}$ is presumably known. Direct evaluation of the damping matrix for a practical structure is not usually possible and assumptions have to be made about the distribution of damping, given that limited experimental data on overall damped response may be available. This subject will be discussed in detail in Chapter 4.

3.4.1 Undamped free vibration

The equation of motion of free vibration, without damping, is

$$\boldsymbol{M}\ddot{\boldsymbol{u}} + \boldsymbol{K}\boldsymbol{u} = 0. \tag{3.98}$$

Using the same procedure adopted for one-degree-of-freedom and distributed parameter systems, we begin by assuming a periodic motion in which all the masses move in phase with each other so that

$$\boldsymbol{u} = \boldsymbol{a}\sin(\omega t + \phi) \tag{3.99}$$

where $\boldsymbol{a}$ is a vector of amplitudes a_1, a_2, etc. Substitution of (3.99) into (3.98) yields

$$[\boldsymbol{K}-\omega^2\boldsymbol{M}]\,\boldsymbol{a}=0. \tag{3.100}$$

This equation is analogous to Equation (3.8) obtained for a two-degree-of-freedom system. The same comments apply, in that it is in the standard form of an eigenvalue problem and that for $\boldsymbol{a}$ to be non zero, the determinant of $\boldsymbol{K}-\omega^2\boldsymbol{M}$ must be zero:

$$|\boldsymbol{K}-\omega^2\boldsymbol{M}|=0. \tag{3.101}$$

However, for all but the smallest problems, solution by expansion of the determinant in (3.101) is impractical. Numerical methods are required for solving (3.100) and several efficient methods are available. Bearing in mind that it is our aim to analyse the vibrations of large structures, we shall consider matrix iteration methods of the kind used in large-scale finite element computer programs.

Basically, the problem is one of finding suitable combinations of the amplitudes a_1, a_2, ... and ω^2 which satisfy Equation (3.100). It turns out that many such combinations are possible, but that for each suitable arrangement of amplitudes there is a corresponding value of ω^2. If the system possesses N degrees of freedom (N masses as shown in Fig. 3.13) there will be N such solutions. A suitable arrangement of amplitudes will be denoted by the vector $\boldsymbol{a}_n$ and its corresponding frequency by ω_n. These are mode shapes and natural frequencies analogous to those found for distributed parameter systems. Since the amplitudes of motion are arbitrary in the case of free vibration, it is common practice to normalize the terms in $\boldsymbol{a}_n$ so that the largest is equal to 1.0 and the rest are factored accordingly. Normalized mode shapes will be denoted by $\boldsymbol{\phi}_n$ so that

$$\boldsymbol{\phi}_n=\begin{Bmatrix}\phi_{1n}\\ \phi_{2n}\\ \vdots\\ \phi_{Nn}\end{Bmatrix}=\frac{1}{a_{\max}}\begin{Bmatrix}a_{1n}\\ a_{2n}\\ \vdots\\ a_{Nn}\end{Bmatrix} \tag{3.102}$$

in which $a_{\max}$ is the largest amplitude in $\boldsymbol{a}_n$.

3.4.2 Orthogonality conditions

First, it is advantageous to demonstrate the existence of orthogonality conditions for normal modes of vibration. The equations of free vibration may be written in the form

$$\boldsymbol{K}\boldsymbol{\phi}_n=\omega_n^2\boldsymbol{M}\boldsymbol{\phi}_n=\boldsymbol{q}_n \tag{3.103}$$

where $\boldsymbol{\phi}_n$ is the nth mode shape. The right-hand side can be interpreted as the inertia forces, $\boldsymbol{q}_n$, arising from harmonic motion in this characteristic mode shape. The left-hand side is the usual stiffness equilibrium condition required when these forces are applied as loads to the structure. Therefore $\boldsymbol{\phi}_n$ are the ensuing deflections. Similarly, in another mode of vibration denoted by m, the inertia forces would be $\boldsymbol{q}_m$ and the corresponding deflections $\boldsymbol{\phi}_m$. According to Betti's law, if a structure is subjected to two alternate load systems the work done by the first set of loads, when moved through the displacements caused by the second set of loads, is identical to the work done by the second set of loads, when moved through the displacements caused by the first set. In equation form this is written

$$\boldsymbol{\phi}_m^{\mathrm{T}}\boldsymbol{q}_n = \boldsymbol{\phi}_n^{\mathrm{T}}\boldsymbol{q}_m. \tag{3.104}$$

Note that the shape vectors must be transposed in order to carry out the appropriate matrix product. Substituting the expressions for inertia force given in (3.103) we obtain

$$\omega_n^2\,\boldsymbol{\phi}_m^{\mathrm{T}}\boldsymbol{M}\boldsymbol{\phi}_n = \omega_m^2\boldsymbol{\phi}_n^{\mathrm{T}}\boldsymbol{M}\boldsymbol{\phi}_m. \tag{3.105}$$

The terms on each side of the equation represent work, which is a scalar quantity. Therefore, transposition of the matrix terms is permissible and the equation can read

$$(\omega_n^2 - \omega_m^2)\,\boldsymbol{\phi}_m^{\mathrm{T}}\,\boldsymbol{M}\boldsymbol{\phi}_n = 0. \tag{3.106}$$

Hence, if any two natural frequencies are not equal we have the orthogonality condition

$$\boldsymbol{\phi}_m^{\mathrm{T}}\boldsymbol{M}\boldsymbol{\phi}_n = 0; \quad \omega_m \neq \omega_n. \tag{3.107}$$

A second orthogonality condition may be obtained by pre-multiplying (3.103) by $\boldsymbol{\phi}_m^{\mathrm{T}}$. Therefore

$$\boldsymbol{\phi}_m^{\mathrm{T}}\boldsymbol{K}\boldsymbol{\phi}_n = \omega_n^2\boldsymbol{\phi}_m^{\mathrm{T}}\boldsymbol{M}\boldsymbol{\phi}_n. \tag{3.108}$$

But because of (3.107) it follows that

$$\boldsymbol{\phi}_m^{\mathrm{T}}\boldsymbol{K}\boldsymbol{\phi}_n = 0; \quad \omega_m \neq \omega_n. \tag{3.109}$$

The two orthogonality conditions given by (3.107) and (3.109) have important implications when we consider eigenvalue solution methods and dynamic response.

3.4.3 Eigenvalue solution by inverse iteration

An effective method of obtaining the natural frequencies of rotating shafts, using hand calculation, was developed by Stodola (1927). It is based upon the idea of assuming a mode shape, evaluating the inertia forces corresponding to the assumed shape, applying these as loads to the structure and finally calculating the deflected shape from the equilibrium equation (3.103). The calculated deflected shape is then a better approximation to the true mode. The method of inverse iteration is based upon this concept and consists of iterative vector operations on the basic eigenvalue equation

$$\boldsymbol{K}\boldsymbol{\phi}=\omega^2\boldsymbol{M}\boldsymbol{\phi}. \tag{3.110}$$

We begin with a trial vector $\boldsymbol{x}_1$ which is a first approximation to $\omega^2\boldsymbol{\phi}$. Hence, we may evaluate the right-hand side which can be interpreted as an inertia force given by

$$\boldsymbol{q}=\boldsymbol{M}\boldsymbol{x}_1. \tag{3.111}$$

An improved displacement vector may then be calculated from the equilibrium equation

$$\boldsymbol{K}\bar{\boldsymbol{x}}_2=\boldsymbol{q}. \tag{3.112}$$

Now if ω^2 were known it would be possible to scale $\bar{\boldsymbol{x}}_2$ and obtain the next trial vector $\boldsymbol{x}_2$ from

$$\boldsymbol{x}_2=\omega^2\,\bar{\boldsymbol{x}}_2. \tag{3.113}$$

A very simple procedure is to assume that the maximum amplitude of $\boldsymbol{x}_2$ is the same as the maximum amplitude of $\boldsymbol{x}_1$ and therefore ω^2 may be calculated from

$$\omega^2\simeq\frac{(x_1)_{\max}}{(\bar{x}_2)_{\max}}. \tag{3.114}$$

$\boldsymbol{x}_2$ is used as the next trial vector in (3.111) and it is assumed that successive iterations converge towards $\omega^2\boldsymbol{\phi}$.

The procedure for obtaining ω^2 using (3.114) is rather crude and really only suitable for small problems. A much better approximation may be obtained by using Rayleigh's quotient as follows:

$$\omega^2=\frac{\bar{\boldsymbol{x}}_2^{\mathrm{T}}\boldsymbol{K}\bar{\boldsymbol{x}}_2}{\bar{\boldsymbol{x}}_2^{\mathrm{T}}\boldsymbol{M}\bar{\boldsymbol{x}}_2}. \tag{3.115}$$

Rayleigh's quotient is obtained by equating the strain energy at maximum displacement to the kinetic energy when the structure passes through the null position. The quotient possesses a stationary property such that its value changes little with quite large variations in $\bar{\boldsymbol{x}}_2$. It is known to be a good method of estimating the frequency from an approximate mode shape.

Clough and Penzien (1975) have shown that inverse iteration converges onto the lowest natural frequency. It is important also to obtain higher frequencies and mode shapes. In order to understand how this is done it should first be realized that any trial vector is a linear combination of the true mode shapes:

$$\boldsymbol{x}=\alpha_1\boldsymbol{\phi}_1+\alpha_2\boldsymbol{\phi}_2+\alpha_3\boldsymbol{\phi}_3+\ \ldots \tag{3.116}$$

Premultiplying by $\boldsymbol{\phi}_1^{\mathrm{T}} \boldsymbol{M}$ we obtain

$$\boldsymbol{\phi}_1^{\mathrm{T}}\boldsymbol{M}\boldsymbol{x}=\alpha_1\boldsymbol{\phi}_1^{\mathrm{T}}\boldsymbol{M}\boldsymbol{\phi}_1+\alpha_2\boldsymbol{\phi}_1^{\mathrm{T}}\boldsymbol{M}\boldsymbol{\phi}_2+\alpha_3\boldsymbol{\phi}_1^{\mathrm{T}}\boldsymbol{M}\boldsymbol{\phi}_3+\ \ldots \tag{3.117}$$

in which all terms on the right-hand side, except the first, are zero on account of the orthogonality condition (3.107). Hence, the coefficient α_1 may be determined:

$$\alpha_1=\frac{\boldsymbol{\phi}_1^{\mathrm{T}}\boldsymbol{M}\boldsymbol{x}}{\boldsymbol{\phi}_1^{\mathrm{T}}\boldsymbol{M}\boldsymbol{\phi}_1}. \tag{3.118}$$

This coefficient can now be used to 'sweep' the first mode component out of the trial vector, yielding a purified vector $\boldsymbol{x}^*$ as follows:

$$\boldsymbol{x}^*=\boldsymbol{x}-\alpha_1\boldsymbol{\phi}_1=\boldsymbol{x}-\frac{\boldsymbol{\phi}_1^{\mathrm{T}}\boldsymbol{M}\boldsymbol{x}}{\boldsymbol{\phi}_1^{\mathrm{T}}\boldsymbol{M}\boldsymbol{\phi}_1}. \tag{3.119}$$

Since errors will tend to accumulate it is wise to repeat the 'sweeping' process at every cycle of iteration. The iterative procedure contained in Equations (3.111) to (3.115) will now converge onto the second mode of vibration. It is clear that the same procedure can be applied to the extraction of higher modes, always sweeping lower-mode components out of the trial vector at every cycle.

EXAMPLE 3.6

Calculate all three natural frequencies and modes of vibration of the building frame shown in Fig. 3.14(a). The flexural stiffness (EI) of the columns is 4.5×10^6 N m^2, and the mass of the structure is concentrated at the three floor levels as shown.

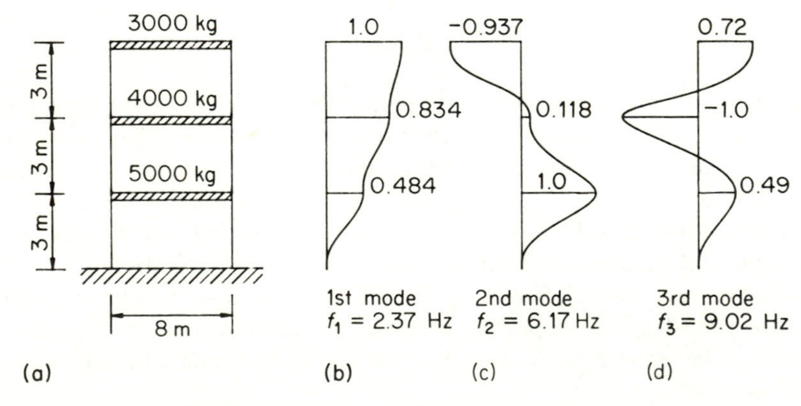

Fig. 3.14 Natural frequencies of a multi-storey frame.

Solution

The structure can be considered to behave like a simple lumped mass system as shown in Fig. 3.13. Only three masses exist, these being

$$m_1 = 5000 \text{ kg}, \quad m_2 = 4000 \text{ kg}, \quad m_3 = 3000 \text{ kg}.$$

The spring stiffnesses are equal to the horizontal shear stiffness in the columns of each storey. These will be

$$k_1 = k_2 = k_3 = k = 4.0 \times 10^6 \text{ N/m}.$$

Note that exactly the same shear stiffness was calculated in Example 2.1.

Therefore, neglecting damping, the equations of free vibration are

$$m_1\ddot{u}_1 + ku_1 + k(u_1 - u_2) = 0$$

$$m_2\ddot{u}_2 + k(u_2 - u_1) + k(u_2 - u_3) = 0$$

$$m_3\ddot{u}_3 + k(u_3 - u_2) = 0$$

Using matrix notation and introducing the numerical data obtained above, we may write

$$10^3 \times \begin{bmatrix} 5 & 0 & 0 \\ 0 & 4 & 0 \\ 0 & 0 & 3 \end{bmatrix} \begin{Bmatrix} \ddot{u}_1 \\ \ddot{u}_2 \\ \ddot{u}_3 \end{Bmatrix} + 10^6 \times \begin{bmatrix} 8 & -4 & 0 \\ -4 & 8 & -4 \\ 0 & -4 & 4 \end{bmatrix} \begin{Bmatrix} u_1 \\ u_2 \\ u_3 \end{Bmatrix} = 0$$

or

$$\boldsymbol{M}\ddot{\boldsymbol{u}} + \boldsymbol{K}\boldsymbol{u} = 0.$$

A good choice of initial trial vector is the unit full vector given by

$$\boldsymbol{x}_1^{\mathrm{T}} = \{1 \ 1 \ 1\}.$$

Evaluating (3.111) we obtain

$$\boldsymbol{q} = \boldsymbol{M}\boldsymbol{x}_1 = 10^3 \times \begin{Bmatrix} 5 \\ 4 \\ 3 \end{Bmatrix}.$$

In order to solve (3.112) we need to calculate the inverse of $\boldsymbol{K}$. Methods of doing this will be discussed in the next chapter. Meanwhile, it can be verified that

$$\boldsymbol{K}^{-1} = 1/(4 \times 10^6) \begin{bmatrix} 1 & 1 & 1 \\ 1 & 2 & 2 \\ 1 & 2 & 3 \end{bmatrix}.$$

Hence from (3.112)

$$\bar{\boldsymbol{x}}_2 = \boldsymbol{K}^{-1}\boldsymbol{q} = 1/(4\times 10^3)\begin{Bmatrix} 12 \\ 19 \\ 22 \end{Bmatrix}.$$

The frequency is obtained approximately using (3.114) giving

$$\omega^2 \simeq 1/(22/4\times 10^3) = 181.8.$$

Hence the next trial vector may be obtained from (3.113), which gives

$$\boldsymbol{x}_2 = \begin{Bmatrix} 0.54 \\ 0.86 \\ 1.00 \end{Bmatrix}.$$

The procedure is repeated until convergence, the results of five iterations being given in Table 3.1.

It is interesting to evaluate Rayleigh's quotient at the end of the first iteration. The numerator of (3.115) is found to be

$$\bar{\boldsymbol{x}}_2^{\mathrm{T}}\boldsymbol{K}\bar{\boldsymbol{x}}_2 = 1/(4\times 10^3)\{12\quad 19\quad 22\}\times 4\times 10^6$$

$$\times\begin{bmatrix} 2 & -1 & 0 \\ -1 & 2 & -1 \\ 0 & -1 & 1 \end{bmatrix}\times 1/(4\times 10^3)\begin{Bmatrix} 12 \\ 19 \\ 22 \end{Bmatrix} = 50.5,$$

while the denominator is

$$\bar{\boldsymbol{x}}_2^{\mathrm{T}}\boldsymbol{M}\bar{\boldsymbol{x}}_2 = 1/(4\times 10^3)\{12\quad 19\quad 22\}\times 10^3$$

$$\times\begin{bmatrix} 5 & & \\ & 4 & \\ & & 3 \end{bmatrix}\times 1/(4\times 10^3)\begin{Bmatrix} 12 \\ 19 \\ 22 \end{Bmatrix} = 226/10^3,$$

yielding

$$\omega^2 = 223.4.$$

It is clear that this is a closer approximation to the true value than was calculated by the simple method.

Second mode

The second mode is obtained using the same procedure but with the application of Equation (3.119) at every cycle to 'sweep' out the first mode

Table 3.1 Extraction of first mode by inverse iteration

Iteration number	Trial vector x_1	x_2	x_3	Frequency ω^2
1	1.0	1.0	1.0	181.8
2	0.54	0.86	1.0	215.3
3	0.492	0.838	1.0	220.2
4	0.485	0.835	1.0	220.9
5	0.484	0.834	1.0	221.1

component from the trial vector. Note that the term $\phi_1^T M \phi_1$ appears as a constant in Equation (3.119). It is the generalized mass, M_1, of the structure vibrating in the fundamental mode and will be evaluated first. Therefore

$$M_1 = \{0.484 \;\; 0.834 \;\; 1.0\} \begin{bmatrix} 5000 & & \\ & 4000 & \\ & & 3000 \end{bmatrix} \begin{Bmatrix} 0.484 \\ 0.834 \\ 1.0 \end{Bmatrix} = 6954 \text{ kg}.$$

Let us begin again with the unit full vector for the first trial so that

$$x_1^T = \{1 \;\; 1 \;\; 1\}.$$

Evaluating α_1 from (3.118) we obtain

$$\alpha_1 = \{0.484 \quad 0.834 \quad 1.0\} \begin{bmatrix} 5000 & & \\ & 4000 & \\ & & 3000 \end{bmatrix} \begin{Bmatrix} 1 \\ 1 \\ 1 \end{Bmatrix} = 1.259.$$

Hence the purified trial vector, given by (3.119), will be

$$x_1^* = \begin{Bmatrix} 1 \\ 1 \\ 1 \end{Bmatrix} - 1.259 \begin{Bmatrix} 0.484 \\ 0.834 \\ 1.0 \end{Bmatrix} = \begin{Bmatrix} 0.391 \\ -0.05 \\ -0.259 \end{Bmatrix} \Rightarrow \begin{Bmatrix} 1.0 \\ -0.128 \\ -0.662 \end{Bmatrix}.$$

The inertia forces are given by (3.111):

$$q = 10^3 \times \begin{Bmatrix} 5.0 \\ -0.512 \\ -1.986 \end{Bmatrix}$$

and therefore the improved displacement vector is obtained from the equilibrium equation (3.112), giving

$$\bar{x}_2 = K^{-1} q = (1/4000) \times \begin{Bmatrix} 2.502 \\ 0.004 \\ -1.982 \end{Bmatrix}.$$

Using (3.114) to evaluate the natural frequency we obtain

$$\omega^2 = 1599,$$

which is substituted into (3.113) to determine the next trial vector,

$$x_2 = \begin{Bmatrix} 1.0 \\ 0.001 \\ -0.792 \end{Bmatrix}.$$

Using (3.118) it may now be verified that $\alpha_1 = 0.0068$. Thus it is clear that our new trial vector is very much closer to the true second-mode shape than the unit full starting vector. Nevertheless, to avoid the accumulation of errors, the trial vector should be purified of any first-mode component at each iteration. The procedure is then repeated until convergence. The results of eight iterations are given in Table 3.2 where it should be noted that the figures in parentheses are the components of the purified trial vector at every iteration.

Table 3.2 Extraction of second mode

Iteration number	*Trial vector* x (x^*) x_1	x_2	x_3	*1st mode coeff.* α_1	*Frequency* ω^2
1	1.0 (1.0)	1.0 (−0.128)	1.0 (−0.662)	1.259	1599
2	1.0 (1.0)	0.001 (−0.005)	−0.792 (−0.801)	0.0068	1552
3	1.0 (1.0)	0.060 (0.059)	−0.872 (−0.873)	0.0006	1528
4	1.0 (1.0)	0.089 (0.091)	−0.911 (−0.908)	−0.0006	1515
5	1.0 (1.0)	0.016 (0.106)	−0.926 (−0.925)	−0.0016	1510
6	1.0 (1.0)	0.112 (0.113)	−0.935 (−0.932)	0.0008	1506
7	1.0 (1.0)	0.117 (0.117)	−0.935 (−0.936)	0.0022	1504
8	1.0 (1.0)	0.120 (0.118)	−0.935 (−0.937)	0	1505

It is clear from Table 3.2 that convergence is slower in the second mode. It is generally true that there is a loss of accuracy in evaluating higher modes. The third mode may be extracted by an identical procedure to the above except that both first and second modes should be swept out at every iteration using

$$x^* = x - \alpha_1 \phi_1 - \alpha_2 \phi_2.$$

The final results are shown in Fig. 3.14(b)–(d), where it can be seen that the natural frequencies are greater for the more distorted mode shapes.

3.4.4 Dynamic response by mode superposition

It is usually necessary to determine the dynamic response of a structure to applied forcing functions. This amounts to finding the solution to the equation of motion of the equivalent multi-degree-of-freedom system as given by (3.96),

$$\boldsymbol{M}\ddot{\boldsymbol{u}} + \boldsymbol{C}\dot{\boldsymbol{u}} + \boldsymbol{K}\boldsymbol{u} = \boldsymbol{P}. \tag{3.96}$$

Finally, the solution for the displacements $\boldsymbol{u}$ can be used to calculate internal forces and stresses at any instant of time.

There are two principle ways of solving (3.96). The first is by direct numerical integration using methods similar to those described in Chapter 2 but generalized to many degrees of freedom. This would be the only practicable way for structures with nonlinear properties. Direct numerical integration is discussed further in Chapter 4. The second method is by mode superposition, which is generally thought to be more efficient for linear structures. We shall now consider the latter method in detail.

Mode superposition is based upon the fact that the deflected shape of the structure may be expressed as a linear combination of all the modes:

$$\boldsymbol{u} = \boldsymbol{\phi}_1 Y_1 + \boldsymbol{\phi}_2 Y_2 + \boldsymbol{\phi}_3 Y_3 + \ldots + \boldsymbol{\phi}_N Y_N = \sum_{n=1}^{N} \boldsymbol{\phi}_n Y_n. \tag{3.120}$$

This is analogous to Equation (3.77) which was used to analyse the response of beams. The coefficients Y_n are the modal amplitudes, which vary with time. In the case of a system of lumped masses, the motion of the ith mass is given by

$$u_i = \phi_{i1} Y_1 + \phi_{i2} Y_2 + \ldots + \phi_{in} Y_n + \ldots + \phi_{iN} Y_N, \tag{3.121}$$

where ϕ_{in} is the displacement of the ith mass in the nth mode of vibration. Equation (3.120) may be written in the more compact matrix notation

$$\boldsymbol{u} = \boldsymbol{\Phi} \boldsymbol{Y} \tag{3.122}$$

where $\boldsymbol{\Phi}$ is the *modal matrix* whose columns are the mode shapes, so that

$$\boldsymbol{\Phi} = \{\boldsymbol{\phi}_1\ \boldsymbol{\phi}_2\ \boldsymbol{\phi}_3\ \ldots\ \boldsymbol{\phi}_N\} \tag{3.123}$$

and $\boldsymbol{Y}$ is a vector of the modal amplitudes

$$\boldsymbol{Y}^{\mathrm{T}} = \{Y_1\ Y_2\ Y_3\ \ldots\ Y_n\ \ldots\ Y_N] \tag{3.124}$$

The modal amplitudes, Y_n, are often referred to as *generalized* coordinates which may be contrasted with the natural coordinates $\boldsymbol{u}$. Modal analysis is a process of decomposing (3.96), using generalized coordinates, so as to obtain a set of differential equations that are uncoupled, each of which may be analysed as a single degree of freedom. This may be achieved by substituting (3.120) into (3.96), which gives

$$\boldsymbol{M}\sum_{n=1}^{N}\boldsymbol{\phi}_n\ddot{Y}_n+\boldsymbol{C}\sum_{n=1}^{N}\boldsymbol{\phi}_n\dot{Y}_n+\boldsymbol{K}\sum_{n=1}^{N}\boldsymbol{\phi}_nY_n=\boldsymbol{P} \tag{3.125}$$

where it is clear that the shape functions $\boldsymbol{\phi}_n$ are independent of time. Now premultiplying (3.125) by $\boldsymbol{\phi}_m$ we obtain

$$\boldsymbol{\phi}_m^{\mathrm{T}}\boldsymbol{M}\sum_{n=1}^{N}\boldsymbol{\phi}_n\ddot{Y}_n+\boldsymbol{\phi}_m^{\mathrm{T}}\boldsymbol{C}\sum_{n=1}^{N}\boldsymbol{\phi}_n\dot{Y}_n+\boldsymbol{\phi}_m^{\mathrm{T}}\boldsymbol{K}\sum_{n=1}^{N}\boldsymbol{\phi}_nY_n=\boldsymbol{\phi}_m^{\mathrm{T}}\boldsymbol{P}. \tag{3.126}$$

The first term in this expression can be expanded to

$$\boldsymbol{\phi}_m^{\mathrm{T}}\boldsymbol{M}\boldsymbol{\phi}_1\ddot{Y}_1+\boldsymbol{\phi}_m^{\mathrm{T}}\boldsymbol{M}\boldsymbol{\phi}_2\ddot{Y}_2+\ \dots\ +\boldsymbol{\phi}_m^{\mathrm{T}}\boldsymbol{M}\boldsymbol{\phi}_n\ddot{Y}_n+\ \dots\ .$$

Referring to the first orthogonality condition (3.107) we can see that every term in this series will be zero except for the case where $n=m$. Therefore,

$$\begin{aligned}\boldsymbol{\phi}_m^{\mathrm{T}}\boldsymbol{M}\sum_{n=1}^{N}\boldsymbol{\phi}_n\ddot{Y}_n&=\boldsymbol{\phi}_m^{\mathrm{T}}\boldsymbol{M}\boldsymbol{\phi}_m\ddot{Y}_m\\&=M_m\ddot{Y}_m,\end{aligned} \tag{3.127}$$

where M_m is the generalized mass of the mth mode. The same reasoning applies to the stiffness term in (3.126) so that

$$\begin{aligned}\boldsymbol{\phi}_m^{\mathrm{T}}\boldsymbol{K}\sum_{n=1}^{N}\boldsymbol{\phi}_nY_n&=\boldsymbol{\phi}_m^{\mathrm{T}}\boldsymbol{M}\boldsymbol{\phi}_mY_m\\&=K_mY_m,\end{aligned} \tag{3.128}$$

where K_m is the generalized stiffness.

Finally, the generalized force will be given by

$$Q_m=\boldsymbol{\phi}_m^{\mathrm{T}}\boldsymbol{P}. \tag{3.129}$$

A problem arises with the damping term in $\boldsymbol{C}$ because we have no convenient orthogonality condition to simplify this part of the expression. Treatment of damping will be considered in more detail in Chapter 4. Meanwhile, we could assume that the damping matrix $\boldsymbol{C}$ was proportional to either $\boldsymbol{M}$ or $\boldsymbol{K}$ or a combination of both. If this were the case, the same effect of decomposition would be achieved so that

$$\boldsymbol{\phi}_m^{\mathrm{T}}\boldsymbol{C}\sum_{n=1}^{N}\boldsymbol{\phi}_n\dot{Y}_n=C_mY_m. \tag{3.130}$$

Hence, Equation (3.126) can be written in the form

$$M_m\ddot{Y}_m+C_m\dot{Y}_m+K_mY_m=Q_m;\quad m=1,2,3,\ \dots,N \tag{3.131}$$

or

$$\ddot{Y}_m + 2\xi_m\omega_m\dot{Y}_m + \omega_m^2 Y_m = Q_m/M_m; \quad m = 1, 2, 3, \ldots N. \qquad (3.132)$$

It is clear that there will be one such equation for every degree of freedom and each equation can be treated independently as a single degree of freedom. Equation (3.132) should be compared with (2.2) and it will be appreciated that subsequent analysis may be achieved using the methods of Chapter 2. Finally, the modal contributions have to be summed using (3.120) to yield the actual displacements of the masses.

Mode superposition is an efficient method of analysis for many problems for two main reasons. The first is because the modal summation, given by (3.120), is usually dominated by the lower modes of vibration, allowing higher modes to be excluded from the analysis without significant error. In fact, there are certain problems, such as the vibration of bridges discussed in Chapter 9, in which sufficient accuracy can often be achieved by taking only the first or fundamental mode into account. Problems involving impact or shock generally require a greater number of modes to be included, with the result that mode superposition is not necessarily more efficient than direct integration. The second reason for the effectiveness of mode superposition is that modal analysis can be done experimentally. Natural frequencies and mode shapes can be obtained by forced excitation of an actual structure and the performance under dynamic loading predicted by the method described in this section (see Ewins, 1984). However, this latter procedure is more appropriate to mechanical and aeronautical prototype structures and less relevant in the field of civil engineering.

EXAMPLE 3.7

Determine the dynamic response of the multi-storey frame of Fig. 3.14 when it is subjected to a dynamic load as shown in Fig. 3.15. Damping may be neglected.

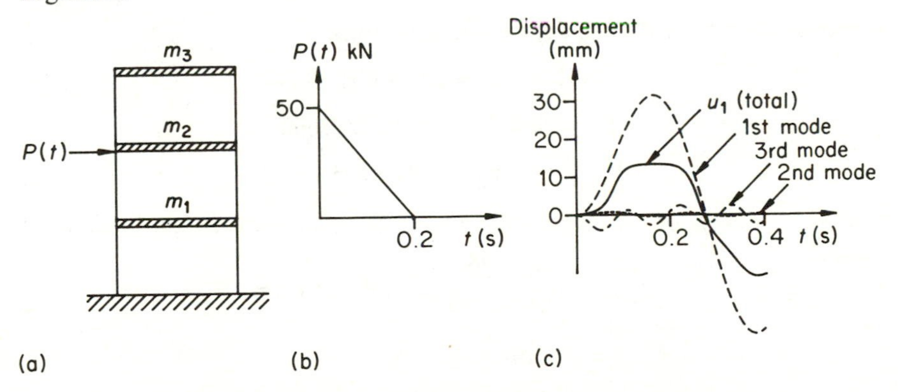

Fig. 3.15 Dynamic loading of a multi-storey frame.

Solution

The natural frequencies and mode shapes of all three modes of vibration are shown in Fig. 3.14. The generalized mass of the first mode was calculated in Example 3.6:

$$M_1 = 6954 \text{ kg.}$$

The generalized masses of the second and third modes may be calculated using (3.118) in the same way, giving

$$M_2 = 7690 \text{ kg}, \quad M_3 = 6756 \text{ kg.}$$

The natural frequencies squared (ω^2) are

$$\omega_1^2 = 221.1, \quad \omega_2^2 = 1505, \quad \omega_3^2 = 3210$$

and the periods are

$$T_1 = 0.422 \text{ s}, \quad T_2 = 0.162 \text{ s}, \quad T_3 = 0.111 \text{ s.}$$

The generalized stiffnesses need not be calculated from (3.128) but follow directly from the relation

$$\omega_n^2 = K_n/M_n.$$

Therefore

$$K_1 = 1.538 \times 10^6, \quad K_2 = 11.573 \times 10^6, \quad K_3 = 21.687 \times 10^6.$$

The generalized forces are evaluated using (3.129). For example,

$$Q_1 = \boldsymbol{\phi}_1^{\mathrm{T}} \boldsymbol{P} = \{0.484 \;\; 0.834 \;\; 1.0\} \begin{Bmatrix} 0 \\ 1 \\ 0 \end{Bmatrix} P(t) = 0.834\, P(t).$$

In the same way

$$Q_2 = 0.118\, P(t), \quad Q_3 = -1.0\, P(t).$$

Therefore, we have three independent equations,

$$\ddot{Y}_1 + 221.1\, Y_1 = 0.834\, P(t)/6954 \tag{a}$$

$$\ddot{Y}_2 + 1505\, Y_2 = 0.118\, P(t)/7690 \tag{b}$$

$$\ddot{Y}_3 + 3210\, Y_3 = -P(t)/6756. \tag{c}$$

The static deflections of each mode, under the maximum 50 kN load, are

$$(Y_1)_{st} = \frac{0.834 \times 50 \times 10^3}{1.538 \times 10^6} \text{ m} = 27.1 \text{ mm},$$

$$(Y_2)_{st} = 0.5 \text{ mm},$$

$$(Y_3)_{st} = -2.3 \text{ mm}.$$

The dominance of the first mode, because of the higher stiffnesses of the other two modes, is already evident.

For an accurate calculation of the dynamic response of the structure, Equations (a), (b) and (c) should be solved using numerical integration or Duhamel's integral. Explicit formulae for the displacement of a one-degree-of-freedom system, subjected to a triangular forcing function, were obtained in Example 2.4. Hence, at every instant of time it is possible to write expressions of the form

$$Y_n(t) = (Y_n)_{st} f_n(t),$$

where $f_n(t)$ is evaluated from (2.36) or (2.38). Then the displacement of the ith mass may be obtained by summing the modal contributions using (3.121). The displacement of the first storey, or m_1, would be given by

$$u_1 = 0.484\ Y_1(t) + 1.0\ Y_2(t) + 0.49\ Y_3(t).$$

The separate modal contributions, together with the total displacement, are shown in Fig. 3.15(c).

The maximum response may be estimated by using dynamic load factors (DLF) to calculate the maximum response in each mode and summing the maxima. For example, in the first mode the ratio of load duration to natural period is

$$T_d/T_1 = 0.2/0.422 = 0.474.$$

Then from Fig. 2.12(b) the DLF is found to be

$$(\mathrm{DLF})_1 = 1.17.$$

In the same way

$$T_d/T_2 = 1.234, \quad T_d/T_3 = 1.802,$$

and therefore

$$(\mathrm{DLF})_2 = 1.62, \quad (\mathrm{DLF})_3 = 1.73.$$

Hence, an estimate of the maximum response is given by

$$\boldsymbol{u}_{max} = |\boldsymbol{\phi}_1| \times 1.17 \times 27.1 + |\boldsymbol{\phi}_2| \times 1.62 \times 0.5 + |\boldsymbol{\phi}_3| \times 1.73 \times 2.3,$$

noting that absolute values of the modal contributions are chosen in order to obtain an upper bound. Therefore, the maximum response of the first storey is

$$(u_1)_{max} = 0.484 \times 1.17 \times 27.1 + 1.0 \times 1.62 \times 0.5 + 0.49 \times 1.73 \times 2.3$$
$$= 18.1 \text{ mm}.$$

This result should be compared with Fig. 3.15(c). The reason that it is greater than the accurate value is because the three modal contributions are not in phase at the instant of overall maximum response. The above procedure gives a conservative estimate of maximum response which might be satisfactory for design calculations. Better methods of combining modal responses are discussed in Chapter 5.

3.5 PROPAGATION OF WAVES IN INFINITE MEDIA

We now turn to the dynamic behaviour of solid bodies of infinite extent. The theory that will be developed has important applications in earthquake engineering, the transmission of vibrations due to pile-driving and blasting, and the dynamic interaction of large structures with their foundations. In all of these applications the structure under consideration is the ground, consisting of soil or geological strata, depending on the scale of the problem.

In order to study such problems it is clearly necessary to analyse simplified models which may be somewhat removed from reality. A basic assumption is that soil or rock can be represented by a uniform elastic medium extending to infinity on one side of a free surface. Such a model is called a semi-infinite elastic solid or an *elastic half space*. Loads are usually applied at the free surface, although internal sources of disturbance are important in studying the nature of earthquakes. Useful results can be obtained by modelling geological strata in the form of elastic layers of different material properties.

An elastic solid is one example of a system with distributed mass and stiffness. The method of analysis is related to the methods discussed in §3.3, in that an equation of motion for a differential element is obtained by considering its dynamic equilibrium, and solution proceeds by separating the time and spatial variables. The main difference is the lack of boundary conditions in the direction of infinite extent. The result of this is that the response is in the form of travelling waves which propagate through and on the surface of the solid.

A rigorous study of wave motion in elastic solids requires the use of some advanced mathematical techniques including Bessel functions, integral transforms and contour integration of complex variables. For a full treatment the reader is referred to Ewing, Jardetzky and Press (1957) or to Graff (1975). The former text discusses many important applications in seismology. The treatment that now follows is limited to a demonstration of the existence of certain types of travelling wave, followed by the presentation of some useful formulae for predicting ground motion caused by dynamic loads applied at the surface.

3.5.1 The wave equation

The phenomenon of wave motion may be explained with reference to the dynamic motion of a stretched string. Consider the differential element shown in Fig. 3.16, in which m is the mass per unit length of the string, and T is the tension, which is assumed to remain constant during small displace-

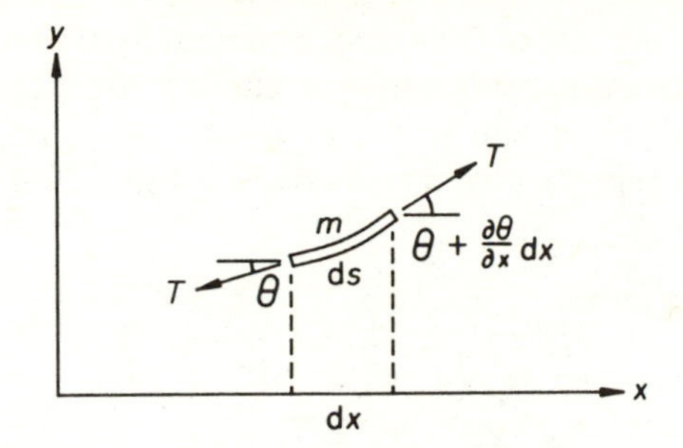

Fig. 3.16 Differential element of a taut string.

ments. The equation of dynamic equilibrium in the y direction is

$$-T\sin\theta + T\sin\left(\theta + \frac{\partial\theta}{\partial x}\mathrm{d}x\right) - m\mathrm{d}s\frac{\partial^2 v}{\partial t^2} = 0. \tag{3.133}$$

In the case of small deflections $\sin\theta = \theta = \partial v/\partial x$ and $\mathrm{d}s \simeq \mathrm{d}x$. Therefore the equation reduces to

$$\frac{\partial^2 v}{\partial x^2} = \frac{1}{c^2}\frac{\partial^2 v}{\partial t^2} \tag{3.134a}$$

where

$$c = \sqrt{(T/m)}. \tag{3.134b}$$

This is known as the one-dimensional wave equation. Solution of the equation may be achieved by separation of variables, as in earlier problems. Assume that there is a solution of the form

$$v = \phi(x)\,Y(t). \tag{3.135}$$

Substituting in (3.134a) we obtain

$$\frac{1}{\phi(x)}\frac{\partial^2\phi}{\partial x^2} = \frac{1}{c^2 Y(t)}\frac{\partial^2 Y}{\partial t^2} = \text{constant} = -k^2. \tag{3.136}$$

By comparing this with (3.36) it may be verified that there is a solution given by

$$\begin{aligned} v &= (A_1 \sin kx + A_2 \cos kx)(A_3 \sin kct + A_4 \cos kct) \\ &= A_1 A_4 \sin kx \cos kct + A_1 A_3 \sin kx \sin kct. \end{aligned} \tag{3.137}$$

This may be written in the alternative form

$$\begin{aligned} v = {} & B_1 \sin k(x+ct) + B_2 \sin k(x-ct) \\ & + B_3 \cos k(x+ct) + B_4 \cos k(x-ct). \end{aligned} \tag{3.138}$$

Consider the behaviour exhibited by a typical term, e.g. $B_2 \sin k(x-ct)$. This term will be zero when $x = ct$. In other words, its null position travels along the string with a velocity of c. The same argument applies to the other terms, and to higher harmonics, from which it can be concluded that Equation (3.138) represents displacement in the form of a wave travelling along the string with a *phase velocity* of c. This is depicted in Fig. 3.17.

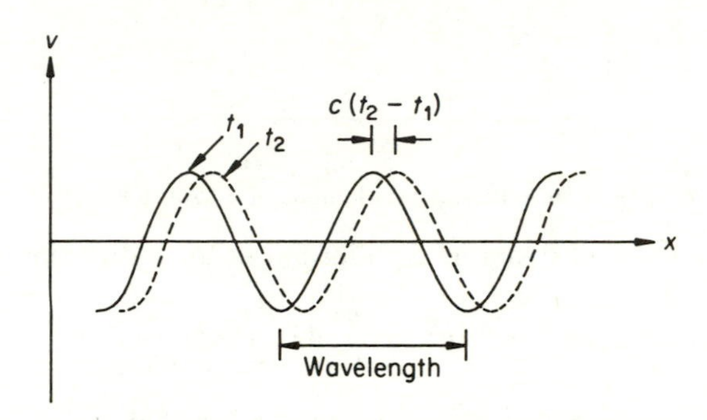

Fig. 3.17 Travelling wave in a taut string.

It should be noted that the factor k^2c^2, found in (3.136), is equivalent to the frequency factor ω^2 in (3.38). Therefore any fixed point on the string appears to vibrate with a frequency of $\omega/2\pi$. Furthermore, a first-harmonic travelling wave must have a wavelength of $2\pi/k$. The frequency number k depends on the boundary and loading conditions.

3.5.2 Shear and compressional waves in elastic solids

The equations of motion of an isotropic elastic solid may be derived by considering the dynamic equilibrium of the differential element shown in Fig. 3.18. The equations of motion, in terms of the displacements u, v and w in the x, y and z directions, are as follows:

$$\left.\begin{aligned}(\lambda+G)\frac{\partial e}{\partial x}+G\nabla^2 u-\rho\frac{\partial^2 u}{\partial t^2}&=0\\(\lambda+G)\frac{\partial e}{\partial y}+G\nabla^2 v-\rho\frac{\partial^2 u}{\partial t^2}&=0\\(\lambda+G)\frac{\partial e}{\partial z}+G\nabla^2 w-\rho\frac{\partial^2 u}{\partial t^2}&=0\end{aligned}\right\}\tag{3.139}$$

in which the volume expansion e is given by

$$e=\varepsilon_x+\varepsilon_y+\varepsilon_z=\frac{\partial u}{\partial x}+\frac{\partial v}{\partial y}+\frac{\partial w}{\partial z},\tag{3.140}$$

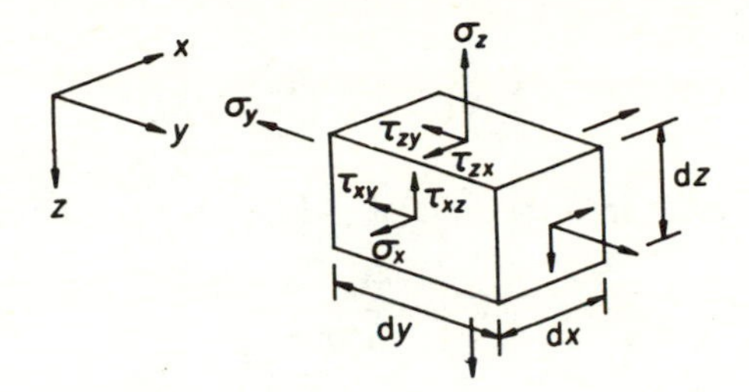

Fig. 3.18 Differential element within an elastic solid.

the operator ∇^2 is defined as

$$\nabla^2 = \frac{\partial^2}{\partial x^2} + \frac{\partial^2}{\partial y^2} + \frac{\partial^2}{\partial z^2} \tag{3.141}$$

and Lamé's elastic constants are

$$G = E/2(1+\nu), \quad \lambda = \nu E/(1+\nu)(1-2\nu). \tag{3.142}$$

The full derivation of the above equations may be found in *Theory of Elasticity* by Timoshenko and Goodier (1951).

Two types of *body waves* occur within the interior of a solid body. The first are associated with shear or distortion of a solid body. If the volume expansion is zero ($e=0$), only distortional waves would occur. Substituting this condition into (3.139) we obtain three equations, the first of which is

$$\nabla^2 u = \frac{1}{\beta^2}\frac{\partial^2 u}{\partial t^2} \tag{3.143a}$$

where

$$\beta = \sqrt{(G/\rho)}. \tag{3.143b}$$

By comparing with (3.134), it is evident that (3.143) is a wave equation with a phase velocity of β. These are often called S waves.

The other type of body wave occurs under the condition of no rotation of the differential element. This condition is fulfilled about the three cartesian axes if

$$\frac{\partial v}{\partial x} - \frac{\partial u}{\partial y} = 0, \quad \frac{\partial w}{\partial y} - \frac{\partial v}{\partial z} = 0, \quad \frac{\partial u}{\partial z} - \frac{\partial w}{\partial x} = 0. \tag{3.144}$$

These conditions are satisfied by a single function ϕ provided that

$$u = \frac{\partial \phi}{\partial x}, \quad v = \frac{\partial \phi}{\partial y}, \quad w = \frac{\partial \phi}{\partial z}. \tag{3.145}$$

Therefore, it follows that

$$e = \nabla^2 \phi \tag{3.146}$$

and hence

$$\frac{\partial e}{\partial x} = \frac{\partial}{\partial x}\nabla^2\phi = \nabla^2\frac{\partial \phi}{\partial x} = \nabla^2 u. \tag{3.147}$$

Substituting this result into the first of (3.139) we find that

$$\nabla^2 u = \frac{1}{\alpha^2}\frac{\partial^2 u}{\partial t^2} \tag{3.148a}$$

where

$$\alpha = \sqrt{\left(\frac{\lambda + 2G}{\rho}\right)}. \tag{3.148b}$$

Since all elements within the body do not rotate under these conditions it is clear that Equation (3.148) represents compressional or *dilatational* (P waves) with a velocity of α. The P waves evidently travel much faster than S waves.

The preceding arguments apply equally well to the second and third equations (3.139).

3.5.3 Surface waves

An early discovery of seismology was that P and S waves, detected by instruments on the surface of the earth, were quickly followed by another tremor of substantially greater magnitude. Furthermore, it was noted that, whereas P and S waves diminished rapidly with distance from the source (proportional to $1/x^2$), the subsequent tremor diminished far less rapidly. This suggested that the energy dissipation was being confined to the surface, in contrast to the volumetric dispersion of the P and S waves. The phenomenon was investigated by Lord Rayleigh (1885) who verified the existence of surface waves and showed that their velocity was slightly less than that of shear waves, and that their amplitudes diminished at the rate of $1/\sqrt{x}$.

In most applications we are interested in waves that radiate outwards from a source of disturbance. On the surface these would be analogous to the ripples on a pond radiating in the form of concentric rings increasing in size with time. Rigorous analysis of this behaviour requires advanced mathematical techniques that are beyond the scope of this book. However, the existence of surface waves can be readily demonstrated if we assume that at great distances from the source of disturbance the wave motion is effectively plane.

Under the special conditions considered in §3.5.2 there were two types of body waves. However, in general it should be expected that displacements arise from the combined effects of both types of wave. This is the basis of the following discussion.

It can be shown that the following expressions for displacement in the x

and z directions satisfy the wave equations:

$$u = u_1 + u_2 = k e^{-\eta z} \sin k(x - ct) + A\zeta e^{-\zeta z} \sin k(x - ct) \tag{3.149a}$$

$$w = w_1 + w_2 = \eta e^{-\eta z} \cos k(x - ct) + Ak e^{-\zeta z} \cos k(x - ct), \tag{3.149b}$$

where

$$\eta^2 = k^2(1 - c^2/\alpha^2), \quad \zeta^2 = k^2(1 - c^2/\beta^2). \tag{3.149c}$$

A and k are constants and c is a velocity. These equations correspond to (3.138) for a taut string. It can easily be shown that u_1 and w_1 satisfy (3.148) while u_2 and w_2 satisfy (3.143). Hence, we have displacements, which are composed of P and S waves, taking the form of waves travelling at a velocity c. It should also be noted that these displacements diminish rapidly with the depth z and are therefore confined to a region close to the surface. The velocity in (3.149) may be determined by introducing the boundary conditions at the free surface. These conditions are that both the shear and normal stress must be zero at $z = 0$. These are expressed respectively by the following:

$$\frac{\partial u}{\partial z} + \frac{\partial w}{\partial x} = 0 \tag{3.150a}$$

$$\lambda e + 2G \frac{\partial w}{\partial z} = 0. \tag{3.150b}$$

Therefore, using Equations (3.149) we obtain

$$2\eta k + A(\zeta^2 + k^2) = 0 \tag{3.151a}$$

$$(\lambda/G)(k^2 - \eta^2) - 2(\eta^2 + Ak\zeta) = 0. \tag{3.151b}$$

Eliminating A, and expressing λ/G in the form

$$\lambda/G = (\alpha^2/\beta^2) - 2, \tag{3.152}$$

it can be shown that the velocity must satisfy the relation

$$\left(2k^2 - \frac{k^2 c^2}{\beta^2}\right)^2 - 4k^2 \eta\zeta = 0. \tag{3.153}$$

This is Rayleigh's equation. When the appropriate forms for η and ζ are introduced it is clear that k^2 will cancel out. The equation has three roots for c^2, only one of which yields positive values for η and ζ, which are necessary for surface waves. Substituting different values of ν into (3.153) the following results may be obtained:

$$\left.\begin{array}{ll} \text{for } \nu = 0, & c_R = 0.8740\,\beta \\ \text{for } \nu = 0.25, & c_R = 0.9194\,\beta \\ \text{for } \nu = 0.5, & c_R = 0.9553\,\beta \end{array}\right\}. \tag{3.154}$$

These results show that Rayleigh waves travel at a speed just less than shear waves. By substituting appropriate values of c into (3.149) it is also possible to show that surface particles trace elliptic retrograde orbits in which the horizontal displacement amplitudes are about two-thirds of the vertical.

Unfortunately, the theoretical derivation of Rayleigh waves fails to explain the observed fact that horizontal ground motions are generally, though not always, greater than the vertical motions during earthquakes. Furthermore, horizontal ground motions are often transverse to the direction of propagation of the waves.

The explanation was discovered by Love (1926) who studied the behaviour of waves in a thin layer of different geological properties near the surface. He modelled this as an elastic layer of thickness H overlying an elastic half space, as shown in Fig. 3.19. He showed that, under certain conditions, horizontally polarized shear waves are trapped within the layer and propagated by multiple reflections. The velocity of propagation of the ensuing composite wave lies between the shear-wave velocities of the two media. Furthermore, since the motion is confined to the thin surface layer (2 to 20 km in geological terms) the waves diminish slowly and are propagated great distances across continents in a similar way to Rayleigh wave dispersion. A detailed discussion is given by Ewing, Jardetzky and Press (1957).

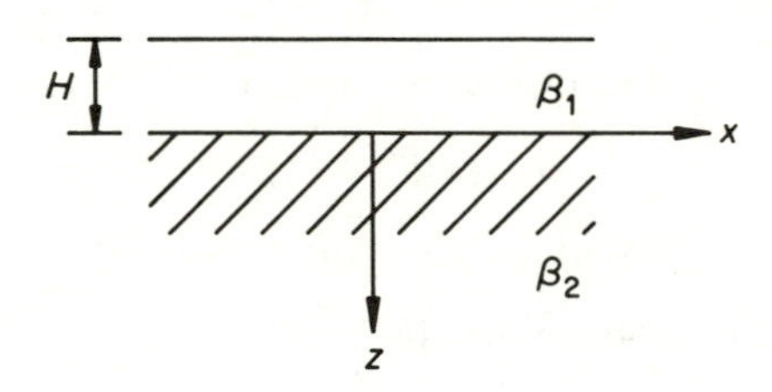

Fig. 3.19 Elastic layer overlying an elastic half-space.

3.5.4 Displacements caused by a surface disturbance

The preceding analysis has been restricted to the identification of different kinds of waves in solid bodies. This is analogous to the study of normal modes of vibration of bodies with finite boundaries. However, we are often more interested in the amplitudes of motion caused by specified dynamic loads or, in other words, the dynamic response. Rigorous analysis of even quite simple loading cases requires some relatively difficult mathematics which is beyond the scope of this book. However, some useful results will now be presented without derivation.

A surface disturbance in the form of a dynamic point load has a number of important civil engineering applications, including the effects of pile-driving and blasting. The configuration is as shown in Fig. 3.20. The problem was

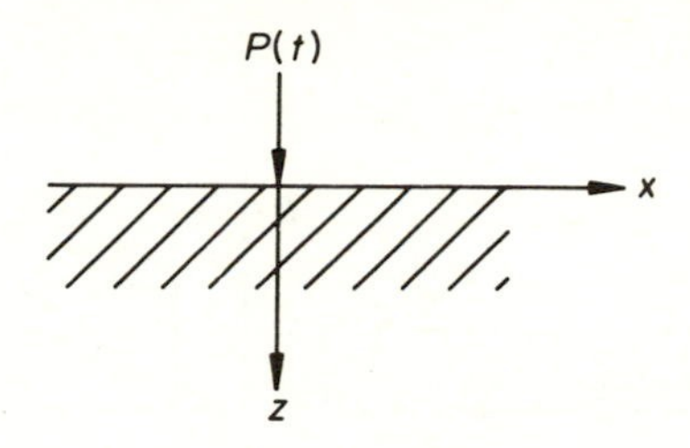

Fig. 3.20 Dynamic point load on surface of an elastic half-space.

first solved by Lamb (1904) and is often known as 'Lamb's problem'. The most basic loading function is a steady-state periodic load given by

$$P(t)=P_0 \cos \omega t. \tag{3.155}$$

We noted in §3.5.1 that wave motion could be interpreted as vibration of a specified point on the solid with a natural frequency related to the phase velocity by

$$\omega = kc \tag{3.156}$$

where k is a frequency number determined by the boundary and loading conditions. Therefore it is reasonable to anticipate a solution in which this frequency is equal to the disturbing frequency. This is found to be true and the expression for vertical displacement of the surface is given approximately by

$$w_0(t)=\frac{AP_0}{G}\sqrt{\left(\frac{\omega}{2\pi c_R x}\right)}\cos(\omega t-\omega x/c_R-\pi/4), \tag{3.157a}$$

where

$$A=-\frac{k_\beta^2\sqrt{(k_c^2-k_\alpha^2)}}{F'(k_c)}, \tag{3.157b}$$

$$k_\beta=\omega/\beta, \quad k_\alpha=\omega/\alpha, \tag{3.157c}$$

and

$$F(k)=(2k^2-k_\beta^2)-4k^2\sqrt{[(k^2-k_\alpha^2)(k^2-k_\beta^2)]}. \tag{3.158}$$

In the above expression $F(k)$ is Rayleigh's function and $k_c=\omega/c_R$ is the root of $F(k)=0$ corresponding to surface waves. $F'(k_c)$ is the first differential of $F(k)$ with respect to k, with the value of k_c substituted.

Lamb also studied the effect of a transient loading function given by

$$P(t)=V_0\tau/\pi(t^2+\tau^2). \tag{3.159}$$

This is the impulse shown in Fig. 3.21(a). The response is given approximately by

$$w_0(t)=\frac{AV_0}{2G\tau\sqrt{(2\tau x c_R)}}\cos^{3/2}\psi\cos(\pi/4-\tfrac{3}{2}\psi) \tag{3.160a}$$

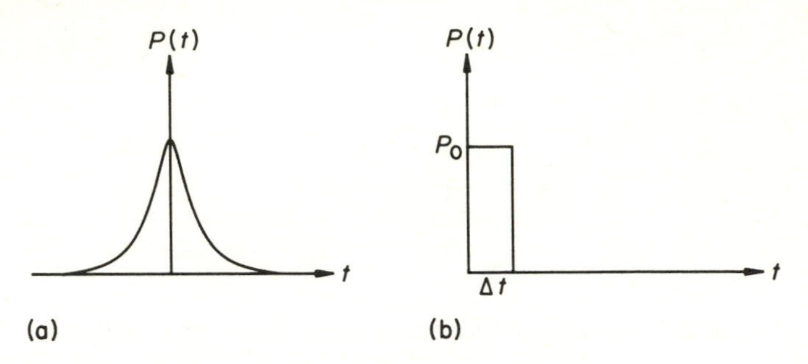

Fig. 3.21 Transient loading functions.

where

$$\cos\psi = \tau/\sqrt{(\tau^2 + t_0^2)} \quad \text{and} \quad t_0 = t - x/c_R. \tag{3.160b}$$

The response to a rectangular impulse, as shown in Fig. 3.21(b), was obtained by Smith (1973a) and is given approximately by

$$w_0(t) = -\frac{AP_0}{\pi G\sqrt{(2xc_R)}}\left(\frac{1 - H(t - x/c_R)}{\sqrt{(x/c_R - t)}} - \frac{1 - H(t - \Delta t - x/c_R)}{\sqrt{(\Delta t + x/c_R - t)}}\right) \tag{3.161}$$

where $H(t-T)$ is a unit step at the instant T. This expression tends to infinity at the beginning and end of the impulse, although in practice nonlinearities in the soil would suppress this effect.

Further terms representing the effects of the shear and compressional waves may be obtained, but have been shown to be of secondary importance and decay much more rapidly with distance. The form of the motions produced by the impulses of Fig. 3.21 are shown in Fig. 3.22. It is evident that the P and S waves are of relatively minor importance as far as likely damage to structures and foundations is concerned. However, the time delay between the arrivals of the P and S waves is used by seismologists for locating the origins of earthquakes.

EXAMPLE 3.8

The impact delivered by a certain pile driver is equivalent to a surface point load with a rectangular impulse of 400 kN magnitude and 0.02 s duration. The properties of the ground may be assumed to be:

$$\rho = 1800\ \text{kg/m}^3, \quad G = 20\ \text{MN/m}^2, \quad \nu = 0.25.$$

Determine the wave velocities and evaluate the surface ground motion at a distance of 20 m from the pile driver.

Solution

Shear wave velocity:

$$\beta = \sqrt{(G/\rho)} = \sqrt{(20 \times 10^6/1800)} = 105.4\ \text{m/s}.$$

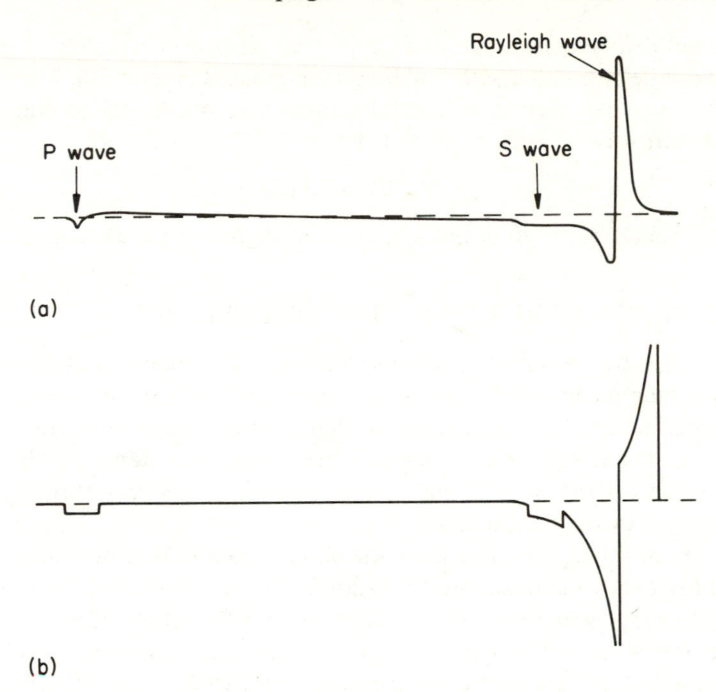

Fig. 3.22 Vertical displacements caused by transient point loads. (a) Effect of Lamb's impulse function; (b) effect of rectangular impulse.

Since $\nu = 0.25$ it follows that $\lambda = G$ and therefore the compressional wave velocity is

$$\alpha = \sqrt{(3G/\rho)} = 182.6 \text{ m/s}.$$

Rayleigh wave velocity:

$$c_R = 0.9194\beta = 96.9 \text{ m/s}.$$

The ground motion may be evaluated from Equation (3.161). First the coefficient A must be obtained from (3.157b). The frequency cancels out of this expression when it is expanded. Omitting the details of the calculation, it may be found that $A = 0.184$ when the above values of α, β and c_R have been substituted. Hence the amplitude factor in (3.161) is given by

$$\frac{AP_0}{\pi G\sqrt{(2xc_R)}} = \frac{0.184 \times 400 \times 10^3}{\pi \times 20 \times 10^6 \sqrt{(2 \times 20 \times 96.9)}} = 18.82 \times 10^{-6}.$$

We shall evaluate the motion just after the leading edge of the Rayleigh wave has arrived. This occurs when $t = x/c_R = 0.206$ s. At this instant the first term within the brackets of (3.161) becomes zero while the second term becomes $-1/\sqrt{(\Delta t)}$. Substituting in (3.161) the vertical displacement at this instant is

$$w_0 = \frac{18.82 \times 10^{-6}}{\sqrt{0.02}} \text{ m} = 0.133 \text{ mm}.$$

Such small displacements are of little interest. However, velocity of ground motion can be correlated with building damage (see Chapter 10). The vertical velocity can be obtained from (3.161) by differentiation. At the leading edge of the Rayleigh wave it may be shown that

$$\dot{w}_0 = w_0/2\Delta t = 3.3\ \text{mm/s}.$$

This is a noticeable motion but not large enough to be a damage risk.

3.6 GROUND–STRUCTURE INTERACTION

It should now be plain that the ground has dynamic properties of its own and therefore dynamic interaction between the ground and a structure should not be ignored. In fact, it is widely accepted that ground–structure interaction can have a significant effect on the natural frequencies and damping characteristics of some structures. A method of assessing the importance of this interaction is evidently required.

The first problem is to determine the effect of foundation flexibility on the natural frequency of structures. In Example 3.6 we considered the vibration of a multi-storey structure assuming that the foundation was absolutely rigid or fixed. But what would happen if the structure were supported by a flexible foundation? It is instructive to consider the extreme case illustrated in Fig. 3.23(a). In this case the structure is assumed to be a rigid vertical cantilever while the foundation flexibility is represented by two springs, one resisting horizontal motion at the base and the other resisting rotation of the cantilever. Vertical flexibility is ignored for the purposes of this discussion. Two ways in which the structure could vibrate are easily identified. The first possibility is a horizontal translation of the entire structure, of mass mL and

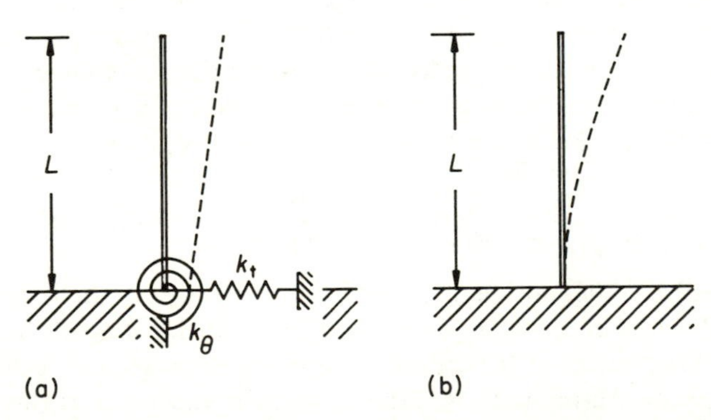

Fig. 3.23 Simple model of ground–structure interaction. (a) Rigid cantilever; (b) rigid foundation.

corresponding stiffness k_t. The natural frequency would be

$$f_t = (1/2\pi)\sqrt{(k_t/mL)}. \tag{3.162}$$

Another possible motion would consist of rotation of the cantilever hingeing at its base. In this case the natural frequency is given by

$$f_\theta = (1/2\pi)\sqrt{(k_\theta/I_\theta)}, \tag{3.163}$$

where k_θ is the rotational stiffness of the equivalent spring at the base and I_θ is the mass moment of inertia of the cantilever about the base. Further discussion of these parameters may be found in Chapter 8.

In effect, the system has two degrees of freedom and the normal modes are combinations of translation and rotation, as shown in Fig. 3.23(a). The other extreme case is shown in Fig. 3.23(b) where the foundation is assumed rigid and the cantilever is flexible. The natural frequency of the fundamental mode in bending is then given by

$$f_f = \frac{(1.875)^2}{2\pi L^2}\sqrt{\left(\frac{EI}{m}\right)}. \tag{3.164}$$

It will now be appreciated that a flexible structure on a flexible foundation should be regarded as having at least three degrees of freedom. The normal modes will be combinations of rigid translation, rigid rotation and fixed base bending. Rigorous analysis of a three-degrees-of-freedom system is possible but it has been suggested by Ellis (1986) that the amount of interaction can be estimated by a much simpler procedure. This makes use of Dunkerley's formula in which the natural frequency of the system may be expressed as

$$1/f^2 = 1/f_t^2 + 1/f_\theta^2 + 1/f_f^2. \tag{3.165}$$

This formula yields the lowest natural frequency of the three-degrees-of-freedom system and inspection of the formula reveals that this frequency will be less than any of the isolated frequencies.

Ellis (1986) used the data obtained from full-scale dynamic tests on four buildings to evaluate their base stiffnesses and compared them with theoretical base stiffnesses. He also examined the results of tests on 37 buildings to quantify the importance of ground–structure interaction. His studies revealed that ground–structure interaction was negligible for tall buildings but became increasingly important for shorter buildings. In one case a 48% reduction in frequency was noted. Ground–structure interaction is usually taken into account when analysing nuclear reactors, which tend to be stiff structures with relatively low ratios of height to width. Ground–structure interaction is also important in the vibration of foundations for industrial machinery (see Chapter 8).

The second problem is to consider the effect of the ground on damping or dissipation of dynamic energy. Ellis (1986) has shown that the energy dissipated in the foundations of buildings can be considerable and cited one case where it accounted for 60% of the input energy.

There are three mechanisms by which the vibration energy of a building is dissipated, these being: (a) damping in the materials and joints of the structure; (b) internal damping in the mass of soil that participates in the vibration of the entire ground–structure system; and (c) energy dissipation by wave motion in the ground which behaves like a semi-infinite solid. The first two mechanisms constitute conventional damping in which interparticle friction dissipates energy as heat. The third mechanism is often known as 'radiation damping' and is analogous to the ripples on a pond which propagate outwards from a source of disturbance. Energy is not lost in an ideal material but is radiated away from the source and is distributed more thinly over a wavefront of increasing size.

It is almost impossible to quantify the energy losses due to these mechanisms by theoretical analysis and recourse must be had to experimental data. Damping will be discussed in more detail in the next chapter.

4

Finite element modelling of vibration problems

4.1 INTRODUCTION

The finite element method first made its appearance in engineering journals in a famous paper by Turner, Clough, Martin and Topp (1956). Although the term 'finite element' was not coined until later, there is no doubt that this paper heralded a completely new era in engineering analysis. The authors had been working on a method of dynamic analysis of aircraft structures and had introduced the concept of subdividing a structure of irregular shape into a large number of simpler geometrical entities or elements. They discovered that if the load–displacement equations for a single element were derived in matrix form it was possible to use matrix algebra to combine the interacting effects of all the elements in a systematic and conceptually straightforward manner. The advent of the digital computer made it possible to implement the procedure on a large enough scale to solve many difficult practical problems. In the abstract to their paper, the authors said that 'considerable extension of the material presented in the paper is possible'. This remarkable understatement can now be seen in the light of the intervening thirty years during which time literally thousands of papers have been published, huge computer programs have been written and hundreds of companies earn their living by solving practical problems by the finite element method. The basic idea of the finite element method is illustrated in Fig. 4.1 where a rockfill dam, seen in cross section, is divided up into a large number of plane triangles. The triangular elements are connected together at nodes each of which is capable of being displaced in the x and y directions, except at the base where the boundary conditions stipulate zero displacements. The nodes at the upstream face are subjected to loads due to water pressure. This is a static problem but there are many important dynamic problems in civil engineering, such as

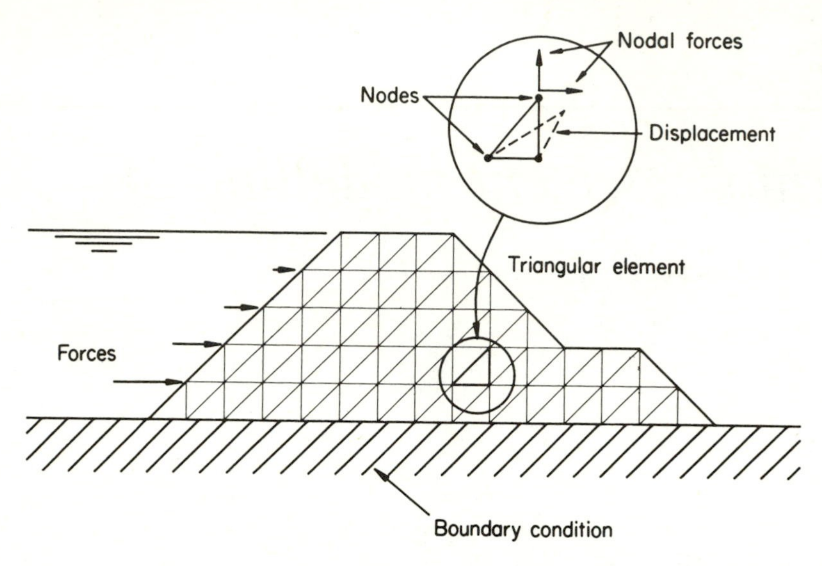

Fig. 4.1 Finite element mesh.

earthquake assessment of large structures, which require solution by finite elements.

In this chapter the theory of the finite element method will be outlined using beam elements in flexure for illustration. Other shapes of element will be reviewed and then the numerical solution procedures employed in dynamic analysis will be explained. The chapter will conclude with a discussion of damping in structures and how it may be included in dynamic analysis.

Readers unfamiliar with the finite element method are referred to an introductory text by Cheung and Yeo (1979). This is restricted to static analysis but includes some elementary problems together with FORTRAN source codes. A comprehensive treatment of the numerical procedures underlying finite element methods by Bathe (1982) is strongly recommended to anyone desiring a deeper fundamental knowledge. A wide range of applications and advanced principles may be found in the textbook by Zienkiewicz (1977).

4.2 THE FINITE ELEMENT METHOD USING 2-*D* BEAM ELEMENTS

Many civil engineering structures, such as building frames and bridges, can be represented by interconnected two-dimensional beam elements. For example, the reinforced concrete bridge shown in Fig. 4.2 is idealized by seven elements rigidly connected at the nodes. The structure is pinned at the

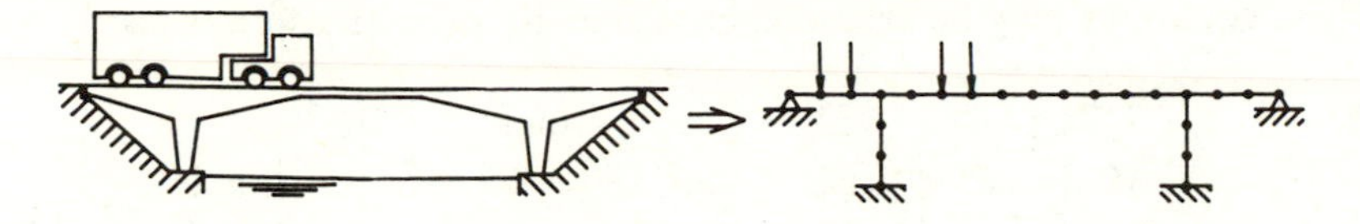

Fig. 4.2 Bridge structure represented by two-dimensional beam elements.

abutments and built in at the column foundations. The weight of the vehicle is represented by point loads applied to the appropriate elements.

The load–deflection behaviour of an individual element is shown in Fig. 4.3. A general relationship between the forces and displacements may be obtained by either of the well known strain energy or slope–deflection methods (Timoshenko and Young, 1965). The full derivation may be found in many textbooks and is not lengthy. The relevant equations are:

$$S_1 = \frac{12EI}{a^3}v_1 + \frac{6EI}{a^2}\theta_1 - \frac{12EI}{a^3}v_2 + \frac{6EI}{a^2}\theta_2$$

$$M_1 = \frac{6EI}{a^2}v_1 + \frac{4EI}{a}\theta_1 - \frac{6EI}{a^2}v_2 + \frac{2EI}{a}\theta_2$$

$$S_2 = -\frac{12EI}{a^3}v_1 - \frac{6EI}{a^2}\theta_1 + \frac{12EI}{a^3}v_2 - \frac{6EI}{a^2}\theta_2$$

$$M_2 = \frac{6EI}{a^2}v_1 + \frac{2EI}{a}\theta_1 - \frac{6EI}{a^2}v_2 + \frac{4EI}{a}\theta_2 \tag{4.1}$$

where S_1, S_2, M_1 and M_2 are the forces and moments at the ends of the member, length a, and v_1, v_2, θ_1 and θ_2 are the corresponding displacements.

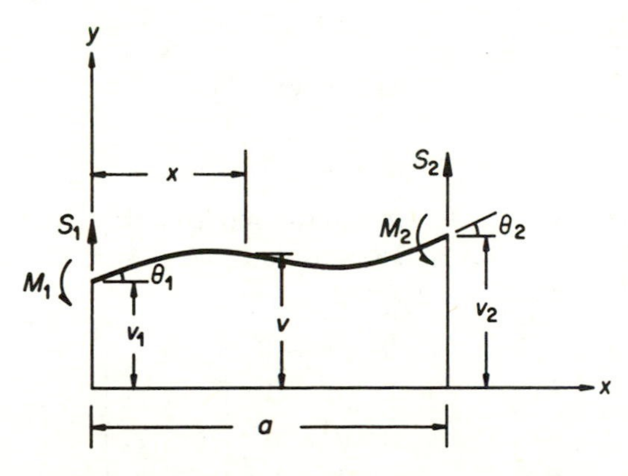

Fig. 4.3 Forces and displacements at the nodes of a two-dimensional beam element.

These equations may be arranged conveniently in matrix form thus:

$$\begin{Bmatrix} S_1 \\ M_1 \\ S_2 \\ M_2 \end{Bmatrix} = \frac{EI}{a^3} \begin{bmatrix} 12 & & \text{symm.} & \\ 6a & 4a^2 & & \\ -12 & -6a & 12 & \\ 6a & 2a^2 & -6a & 4a^2 \end{bmatrix} \begin{Bmatrix} v_1 \\ \theta_1 \\ v_2 \\ \theta_2 \end{Bmatrix} \tag{4.2}$$

The flexural stiffness of the beam element is denoted by EI.

In compact matrix notation the equations may be written

$$\boldsymbol{f} = \boldsymbol{k}\boldsymbol{u} \tag{4.3}$$

where $\boldsymbol{f}$ is a load vector, $\boldsymbol{k}$ is the stiffness matrix for the element and $\boldsymbol{u}$ is a displacement vector. The stiffness terms, $12EI/a^3$ etc., are, in effect, the forces occurring at the ends of the element when unit displacements are imposed.

4.2.1 General method for derivation of stiffness matrix

The most versatile method of deriving the stiffness matrices of finite elements is by utilizing the principle of virtual displacements. It has the great advantage of being equally valid in both linear and nonlinear ranges and is therefore more general than the strain energy method.

The first, and possibly the most important, step in the finite element formulation is to choose a suitable displacement function. This function should be reasonably faithful to the deflection curve and should satisfy continuity requirements at the nodes. The most convenient functions for this purpose are polynomials with as many terms as are compatible with the nodal displacements. In the case of a beam element a third-order or cubic polynomial is appropriate. The displacement at a point x from the left node is given by

$$v = \alpha_1 + \alpha_2 x + \alpha_3 x^2 + \alpha_4 x^3 = \boldsymbol{X}\boldsymbol{\alpha} \tag{4.4a}$$

where

$$\boldsymbol{X} = [1 \;\; x \;\; x^2 \;\; x^3] \tag{4.4b}$$

and

$$\boldsymbol{\alpha}^{\mathrm{T}} = [\alpha_1 \;\; \alpha_2 \;\; \alpha_3 \;\; \alpha_4]. \tag{4.4c}$$

The coefficients $\alpha_1, \alpha_2, \ldots$ etc. are determined from the nodal displacements. For example at $x=0$, $v_1 = \alpha_1$ and $\theta_1 = (\mathrm{d}v/\mathrm{d}x)_0 = \alpha_2$. Thus the nodal displacements may be written as

$$\boldsymbol{u} = \begin{Bmatrix} v_1 \\ \theta_1 \\ v_2 \\ \theta_2 \end{Bmatrix} = \begin{bmatrix} 1 & 0 & 0 & 0 \\ 0 & 1 & 0 & 0 \\ 1 & a & a^2 & a^3 \\ 0 & 1 & 2a & 3a^2 \end{bmatrix} \begin{Bmatrix} \alpha_1 \\ \alpha_2 \\ \alpha_3 \\ \alpha_4 \end{Bmatrix} = \boldsymbol{A}\boldsymbol{\alpha}. \tag{4.5}$$

The coefficients may be determined by inversion of the A matrix.

$$\boldsymbol{\alpha}=\boldsymbol{A}^{-1}\boldsymbol{u}=\begin{bmatrix} 1 & 0 & 0 & 0 \\ 0 & 1 & 0 & 0 \\ -3/a^2 & -2/a & 3/a^2 & -1/a \\ 2/a^3 & 1/a^2 & -2/a^3 & 1/a^2 \end{bmatrix}\begin{Bmatrix} v_1 \\ \theta_1 \\ v_2 \\ \theta_2 \end{Bmatrix}. \tag{4.6}$$

The principle of virtual displacements states that when a system of forces, which is in equilibrium, undergoes any set of arbitrary virtual displacements the total work done is zero. It can therefore be looked upon as an alternative expression of the equilibrium of a structure. The displacements are described as virtual because they need not actually occur. They may be chosen to suit the particular problem in question. Therefore, applying this principle to the beam element in Fig. 4.3, which is in equilibrium under the loads shown, we now subject it to virtual displacements of the form shown in Fig. 4.4. These need not be equal to the actual displacements, but for generality a cubic curve of similar shape is shown.

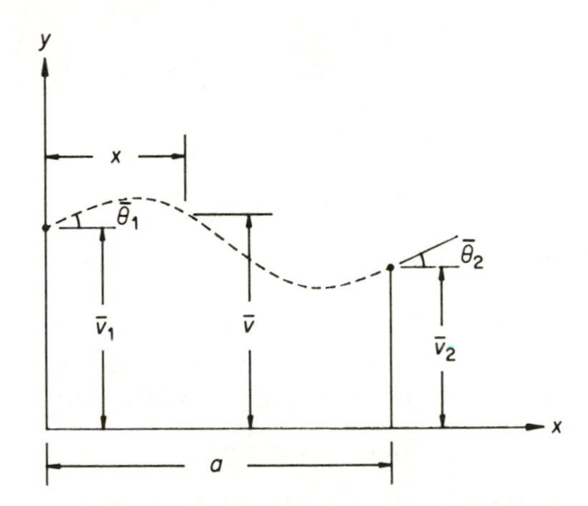

Fig. 4.4 Virtual displacements of beam.

The external virtual work done as a result of this displacement is the product of the nodal forces and the corresponding virtual displacements. Thus

$$W_e=\{\bar{v}_1\ \bar{\theta}_1\ \bar{v}_2\ \bar{\theta}_2\}\begin{Bmatrix} S_1 \\ M_1 \\ S_2 \\ M_2 \end{Bmatrix}=\bar{\boldsymbol{u}}^{\mathrm{T}}\boldsymbol{f} \tag{4.7}$$

where $\bar{v}_1, \bar{\theta}_1, \ldots$ are virtual displacements.

The internal virtual work must be considered rather carefully. Referring to Fig. 4.5(a), a member subjected to a bending moment may be regarded as being composed of a large number of infinitesimal lengths dx. The moments internal to the shaded length act in directions opposite to the rotations at its boundaries (Fig. 4.5(b)). Hence the work done by the internal forces within the shaded length is

$$\mathrm{d}W_{\mathrm{i}} = -M\mathrm{d}\bar{\theta} = -M\,\mathrm{d}x/\bar{R} = -M\,\mathrm{d}^2\bar{v}/\mathrm{d}x^2\,\mathrm{d}x \tag{4.8}$$

where the rotation $\bar{\theta}$ and radius of curvature $\bar{R}$ correspond to the imposed virtual displacement $\bar{v}$.

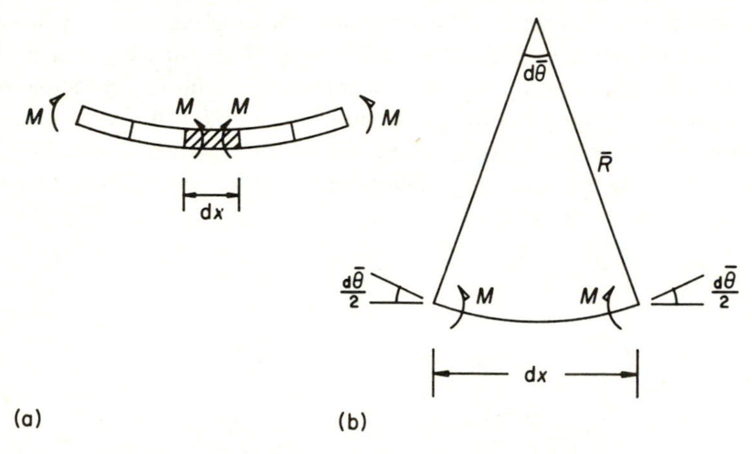

Fig. 4.5 Internal moments and virtual displacements.

In general the bending moment in the flexural element (Fig. 4.3) varies with x and is denoted by $M(x)$. It can be written in terms of the actual displacements of the beam from elementary bending theory, thus:

$$M(x) = EI\,\mathrm{d}^2v/\mathrm{d}x^2. \tag{4.9}$$

Therefore the expression for internal virtual work becomes

$$\mathrm{d}W_{\mathrm{i}} = -EI\frac{\mathrm{d}^2v}{\mathrm{d}x^2}\frac{\mathrm{d}^2\bar{v}}{\mathrm{d}x^2}\,\mathrm{d}x. \tag{4.10}$$

Introducing the displacement function of Equation (4.4),

$$\frac{\mathrm{d}^2v}{\mathrm{d}x^2} = [0\;\;0\;\;2\;\;6x]\boldsymbol{\alpha} = \boldsymbol{B}\boldsymbol{\alpha} = \boldsymbol{B}\boldsymbol{A}^{-1}\boldsymbol{u}. \tag{4.11}$$

A similar expression holds for $\mathrm{d}^2\bar{v}/\mathrm{d}x^2$.

Substituting in (4.10)

$$\mathrm{d}W_{\mathrm{i}} = -EI(\boldsymbol{B}\boldsymbol{A}^{-1}\boldsymbol{u})(\boldsymbol{B}\boldsymbol{A}^{-1}\bar{\boldsymbol{u}}). \tag{4.12}$$

The expressions within parentheses represent scalar quantities and may be validly interchanged or transposed. Integrating over the length of the beam and rearranging we obtain

$$W_i = -EI\bar{\boldsymbol{u}}^T(\boldsymbol{A}^{-1})^T \int_0^a \boldsymbol{B}^T\boldsymbol{B}\,dx\,\boldsymbol{A}^{-1}\,\boldsymbol{u}. \tag{4.13}$$

The principle of virtual displacements states that the total virtual work should be zero if the system is in equilibrium. Therefore

$$W_e + W_i = 0. \tag{4.14}$$

Accordingly we may combine (4.7) and (4.13) and cancel out $\bar{\boldsymbol{u}}^T$, giving

$$\boldsymbol{f} = EI(\boldsymbol{A}^{-1})^T \int_0^a \boldsymbol{B}^T\boldsymbol{B}\,dx\,\boldsymbol{A}^{-1}\,\boldsymbol{u} = \boldsymbol{k}\boldsymbol{u}, \tag{4.15}$$

from which the stiffness matrix is given by

$$\boldsymbol{k} = EI(\boldsymbol{A}^{-1})^T \int_0^a \boldsymbol{B}^T\boldsymbol{B}\,dx\,\boldsymbol{A}^{-1}. \tag{4.16}$$

By utilizing the expressions for $\boldsymbol{A}^{-1}$ and $\boldsymbol{B}$, already obtained in (4.6) and (4.11), it is quite easy to verify that $\boldsymbol{k}$ is the same as that given by Equation (4.2).

The procedure we have just described appears to be a lengthy way of deriving the stiffness matrix of a beam element. However, it is the most general method available and can be applied to the derivation of stiffness matrices for more sophisticated elements, non-linear materials, or large displacements without further conceptual difficulty.

4.2.2 External load vector for distributed loads

Equation (4.3) represents the relation between the forces and displacements at the ends of a flexural member. We now consider the effect of an external distributed load, as shown in Fig. 4.6. In this figure the loads S_1, M_1, . . . etc. are to be regarded as the internal forces required to connect the element to adjacent elements. In addition, we now require nodal forces that are equivalent to the distributed load of intensity $w(x)$. This is done by including the distributed load in the expression for external virtual work when the virtual displacements shown in Fig. 4.4 are imposed on the element. Thus Equation (4.7) now becomes

$$W_e = \{\bar{v}_1\ \bar{\theta}_1\ \bar{v}_2\ \bar{\theta}_2\} \begin{Bmatrix} S_1 \\ M_1 \\ S_2 \\ M_2 \end{Bmatrix} + \int_0^a \bar{v}\,w(x)\,dx, \tag{4.17}$$

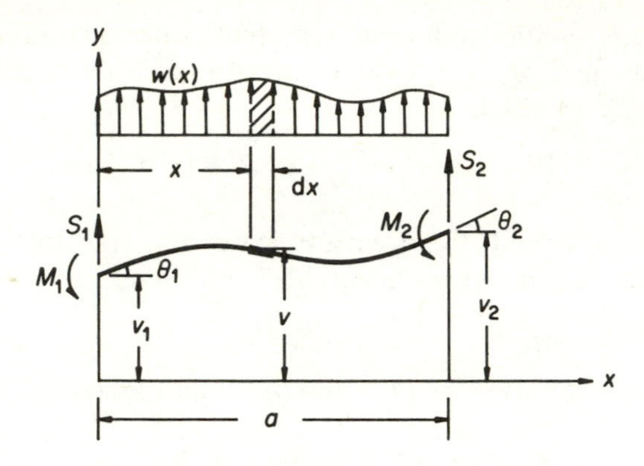

Fig. 4.6 External distributed load on beam element.

where it can be seen that the second term is the product of the load, $w(x)\mathrm{d}x$, and the corresponding virtual displacement, $\bar{v}$, integrated over the length of the element. Using Equation (4.4) it can be seen that

$$\bar{v}=\boldsymbol{X}\bar{\boldsymbol{\alpha}}=\boldsymbol{X}\boldsymbol{A}^{-1}\bar{\boldsymbol{u}}=\bar{\boldsymbol{u}}^{\mathrm{T}}(\boldsymbol{A}^{-1})^{\mathrm{T}}\boldsymbol{X}^{\mathrm{T}}. \tag{4.18}$$

Substituting this in (4.17) we obtain

$$W_{\mathrm{e}}=\bar{\boldsymbol{u}}^{\mathrm{T}}\boldsymbol{f}+\bar{\boldsymbol{u}}^{\mathrm{T}}(\boldsymbol{A}^{-1})^{\mathrm{T}}\int_0^a \boldsymbol{X}^{\mathrm{T}}w(x)\,\mathrm{d}x. \tag{4.19}$$

When the external virtual work is combined with the internal virtual work it is easy to see that the matrix expression becomes

$$\boldsymbol{f}+\boldsymbol{f}_{\mathrm{e}}=\boldsymbol{k}\,\boldsymbol{u} \tag{4.20}$$

where $\boldsymbol{f}_{\mathrm{e}}$ is the external load vector, being the vector of nodal forces equivalent to the external distributed load, and is given by

$$\boldsymbol{f}_{\mathrm{e}}=(\boldsymbol{A}^{-1})^{\mathrm{T}}\int_0^a \boldsymbol{X}^{\mathrm{T}}w(x)\,\mathrm{d}x. \tag{4.21}$$

The vector $\boldsymbol{f}$ contains only the forces which connect adjacent elements together.

EXAMPLE 4.1

The load vector for a uniformly distributed load may be obtained by substituting from (4.4) and (4.6) for $\boldsymbol{X}$ and $\boldsymbol{A}^{-1}$, and by taking w as constant.

Therefore

$$f_e = w \begin{bmatrix} 1 & 0 & -3/a^2 & 2/a^3 \\ 0 & 1 & -2/a & 1/a^2 \\ 0 & 0 & 3/a^2 & -2/a^3 \\ 0 & 0 & -1/a & 1/a^2 \end{bmatrix} \int_0^a \begin{Bmatrix} 1 \\ x \\ x^2 \\ x^3 \end{Bmatrix} dx = w \begin{Bmatrix} a/2 \\ a^2/12 \\ a/2 \\ -a^2/12 \end{Bmatrix}. \tag{4.22}$$

It may be seen that the shear forces applied at each end of the element are both equal to half of the total load. Furthermore, there are moments acting at each node. The moment at node 2 acts negatively, or clockwise, according to the directions indicated in Fig. 4.6.

4.2.3 Mass matrix for beam element

In a dynamic problem there will be, in addition, inertia forces arising from the acceleration of parts of the element. Fig. 4.7 depicts a beam element with distributed mass of intensity $m(x)$ per unit length. If an infinitesimal portion of the mass, $m(x)\mathrm{d}x$, moves in the y direction with an acceleration of $\ddot{v}$, there will be an inertia force, acting in opposition to the motion, given by

$$w(x)\mathrm{d}x = -\ddot{v}m(x)\mathrm{d}x. \tag{4.23}$$

This is equivalent to an external distributed load whose form is determined by the acceleration and the mass distribution of the element. The external load vector is therefore given by

$$f_e = -(A^{-1})^{\mathrm{T}} \int_0^a X^{\mathrm{T}} \ddot{v} m(x)\mathrm{d}x. \tag{4.24}$$

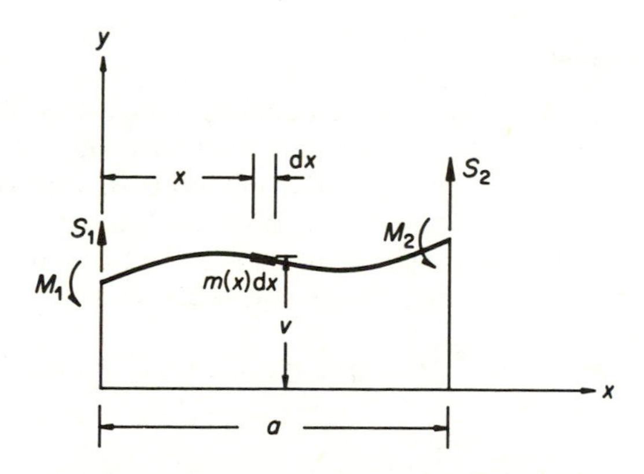

Fig. 4.7 Beam element with distributed mass.

However, from (4.4) and (4.6) it is evident that

$$\ddot{v} = \boldsymbol{X}\boldsymbol{A}^{-1}\ddot{\boldsymbol{u}}. \tag{4.25}$$

Substituting (4.25) into (4.24) and using (4.20) we obtain

$$\begin{aligned} \boldsymbol{f} &= \boldsymbol{k}\boldsymbol{u} + (\boldsymbol{A}^{-1})^{\mathrm{T}} \int_0^a \boldsymbol{X}^{\mathrm{T}} \boldsymbol{X}\, m(x) \mathrm{d}x\, \boldsymbol{A}^{-1}\ddot{\boldsymbol{u}} \\ &= \boldsymbol{k}\boldsymbol{u} + \boldsymbol{m}\ddot{\boldsymbol{u}}. \end{aligned} \tag{4.26}$$

This is the familiar equation of motion for a multi-degree-of-freedom system where $\boldsymbol{m}$ is the mass matrix of the element. In the case of uniformly distributed mass, where m is constant, this is given by

$$\begin{aligned} \boldsymbol{m} &= m(\boldsymbol{A}^{-1})^{\mathrm{T}} \int_0^a \boldsymbol{X}^{\mathrm{T}} \boldsymbol{X} \mathrm{d}x\, \boldsymbol{A}^{-1} \\ &= \frac{ma}{420} \begin{bmatrix} 156 & & \text{symm.} & \\ 22a & 4a^2 & & \\ 54 & 13a & 156 & \\ -13a & -3a^2 & -22a & 4a^2 \end{bmatrix}. \end{aligned} \tag{4.27}$$

This is called a 'consistent' mass matrix because it is derived using the same shape functions as were used in deriving the stiffness matrix. In a simplified lumped-mass calculation there would only be diagonal terms representing masses associated with the linear motions v_1 and v_2 and mass moments of inertia associated with rotations θ_1 and θ_2. The consistent mass matrix is a more accurate representation since it takes into account the dynamic coupling between the rotational and linear motions.

4.2.4 Matrix assembly

The stiffness matrix for a complete structure is obtained by simply adding stiffness terms of adjacent elements wherever they have a degree of freedom in common. This may be illustrated with the uniform beam loaded as shown in Fig. 4.8(a). Considering the beam as two separate elements, as in Fig. 4.8(b), the expressions for the shears and moments at node 2 may be written down by making use of Equations (4.2), (4.20) and (4.22). Thus

$$S_{2L} - \frac{w_1 a_1}{2} = \frac{EI}{a_1^3}(-12v_1 - 6a_1\theta_1 + 12v_2 - 6a_1\theta_2) \tag{4.28a}$$

$$S_{2R} - \frac{w_2 a_2}{2} = \frac{EI}{a_2^3}(12v_2 + 6a_2\theta_2 - 12v_3 + 6a_2\theta_3) \tag{4.28b}$$

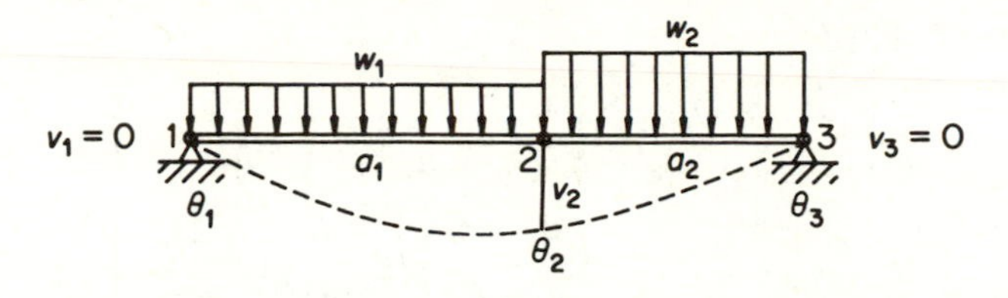

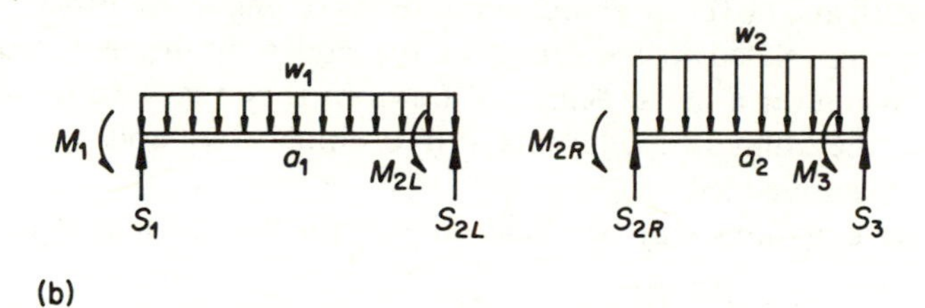

Fig. 4.8 Simple addition of stiffnesses.

$$M_{2L}+\frac{w_1a_1^2}{12}=\frac{EI}{a_1^3}(6a_1v_1+2a_1^2\theta_1-6a_1v_1+4a_1^2\theta_2) \tag{4.28c}$$

$$M_{2R}-\frac{w_2a_2^2}{12}=\frac{EI}{a_2^3}(6a_2v_2+4a_2^2\theta_2-6a_2v_3+2a_2^2\theta_3). \tag{4.28d}$$

The stiffnesses depend on the lengths, a_1 and a_2, of the left- and right-hand elements respectively and the distributed loads are acting in the negative y direction.

It will be noted that for equilibrium at the common node, number 2, the shear forces on either side must take values consistent with the condition

$$S_{2L}+S_{2R}=0. \tag{4.29}$$

Therefore by adding Equations (4.28a) and (4.28b) the unknown shears may be eliminated and a single equation obtained thus:

$$-\frac{w_1a_1}{2}-\frac{w_2a_2}{2}=EI\left[-\frac{12}{a_1^3}v_1-\frac{6}{a_1^2}\theta_1+\left(\frac{12}{a_1^3}+\frac{12}{a_2^3}\right)v_2 + \left(-\frac{6}{a_1^2}+\frac{6}{a_2^2}\right)\theta_2-\frac{12}{a_2^3}v_3+\frac{6}{a_2^2}\theta_3\right]. \tag{4.30}$$

In the same way, equilibrium of the moments at node 2 requires that

$$M_{2L}+M_{2R}=0, \tag{4.31}$$

and therefore a single equation may be obtained by adding Equations

(4.28c) and (4.28d):

$$+\frac{w_1 a_1^2}{12}-\frac{w_2 a_2^2}{12}=EI\left[\frac{6}{a_1^2}v_1+\frac{2}{a_1}\theta_1+\left(-\frac{6}{a_1^2}+\frac{6}{a_2^2}\right)v_2 + \left(\frac{4}{a_1}+\frac{4}{a_2}\right)\theta_2-\frac{6}{a_2^2}v_3+\frac{2}{a_2}\theta_3\right]. \tag{4.32}$$

Equations (4.30) and (4.32) are expressed in terms of unknown displacements only, these being related to the forces at the nodes by means of stiffness coefficients. Wherever a nodal degree of freedom is common to more than one element, the stiffness coefficients (for example $4/a_1$ and $4/a_2$ corresponding to θ_2) are added.

Two further equations may be obtained, one for the moment at node 1,

$$M_1-\frac{w_1 a_1^2}{12}=\frac{EI}{a_1^3}(6a_1 v_1+4a_1^2\theta_1-6a_1 v_2+2a_1^2\theta_2), \tag{4.33}$$

and the other for moment at node 3,

$$M_3+\frac{w_2 a_2^2}{12}=\frac{EI}{a_2^3}(6a_2 v_2+2a_2^2\theta_2-6a_2 v_3+4a_2^2\theta_3). \tag{4.34}$$

At the simply supported ends the moments are zero ($M_1=0$ and $M_3=0$), and the displacements are zero ($v_1=0$ and $v_3=0$). Therefore, the four equations (4.30), (4.32), (4.33) and (4.34) contain the four unknown displacements (θ_1, v_2, θ_2 and θ_3) for which they may be solved. This is the reason that the finite element method is called a *displacement* method: the primary variables are displacements. Note that there is no need for the equations in S_1 and S_3 because they correspond to zero displacements.

The general procedure for assembly is demonstrated in the following matrix equation for a complete structure.

$$\left\{\begin{matrix} \uparrow \\ \boldsymbol{fe}_1 \\ \downarrow\uparrow \\ \boldsymbol{fe}_2 \\ \uparrow\downarrow \\ \boldsymbol{fe}_3 \\ \downarrow \end{matrix}\right\} = \left[\begin{matrix} \boldsymbol{k}_1 & & \\ & \boldsymbol{k}_2 & \\ & & \boldsymbol{k}_3 \end{matrix}\right] \left\{\begin{matrix} v_1 \\ \theta_1 \\ v_2 \\ \theta_2 \\ v_3 \\ \theta_3 \\ v_4 \\ \theta_4 \end{matrix}\right\}. \tag{4.35}$$

This may be written compactly in the form

$$\boldsymbol{F}=\boldsymbol{K}\boldsymbol{U}, \tag{4.36}$$

where $\boldsymbol{F}$ is the load vector, $\boldsymbol{K}$ is the structure stiffness matrix and $\boldsymbol{U}$ is the displacement vector.

It may be seen that the stiffness coefficients of the individual elements add together wherever the element matrices $\boldsymbol{k}_1$, $\boldsymbol{k}_2$ overlap within the complete structure stiffness matrix. The same rule applies to the external load vectors $\boldsymbol{fe}_1, \boldsymbol{fe}_2, \ldots$. In finite element programs there is an automatic procedure for assembly of element matrices which involves an 'equation vector' which is simply a book-keeping device for allocating each possible nodal displacement to a particular equation in the final set of equations. Where there are supports or other restraints, such as nodes 1 and 3 in Fig. 4.8 the non-existent displacement is 'deflated' and does not have an equation in the final assembly.

In a frame structure there will also be members that are vertical or inclined. Before assembly into the final equations, the stiffnesses will need to be corrected so that force–displacement relations in local axes are transformed into the corresponding values required in the global axes of the complete structure. This can be done by inspection in the case of vertical members. Reference should be made to specialist textbooks on finite elements for the general case.

The mass matrices for individual elements may be assembled into a structure mass matrix by essentially the same procedure. This can be understood by referring to Equation (4.26) which, if fully expanded, could be treated in the same way as we have described for the stiffness matrix on its own. Thus it is evident that the assembled dynamic equation of motion will be given by

$$\boldsymbol{M}\ddot{\boldsymbol{U}} + \boldsymbol{K}\boldsymbol{U} = \boldsymbol{F} \tag{4.37}$$

where $\boldsymbol{M}$ is the structure mass matrix and $\ddot{\boldsymbol{U}}$ is the vector of second derivatives of the degrees of freedom with respect to time.

Equation (4.37) is of the same form as the equation of motion of a lumped parameter system given in the previous chapter. It should be obvious, therefore, that the solution will proceed by methods already discussed in Chapter 3.

4.2.5 Solution routines and back substitution for stresses

The flow diagram in Fig. 4.9 illustrates the successive stages in a finite element analysis. There are basically three stages, these being: (a) formation and assembly of matrix equations; (b) equation solution; and (c) back-substitution of displacements for stresses.

We have already dealt with the theory of the first stage, though it must be said that a substantial amount of computer code is required so that the problem can be described in a standard, easy to use, numerical format.

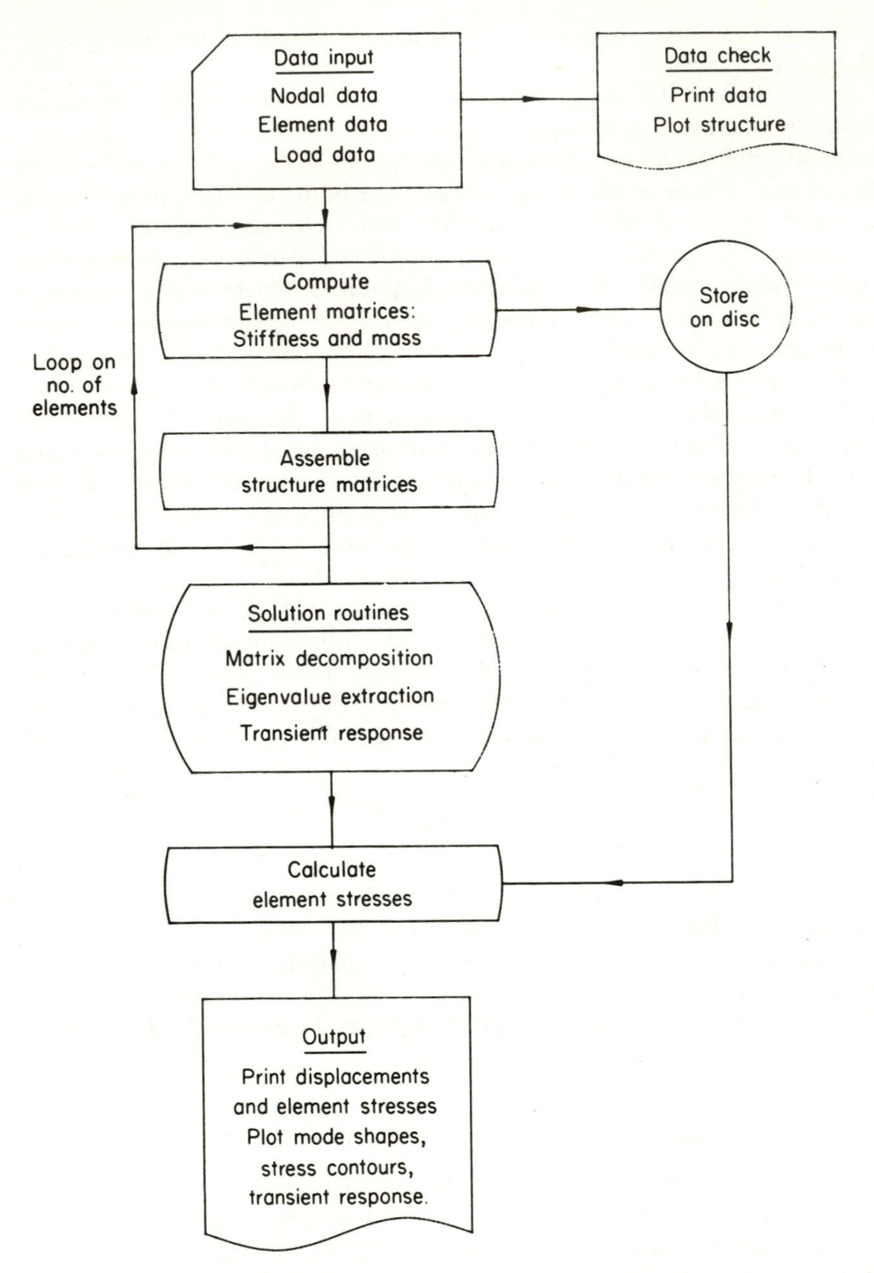

Fig. 4.9 Flow diagram of finite element analysis.

Commercial finite elements programs make effective use of modern interactive computer hardware so that there is a friendly interface between the engineering user and the analysis program. In particular, there would be automatic generation of nodal geometry for regularly shaped parts of a structure, internal data checks for self consistency and optional printouts or graphical plots of the structure as described. Graphical displays are helpful for eliminating mistakes in the data entry of a complicated structure.

It will be noted in Fig. 4.9 that the individual element stiffness matrices are temporarily stored for later use in the calculation of internal stresses. Since the main memory of a computer is limited, it is worthwhile storing this data on secondary media such as magnetic disc or tape. This operation is normally transparent to the user. Meanwhile, the stiffness and mass matrices are assembled in the main memory, though even then there is usually some temporary storage activity going on for large problems.

The main types of equation solution routine are: matrix decomposition; eigenvalue and eigenvector extraction; and step-by-step transient response. The first is conceptually equivalent to matrix inversion though, in fact, it consists of decomposing a large matrix into upper and lower triangular matrices. The normal modes and natural frequencies are obtained by extracting the eigenvalues and eigenvectors from the equations of free vibration, usually employing methods based on inverse iteration. Finally, in many dynamic problems the transient dynamic response to non-periodic external loads is required. This usually involves evaluation of the Duhamel integral, or step-by-step integration of the modal equations using a direct integration procedure. For nonlinear problems direct integration of the original set of coupled matrix equations cannot be avoided. Computer implementation of the principal solution routines are discussed later in this chapter.

The output of the equation solution stage will be in the form of nodal displacements. These may be useful in assessing the performance of a structure from the point of view of human sensitivity to vibration (see Chapter 11). The dynamic stresses, especially the maximum values induced during transient response, are of considerable importance. These must be obtained, element by element, by back-substitution of the displacements into the element stiffness matrix equations (see Equation (4.2)). Note that in these equations M_1, S_1 . . . etc. are the forces required to connect members together, i.e. the internal forces from which stresses must be calculated.

EXAMPLE 4.2

A portal frame, with members of uniform stiffness and mass, is shown in Fig. 4.10(a). It is rigidly jointed at B and C, is pinned to the foundation at A and D, and is loaded with a uniform intensity of w over the transverse beam. Determine the bending moments in the frame when $h = 3$ m.

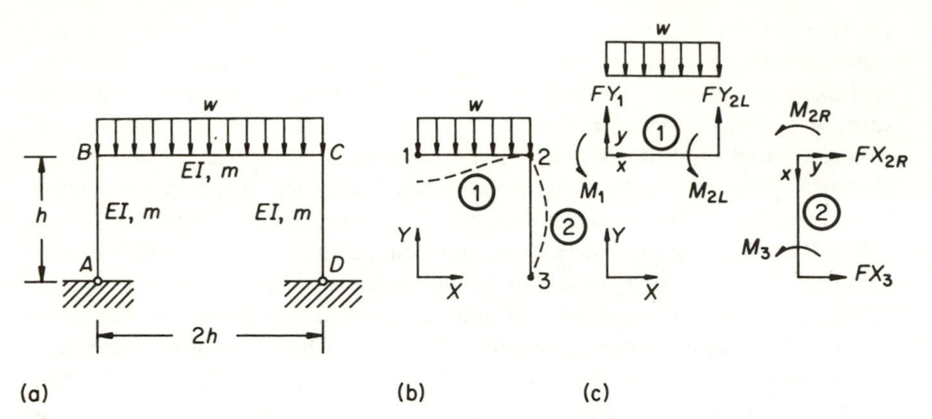

Fig. 4.10 Finite element representation of a portal frame.

Solution

Since both the structure and the loading are symmetrical, only one half need be considered. However, the deflection conditions at midspan have to be chosen carefully. For ease of calculation only two elements will be used, one for the right half of the beam and one for the column as shown in Fig. 4.10(b). A finite element program stores element and nodal data in the form of lists or arrays. The elements are numbered 1 and 2, and the nodes are numbered 1, 2 and 3.

First the stiffness equations for the individual elements will be obtained by referring to the system of forces shown in Fig. 4.10(c). The equations for element 1, using (4.2), (4.20) and (4.22), are given by

$$\begin{Bmatrix} FY_1 \\ M_1 \\ FY_{2L} \\ M_{2L} \end{Bmatrix} - w \begin{Bmatrix} h/2 \\ h^2/12 \\ h/2 \\ -h^2/12 \end{Bmatrix} = \frac{EI}{h^3} \begin{bmatrix} 12 & & \text{symm.} & \\ 6h & 4h^2 & & \\ -12 & -6h & 12 & \\ 6h & 2h^2 & -6h & 4h^2 \end{bmatrix} \begin{Bmatrix} v_1 \\ \theta_1 \\ v_2 \\ \theta_2 \end{Bmatrix}. \tag{4.38}$$

Note that the applied load is in the negative y direction. Because of symmetry there will be no rotation at midspan and therefore $\theta_1 = 0$. Also, if the elements are effectively inextensional (because of their great stiffness longitudinally compared with flexure) then v_2 may also be assumed to be zero. Consequently θ_1 and v_2 will not be amongst the unknown variables and their equations will not be required in the analysis at this stage. Therefore, (4.38) may be reduced to

$$\begin{Bmatrix} FY_1 \\ M_{2L} \end{Bmatrix} - w \begin{Bmatrix} h/2 \\ -h^2/12 \end{Bmatrix} = \frac{EI}{h^3} \begin{bmatrix} 12 & 6h \\ 6h & 4h^2 \end{bmatrix} \begin{Bmatrix} v_1 \\ \theta_2 \end{Bmatrix}. \quad (4.39)$$

Similarly, the stiffness equations for element 2, in the local coordinates (x, y), are given by

$$\begin{Bmatrix} FX_{2R} \\ M_{2R} \\ FX_3 \\ M_3 \end{Bmatrix} = \frac{EI}{h^3} \begin{bmatrix} 12 & & \text{symm.} & \\ 6h & 4h^2 & & \\ -12 & -6h & 12 & \\ 6h & 2h^2 & -6h & 4h^2 \end{bmatrix} \begin{Bmatrix} u_2 \\ \theta_2 \\ u_3 \\ \theta_3 \end{Bmatrix}. \quad (4.40)$$

There will be no displacement, u_3, at the base and, because of symmetry, $u_2=0$ also. Therefore, the effective equations are

$$\begin{Bmatrix} M_{2R} \\ M_3 \end{Bmatrix} = \frac{EI}{h^3} \begin{bmatrix} 4h^2 & 2h^2 \\ 2h^2 & 4h^2 \end{bmatrix} \begin{Bmatrix} \theta_2 \\ \theta_3 \end{Bmatrix}. \quad (4.41)$$

The moment–rotation relationship will be the same regardless of the orientation of element 2. Therefore, no transformation of (4.41) will be required in this instance.

The degrees of freedom of the half structure are v_1, θ_2 and θ_3. It will be noted that u_1, u_2 and θ_1 will be zero because of symmetry, while v_2, u_3 and v_3 are zero because of the boundary and inextensional conditions. Hence, the stiffness equations for the complete structure may be assembled as shown in Equation (4.35). Observing that $FY_1=0$, because of symmetry, M_{2L} and M_{2R} cancel out to satisfy equilibrium, and $M_3=0$, it may be verified that the final equations are given by

$$w \begin{Bmatrix} -h/2 \\ h^2/12 \\ 0 \end{Bmatrix} = \frac{EI}{h^3} \begin{bmatrix} 12 & 6h & 0 \\ 6h & 4h^2+4h^2 & 2h^2 \\ 0 & 2h^2 & 4h^2 \end{bmatrix} \begin{Bmatrix} v_1 \\ \theta_2 \\ \theta_3 \end{Bmatrix}. \quad (4.42)$$

It should be noted that the stiffnesses of the separate elements are added where there is a common degree of freedom, such as θ_2 in this example. Furthermore, note the existence of null terms leaving a symmetric and banded form to the stiffness matrix. This is a common feature of stiffness matrices of large structures, which can be utilized advantageously in the numerical solution of the equations.

Substituting $h=3$, the equations to be solved are

$$w\begin{Bmatrix} -1.5 \\ 0.75 \\ 0 \end{Bmatrix} = \frac{EI}{27}\begin{bmatrix} 12 & 18 & 0 \\ 18 & 72 & 18 \\ 0 & 18 & 36 \end{bmatrix}\begin{Bmatrix} v_1 \\ \theta_2 \\ \theta_3 \end{Bmatrix}. \tag{4.43}$$

Solution by direct inversion of the stiffness matrix is not practicable for large problems. The most efficient solution routines for linear equations are based upon Gauss elimination, a procedure developed in the 19th century. This method will now be employed to solve the above equations. It consists of a simple process of eliminating the unknown displacements from the equations, one at a time, until the final displacement can be calculated directly. Finally all the displacements may be obtained by successive back-substitution.

Considering Equations (4.43), the first step is to eliminate v_1 by subtracting a multiple of the first equation from the remainder. The resulting equations are

$$\begin{bmatrix} 12 & 18 & 0 \\ 0 & 45 & 18 \\ 0 & 18 & 36 \end{bmatrix}\begin{Bmatrix} v_1 \\ \theta_2 \\ \theta_3 \end{Bmatrix} = \frac{27w}{EI}\begin{Bmatrix} -1.5 \\ 3.0 \\ 0 \end{Bmatrix}. \tag{4.44}$$

Next a multiple of the second equation is subtracted from the third yielding

$$\begin{bmatrix} 12 & 18 & 0 \\ 0 & 45 & 18 \\ 0 & 0 & 28.8 \end{bmatrix}\begin{Bmatrix} v_1 \\ \theta_2 \\ \theta_3 \end{Bmatrix} = \frac{27w}{EI}\begin{Bmatrix} -1.5 \\ 3.0 \\ -1.2 \end{Bmatrix}. \tag{4.45}$$

Note that the operation was carried out on the load vector also. We are now in a position to determine θ_3 directly from the third equation:

$$\theta_3 = -1.125w/EI. \tag{4.46a}$$

The remaining variables may be obtained by back-substitution of (4.46a), giving

$$\theta_2 = \frac{1}{45}\left(\frac{27w}{EI}\times 3.0 - 18\,\theta_3\right) = 2.25\,w/EI \tag{4.46b}$$

$$v_1 = \frac{1}{12}\left(-\frac{27w}{EI}\times 1.5 - 18\,\theta_2\right) = -6.75\,w/EI. \tag{4.46c}$$

The analysis just described is the basis of efficient matrix decomposition procedures as used in large-scale finite element programs. These will be discussed in more detail later in this chapter. To complete the solution to a practical problem the primary variables v_1, θ_2 and θ_3 have to be back-substituted into the element equations to obtain the internal forces and, hence, the stresses. Therefore, substituting (4.46) into (4.38) and (4.40) we obtain

$$\begin{aligned} FY_1 &= 0 \\ M_1 &= -2.25\,w \\ FY_{2L} &= 3\,w \\ M_{2L} &= -2.25\,w \\ FX_{2R} &= 0.75\,w \\ M_{2R} &= 2.25\,w \\ FX_3 &= -0.75\,w \\ M_3 &= 0. \end{aligned}$$

Note how the conditions of equilibrium are satisfied at all of the nodes. The moments and forces calculated are in agreement with 'exact' values.

4.3 OTHER ELEMENT TYPES

In order to subdivide complex structural geometries into finite elements, it is necessary for there to be a range of finite elements, with very simple shapes, capable of making up line segments, areas and volumes. Some examples of typical element shapes are shown in Fig. 4.11. They fall into three basic classes, these being line elements, area elements and volume elements. We have looked at a two-dimensional beam element but there is obviously greater practical scope for a beam element in three dimensions. A broad class of solid mechanics problems can be solved using two-dimensional flat elements. Triangular and isoparametric shapes are illustrated, the latter being a most versatile and effective element because of its accuracy and its ability to be fitted into almost any geometry. Flat elements loaded perpendicular to their planes are useful for plate flexure problems. In three dimensions a flat element with the characteristics of a shell structure may be derived by combining a plate flexure element with a plane stress element. Volume or solid elements are three-dimensional by their very nature. A tetrahedron and a twenty-noded isoparametric element are illustrated. Other types are possible, including an eight-noded brick and a six-noded wedge. Axisymmetric elements are derivatives of two-dimensional elements and are highly efficient for analysing solid structures which are symmetrical about an axis. Axisymmetric shell elements also exist.

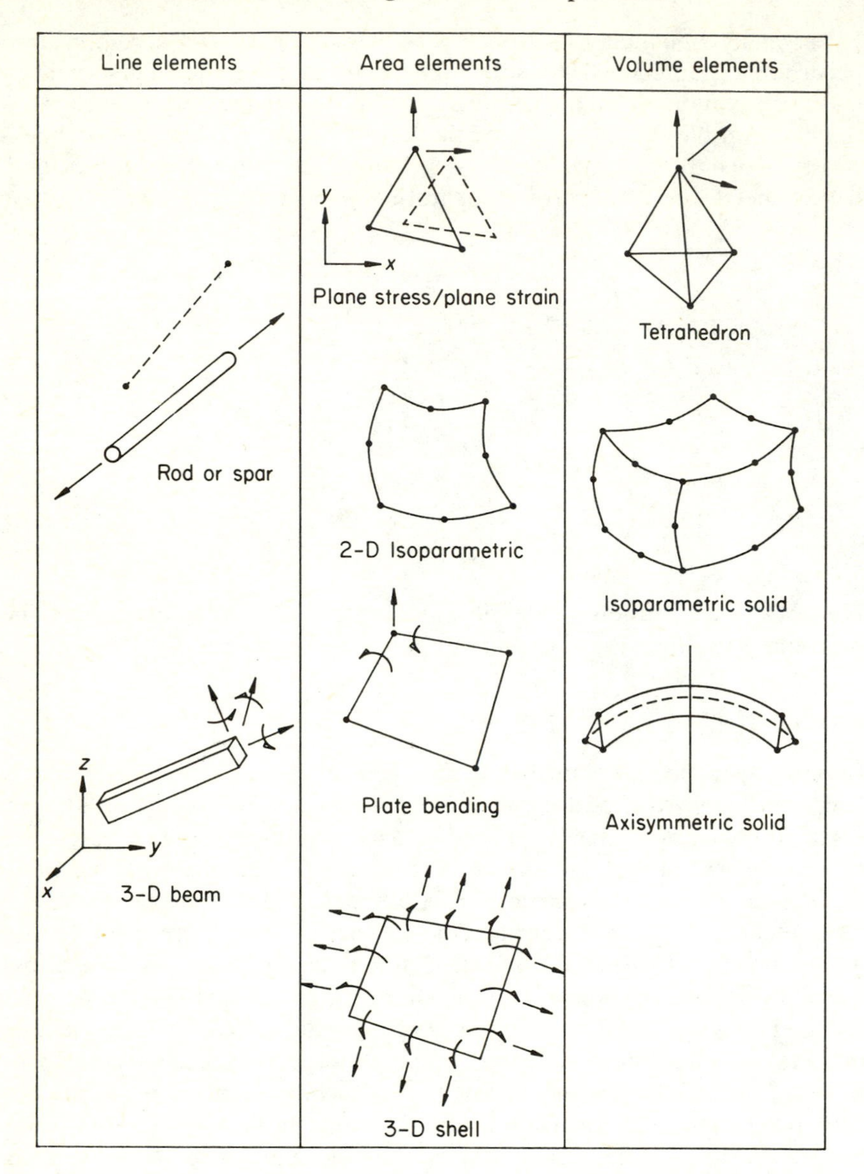

Fig. 4.11 Typical finite element shapes.

Before discussing some important element types in detail it is useful to consider the general form of the formula for a stiffness matrix. This is given by

$$\boldsymbol{k}=(\boldsymbol{A}^{-1})^{\mathrm{T}}\int \boldsymbol{B}^{\mathrm{T}}\boldsymbol{D}\boldsymbol{B}\,\mathrm{d}V(\boldsymbol{A}^{-1}). \tag{4.47}$$

This should be compared carefully with (4.16) and it will be seen that there are two obvious differences. The first is that the integration is taken over a volume. This is because the internal virtual work has to be evaluated over a three-dimensional solid in the general case. The second difference is that the stiffness, EI, of the flexural member has been replaced by a matrix $\boldsymbol{D}$ within the integral. This matrix is required in the general constitutive relation of the material, namely

$$\boldsymbol{\sigma} = \boldsymbol{D}\boldsymbol{\varepsilon} \tag{4.48}$$

where $\boldsymbol{\sigma}$ and $\boldsymbol{\varepsilon}$ are components of stress and strain at a point. The matrix $\boldsymbol{D}$ is usually an elasticity relationship but it could also be a nonlinear material law. The matrices $\boldsymbol{B}$ and $\boldsymbol{A}^{-1}$ are required in the relations between strains and nodal displacements:

$$\boldsymbol{\varepsilon} = \boldsymbol{B}\boldsymbol{A}^{-1}\boldsymbol{u}. \tag{4.49}$$

The forms of $\boldsymbol{B}$ and $\boldsymbol{A}^{-1}$ depend entirely on the polynomial shape functions chosen to express displacement at any point. A cubic polynomial, (4.4), was used in the case of beam flexure.

The general expression for a consistent mass matrix is given by

$$\boldsymbol{m} = (\boldsymbol{A}^{-1})^{\mathrm{T}} \int \boldsymbol{X}^{\mathrm{T}} \boldsymbol{X} m \, \mathrm{d}V (\boldsymbol{A}^{-1}) \tag{4.50}$$

where m is the mass per unit volume and $\boldsymbol{X}$ is the vector of coordinates required in the polynomial shape function.

4.3.1 Three-dimensional rods and beams

A simple and useful element for the analysis of pin-jointed frameworks is the rod element shown in Fig. 4.11. It is capable only of carrying tension or compression along its length. Triangulated lattice girders, trusses and space frames are usually assumed to behave as if pin-jointed, even though secondary stresses exist due to bending at nodes possessing some degree of fixity.

Extensional behaviour is usually ignored in rigid-jointed frame structures, it being assumed that flexure is predominant. However, it is possible to combine both flexural and extensional behaviour in a beam element by including transverse, axial and rotational displacements as nodal degrees of freedom. A fully three-dimensional beam element should include axial forces, bending about perpendicular axes and torsion as shown in Fig. 4.11. Thus there would be six degrees of freedom at each node. A full treatment is given by Przemieniecki (1968). Calculation of stresses is straightforward only for simple sections such as rectangles and circles. In the case of thin-walled sections such as channels there is the added complication of the offset between shear centre and centroid.

4.3.2 Plane stress and plane strain triangle

A frequently used element in two-dimensional stress analysis is the constant-strain triangle shown in Fig. 4.11. It is assumed to have only two possible displacements at each node, with no rotations, and therefore its displaced form will be as shown. Since there are six degrees of freedom in total, internal displacements would have to be given by the expressions

$$u=\alpha_1+\alpha_2 x+\alpha_3 y, \quad v=\alpha_4+\alpha_5 x+\alpha_6 y. \tag{4.51}$$

Note that x and y coordinates should appear in both expressions and therefore the displacement form is linear. In matrix notation this may be written

$$\begin{Bmatrix} u \\ v \end{Bmatrix}=\begin{bmatrix} 1 & x & y & 0 & 0 & 0 \\ 0 & 0 & 0 & 1 & x & y \end{bmatrix}\begin{Bmatrix} \alpha_1 \\ \alpha_2 \\ \alpha_3 \\ \alpha_4 \\ \alpha_5 \\ \alpha_6 \end{Bmatrix}=\boldsymbol{X\alpha}. \tag{4.52}$$

Internal strains may be calculated from Equation (4.52) as follows:

$$\boldsymbol{\varepsilon}=\begin{Bmatrix} \varepsilon_x \\ \varepsilon_y \\ \gamma_{xy} \end{Bmatrix}=\begin{Bmatrix} \dfrac{\partial u}{\partial x} \\ \dfrac{\partial v}{\partial y} \\ \dfrac{\partial u}{\partial y}+\dfrac{\partial v}{\partial x} \end{Bmatrix}=\begin{bmatrix} 0 & 1 & 0 & 0 & 0 & 0 \\ 0 & 0 & 0 & 0 & 0 & 1 \\ 0 & 0 & 1 & 0 & 1 & 0 \end{bmatrix}\boldsymbol{\alpha}=\boldsymbol{B\alpha}. \tag{4.53}$$

It is evident that the strains are independent of x and y. This is why the element is called the 'constant-strain triangle'. Because the strain, and therefore the stress, is constant within each element, it is necessary to use a large number of small elements in regions of stress concentrations so that steep stress gradients can be modelled accurately.

Stress may be calculated from the strains by using the constitutive equations of two-dimensional elasticity (Timoshenko and Goodier, 1951). There are two possible formulations, these being plane stress or plane strain. The former is relevant to thin two-dimensional shapes in which the stress perpendicular to the plane is zero. The constitutive equation for an isotropic material is

$$\boldsymbol{\sigma}=\begin{Bmatrix}\sigma_x\\ \sigma_y\\ \tau_{xy}\end{Bmatrix}=\frac{E}{(1-v^2)}\begin{bmatrix}1 & v & 0\\ v & 1 & 0\\ 0 & 0 & (1-v)/2\end{bmatrix}\begin{Bmatrix}\varepsilon_x\\ \varepsilon_y\\ \gamma_{xy}\end{Bmatrix}=\boldsymbol{D}\boldsymbol{\varepsilon} \qquad (4.54)$$

where v is Poisson's ratio.

A plane-strain form of the stiffness matrix is required for structures in which the z dimension is very large and where the loading and goemetry do not change significantly with z. A gravity dam is a good example in civil engineering. Strain in the z direction is prevented and therefore a different constitutive equation is required in place of Equation (4.54), but in other respects the formulation is identical.

4.3.3 Plate bending

Plate elements are useful for analysis of slab bridges and other flat structures loaded perpendicular to the surface. A rectangular plate element is shown in Fig. 4.12 where it may be seen that there are two rotations and one transverse displacement at each node, giving a total of twelve degrees of freedom. The shape function describing the transverse displacements at any internal point must have twelve coefficients in it and should be a function of x and y. The most suitable function, proposed by Zienkiewicz and Cheung (1964), is as follows:

$$w=\alpha_1+\alpha_2 x+\alpha_3 y+\alpha_4 x^2+\alpha_5 xy+\alpha_6 y^2+\alpha_7 x^3+\alpha_8 x^2 y$$
$$+\alpha_9 xy^2+\alpha_{10} y^3+\alpha_{11} x^3 y+\alpha_{12} xy^3. \qquad (4.55)$$

The function is symmetrical and cubic in x and y. Thus it is capable of describing a deflected profile analogous to the beam element. It should be noted that in plate bending the moment curvature relationship for an

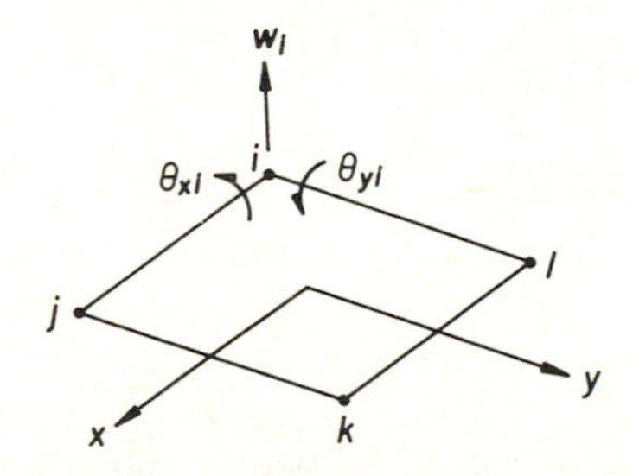

Fig. 4.12 Rectangular plate element.

isotropic plate is given by

$$\boldsymbol{\sigma} = \begin{Bmatrix} M_x \\ M_y \\ M_{xy} \end{Bmatrix} = \frac{Et^3}{12(1-v^2)} \begin{bmatrix} 1 & v & 0 \\ v & 1 & 0 \\ 0 & 0 & \dfrac{1-v}{2} \end{bmatrix} \begin{Bmatrix} -\dfrac{\partial^2 w}{\partial x^2} \\ -\dfrac{\partial^2 w}{\partial y^2} \\ 2\dfrac{\partial^2 w}{\partial x \partial y} \end{Bmatrix} = \boldsymbol{DB\alpha} \qquad (4.56)$$

where t is the plate thickness.

4.3.4 Thin-shell elements

A curved shell may be considered as an assembly of flat elements as shown in Fig. 4.13. The characteristic behaviour of a thin shell is a combination of in-plane or membrane forces combined with bending action as in a thin plate (Timoshenko and Woinowsky-Krieger, 1959). Thus a thin-shell finite element

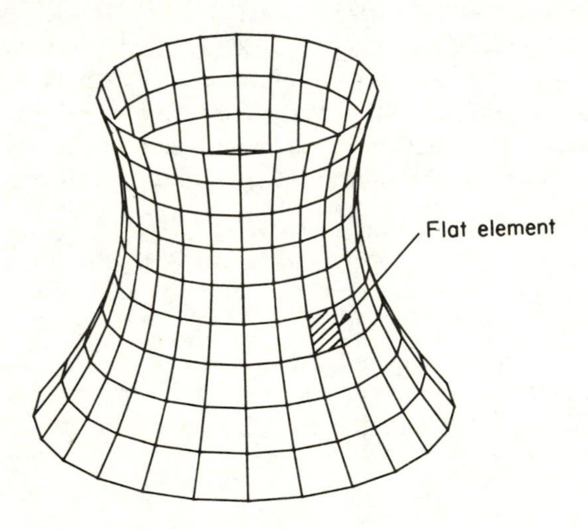

Fig. 4.13 Shell as an assembly of flat elements.

may be obtained by combining the stiffness matrices of a plane stress element with a plate bending element. The forces and nodal displacements of a rectangular shell element are shown in Fig. 4.14. It should be noted that the membrane force, N_x or N_y, is the product of membrane stress, σ_x or σ_y, and the element thickness, t. The moments, M_x and M_y, correspond to the moments in plate flexure. In general there will also be shear forces, N_{xy} (equivalent to $t \times \tau_{xy}$) and twisting moments, M_{xy}.

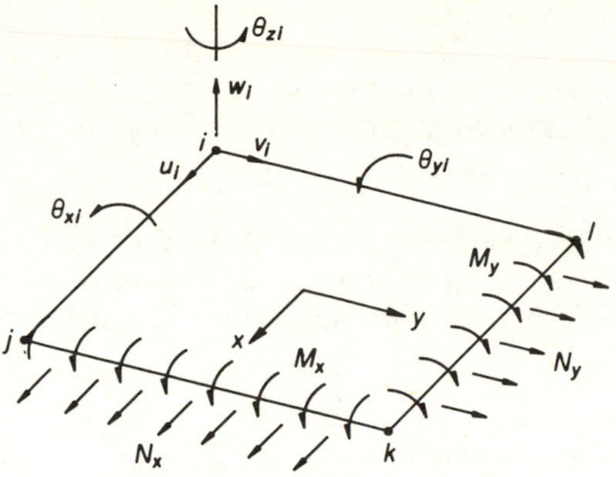

Fig. 4.14 Rectangular shell element.

Because a shell element is fully three-dimensional there should be six degrees of freedom at each node, as shown in Fig. 4.14. However, plane stress elements do not normally contain an in-plane rotational stiffness associated with θ_z. This lack of stiffness could result in the structure stiffness matrix being ill conditioned where elements are coplanar or meet at a shallow angle. The accepted way of avoiding the consequent numerical difficulties is to include a small rotational stiffness proportional to the direct membrane stiffness (Zienkiewicz, Parekh and King, 1968). Alternatively, a higher-order membrane element possessing in-plane rotational stiffness, as suggested by Sisodiya and Cheung (1971), may be used.

A quadrilateral version of the above shell element is generally available, and so too are triangular shell elements. Thus it is possible to mesh irregular geometries without difficulty.

4.3.5 Isoparametric elements

It was shown in §4.2.1 that a fundamental step in the derivation of a stiffness matrix is the choice of polynomial shape function for interpolating element displacements from the nodal displacements. The polynomial coefficients $\boldsymbol{\alpha}$ had to be obtained by evaluating a transformation matrix $\boldsymbol{A}^{-1}$ which was a function of the local coordinates of the element. In contrast, the isoparametric concept, developed by Irons (1966) and Ergatoudis, Irons and Zienkiewicz (1968), is based upon the use of 'natural' coordinate systems for elements which enable the geometry to be expressed in dimensionless form. This approach not only removes the need to evaluate $\boldsymbol{A}^{-1}$ but also leads to a general form of interpolation function from which whole families of continuum elements may be derived. Bathe (1982) maintains that isoparametric elements are more efficient than the conventional type in most practical applications.

The fundamental step is to introduce a system of 'natural' coordinates which are related to the local cartesian coordinates by the transformation

$$x = \boldsymbol{N}\boldsymbol{x} \tag{4.57}$$

where x is any point, $\boldsymbol{x}$ are the nodes in local coordinates, and $\boldsymbol{N}$ is an interpolation function in terms of the natural coordinates. It may be shown that a similar relation holds for displacement at a point in terms of nodal displacements, so that

$$u = \boldsymbol{N}\boldsymbol{u} \tag{4.58}$$

where $\boldsymbol{u}$ are the nodal displacements.

This may be illustrated with the simple rod element in Fig. 4.15 which has nodes located at x_1 and x_2 in the local coordinate system and has natural coordinates, r, whose origin is at mid-length of the bar, ranging from -1 to $+1$. Thus any internal point x is given by

$$x = \{\tfrac{1}{2}(1-r) \quad \tfrac{1}{2}(1+r)\}\begin{Bmatrix} x_1 \\ x_2 \end{Bmatrix} = \boldsymbol{N}\boldsymbol{x}. \tag{4.59}$$

Displacements within the rod are related to nodal displacements in the same way. Thus:

$$u = \{\tfrac{1}{2}(1-r) \quad \tfrac{1}{2}(1+r)\}\begin{Bmatrix} u_1 \\ u_2 \end{Bmatrix} = \boldsymbol{N}\boldsymbol{u}. \tag{4.60}$$

Internal strain is determined by

$$\varepsilon = \frac{\mathrm{d}u}{\mathrm{d}x} = \frac{\mathrm{d}u}{\mathrm{d}r}\frac{\mathrm{d}r}{\mathrm{d}x} = \frac{1}{a}\{-1 \quad -1\}\begin{Bmatrix} u_1 \\ u_2 \end{Bmatrix} = \boldsymbol{B}\boldsymbol{u} \tag{4.61}$$

in which the strain displacement matrix $\boldsymbol{B}$ is obtained directly without the need for transformation using $\boldsymbol{A}^{-1}$.

The stiffness matrix is obtained by evaluation of the integral

$$\boldsymbol{k} = AE\int_0^a \boldsymbol{B}^{\mathrm{T}}\boldsymbol{B}\,\mathrm{d}x \tag{4.62}$$

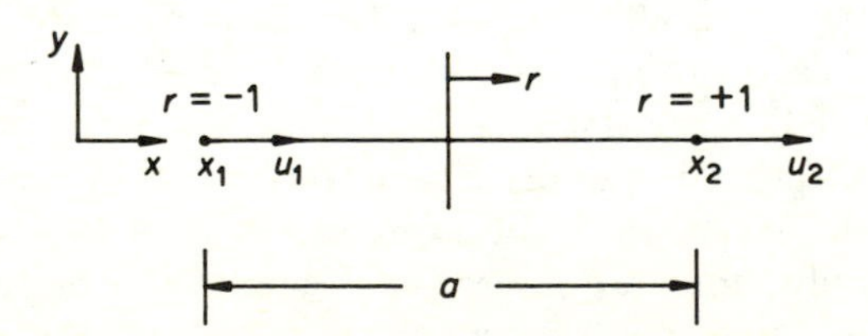

Fig. 4.15 Isoparametric rod element.

which in natural coordinates becomes

$$\boldsymbol{k}=AE\int_{-1}^{+1}\boldsymbol{B}^{\mathrm{T}}\boldsymbol{B}\frac{\mathrm{d}x}{\mathrm{d}r}\,\mathrm{d}r=AE\int_{-1}^{1}\boldsymbol{B}^{\mathrm{T}}\boldsymbol{B}J\,\mathrm{d}r \tag{4.63}$$

where $J=\mathrm{d}x/\mathrm{d}r$ is the Jacobian transforming local to natural coordinates. Hence

$$\boldsymbol{k}=\frac{AE}{a}\begin{bmatrix}1 & -1\\ -1 & 1\end{bmatrix}. \tag{4.64}$$

The interpolation functions for the four-node quadrilateral isoparametric element shown in Fig. 4.16 are given by

$$x=\{n_1\ n_2\ n_3\ n_4\}\begin{Bmatrix}x_1\\ x_2\\ x_3\\ x_4\end{Bmatrix}=\boldsymbol{N}\boldsymbol{x},\quad y=\boldsymbol{N}\boldsymbol{y} \tag{4.65}$$

where

$$\begin{aligned}n_1&=\tfrac{1}{4}(1+r)(1+s)\\ n_2&=\tfrac{1}{4}(1-r)(1+s)\\ n_3&=\tfrac{1}{4}(1-r)(1-s)\\ n_4&=\tfrac{1}{4}(1+r)(1-s).\end{aligned} \tag{4.66}$$

Curved element boundaries may be obtained by including additional mid-side nodes as shown in Fig. 4.11. The interpolation functions of Equation (4.66) may be modified to allow for the additional nodes.

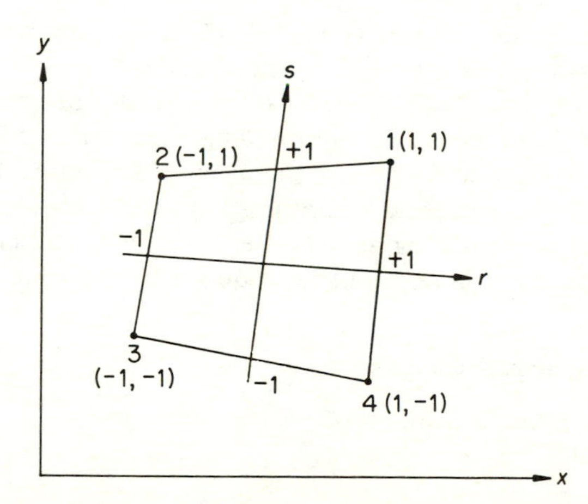

Fig. 4.16 Quadrilateral isoparametric plane element.

The same principles apply to three-dimensional volume elements, a generally available version being the curved solid with eight to twenty variable number nodes curved solid also shown in Fig. 4.11. Interpolation functions are given by Bathe (1982). There are many applications where triangular, wedge or tetrahedral elements can be fitted more conveniently into the geometry. Such elements may be obtained as degenerate versions of plane quadrilaterals or eight-noded solids. This is achieved by collapsing one or more of the sides and thus making some of the nodes coincident as shown in Fig. 4.17.

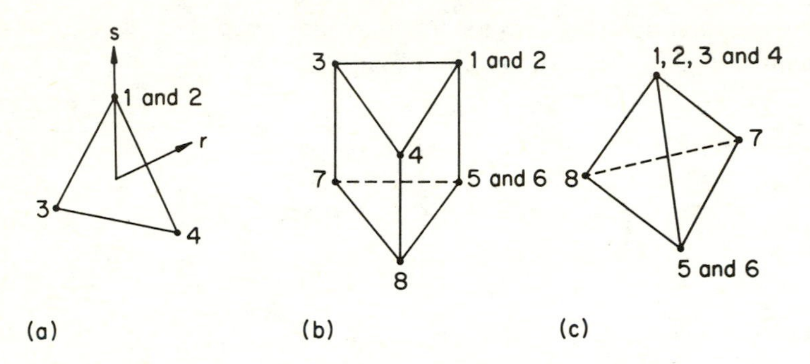

Fig. 4.17 Degenerate versions of isoparametric elements.

4.4 SOLUTION OF EQUATIONS

Finite element analysis is essentially a systematic way of discretizing a structure, having distributed mass and stiffness, and thereby reducing it to a multi-degree-of-freedom system described by the matrix equation

$$\boldsymbol{M}\ddot{\boldsymbol{u}} + \boldsymbol{C}\dot{\boldsymbol{u}} + \boldsymbol{K}\boldsymbol{u} = \boldsymbol{F}. \tag{4.67}$$

Therefore the methods of analysis outlined in Chapter 3 apply equally well to finite element systems. However, it is clear that the order of the matrices in Equation (4.67) will be very large for most practical problems. A few thousand degrees of freedom, with corresponding equations, would not be unreasonable in the case of many civil engineering structures. Consequently, some highly efficient numerical algorithms have been developed and implemented in finite element programs (Bathe, 1982). We shall now review the most generally useful solution routines required for dynamic analysis.

4.4.1 Stiffness matrix decomposition

The matrix equation for static problems is given by

$$\boldsymbol{K}\boldsymbol{u} = \boldsymbol{F}. \tag{4.68}$$

Solution of the equations by direct inversion of the stiffness matrix is inefficient and generally impractical for large problems. The most efficient static solution routines are based on the Gauss elimination procedure. This method, which was introduced in Example 4.2, is merely a systematic way of solving a set of simultaneous equations by successive elimination of variables.

Consider the following equations:

$$\begin{bmatrix} k_{11} & k_{12} & k_{13} & k_{14} \\ k_{21} & k_{22} & k_{23} & k_{24} \\ k_{31} & k_{32} & k_{33} & k_{34} \\ k_{41} & k_{42} & k_{43} & k_{44} \end{bmatrix} \begin{Bmatrix} u_1 \\ u_2 \\ u_3 \\ u_4 \end{Bmatrix} = \begin{Bmatrix} F_1 \\ F_2 \\ F_3 \\ F_4 \end{Bmatrix}. \tag{4.69}$$

Variable u_1 is first eliminated from the second equation by multiplying all coefficients of the first equation by k_{21}/k_{11} and subtracting from the second equation. The same is done for the third equation, using a factor of k_{31}/k_{11}, and similarly for the fourth equation, yielding

$$\left[\begin{array}{c|ccc} k_{11} & k_{12} & k_{13} & k_{14} \\ \hline 0 & k'_{22} & k'_{23} & k'_{24} \\ 0 & k'_{32} & k'_{33} & k'_{34} \\ 0 & k'_{42} & k'_{43} & k'_{44} \end{array}\right] \begin{Bmatrix} u_1 \\ u_2 \\ u_3 \\ u_4 \end{Bmatrix} = \begin{Bmatrix} F_1 \\ F'_2 \\ F'_3 \\ F'_4 \end{Bmatrix}, \tag{4.70a}$$

where

$$k'_{ij} = k_{ij} - \frac{k_{i1}}{k_{11}} k_{ij} \tag{4.70b}$$

and

$$F'_i = F_i - \frac{k_{i1}}{k_{11}} F_1. \tag{4.70c}$$

We now have three simultaneous equations from which to solve for u_2, u_3 and u_4. Thus we can repeat the above procedure, eliminating u_2 by using k'_{22} as the *pivot*. The process is continued until we are left with only one equation from which to obtain the final unknown. The decomposed equations will now be in the form

$$\boldsymbol{S}\boldsymbol{u} = \begin{bmatrix} S_{11} & S_{12} & S_{13} & S_{14} \\ & S_{22} & S_{23} & S_{24} \\ & & S_{33} & S_{34} \\ & & & S_{44} \end{bmatrix} \begin{Bmatrix} u_1 \\ u_2 \\ u_3 \\ u_4 \end{Bmatrix} = \begin{Bmatrix} \phi_1 \\ \phi_2 \\ \phi_3 \\ \phi_4 \end{Bmatrix} = \boldsymbol{\phi}. \tag{4.71}$$

This process is known as *factorization* or *decomposition*. The matrix $\boldsymbol{S}$ may be regarded as the product of a diagonal matrix $\boldsymbol{D}$, whose elements are the pivots in the Gauss elimination, and an upper triangular matrix $\boldsymbol{H}^{\mathrm{T}}$, having a unit leading diagonal thus:

$$\boldsymbol{S}=\boldsymbol{D}\boldsymbol{H}^{\mathrm{T}}=\begin{bmatrix} d_{11} & & & \\ & d_{22} & & \\ & & d_{33} & \\ & & & d_{44} \end{bmatrix}\begin{bmatrix} 1 & h_{12} & h_{13} & h_{14} \\ & 1 & h_{23} & h_{24} \\ & & 1 & h_{34} \\ & & & 1 \end{bmatrix} \tag{4.72}$$

where $d_{ii}=s_{ii}$ and $h_{ij}=s_{ij}$.

It is now clear that having factorized $\boldsymbol{K}$ it is easy to obtain the displacements by successive back-substitution in Equation (4.71) beginning with u_4. Gauss elimination may therefore be summed up in the following three steps:

(a) factorization of the stiffness matrix so that

$$\boldsymbol{K}=\boldsymbol{H}\boldsymbol{D}\boldsymbol{H}^{\mathrm{T}}; \tag{4.73}$$

(b) reduction of the load vector using

$$\boldsymbol{H}\boldsymbol{\phi}=\boldsymbol{F}; \tag{4.74}$$

(c) back-substitution to obtain $\boldsymbol{u}$ using

$$\boldsymbol{D}\boldsymbol{H}^{\mathrm{T}}\boldsymbol{u}=\boldsymbol{\phi}, \tag{4.75}$$

this equation being identical to (4.71).

Equation (4.73) is valid provided that the matrix $\boldsymbol{K}$ is symmetrical. This is always true of stiffness matrices. The equations may be reduced in any order but to retain numerical accuracy it is best to use the largest current diagonal element as the pivot.

In a practical structural problem the stiffness matrix is usually of banded form so that all the non-zero terms are clustered around the leading diagonal thus:

$$\boldsymbol{K} = \tag{4.76}$$

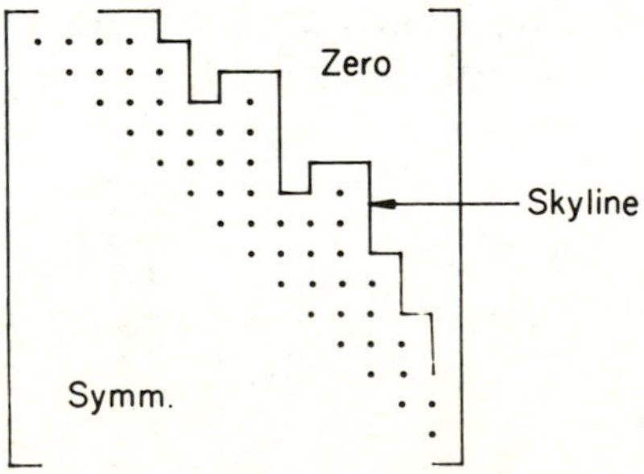

Zero
Skyline
Symm.

An efficient storage scheme makes use of this fact and it is therefore only necessary to operate on the non-zero elements of one half of the matrix.

An important variation of the above scheme is known as Cholesky factorization. In this method the stiffness matrix is decomposed into upper and lower triangular matrices as follows:

$$\boldsymbol{K} = \boldsymbol{L}\boldsymbol{L}^{\mathrm{T}} \tag{4.77a}$$

where

$$\boldsymbol{L} = \boldsymbol{H}\boldsymbol{D}^{1/2}. \tag{4.77b}$$

4.4.2 Eigenvalue solution

We now turn to the problem of obtaining natural frequencies and modes of vibration from the equations of undamped free vibration:

$$\boldsymbol{M}\ddot{\boldsymbol{u}} + \boldsymbol{K}\boldsymbol{u} = 0. \tag{4.78}$$

In Chapter 3 it was shown that this could be transformed into a standard eigenvalue problem:

$$\boldsymbol{K}\boldsymbol{\phi} = \omega^2 \boldsymbol{M}\boldsymbol{\phi}. \tag{4.79}$$

The equations are satisfied by eigenvalues, ω_i^2, and corresponding eigenvectors $\boldsymbol{\phi}_i$. These represent the natural frequencies and modes of vibration, and there are as many combinations as there are degrees of freedom in the finite element model.

Solution by inverse iteration was described in §3.4.3 in which a trial vector is introduced into the right-hand side of (4.79) and an improved vector extracted from the ensuing equilibrium equation. Iteration results in an eigenvalue and corresponding eigenvector converging onto the fundamental mode of vibration. Higher modes may then be obtained by using the property of orthogonality to sweep lower modes out of the trial vector and thereby converging onto the next lowest mode. However, some large-scale finite element problems may possess a few thousand degrees of freedom with perhaps fifty or a hundred significant modes. On this scale, it is not feasible to contain all of the data in a computer's main memory at one time and therefore transfers into and out of secondary storage, e.g. disc, are inevitable. Consequently, some highly efficient variations of inverse iteration have been developed in recent years.

Many of these derive from the method of simultaneous iteration suggested by Jennings (1967), in which inverse iteration is performed using a group of vectors simultaneously. The first step is to choose a set of q starting vectors, when $q > p$ and p is the number of required modes. The starting vectors should be initially orthogonal and a suitable choice has been suggested by Bathe (1982). Simultaneous iteration is then performed as follows:

$$\boldsymbol{K}\bar{\boldsymbol{X}}_{k+1} = \boldsymbol{M}\boldsymbol{X}_k \tag{4.80}$$

where $\boldsymbol{X}_k=\{x_1\ x_2\ \ldots\ x_n\ \ldots\ ]_k$ is the set of q trial vectors at the kth iteration. New trial vectors are obtained by extracting the vectors $\bar{\boldsymbol{x}}_{k+1}$ from the ensuing equilibrium equation, as in inverse iteration, and normalizing them by

$$\boldsymbol{x}_{k+1}=\frac{\bar{\boldsymbol{x}}_{k+1}}{(\bar{\boldsymbol{x}}_{k+1}^{\mathrm{T}}\boldsymbol{M}\bar{\boldsymbol{x}}_{k+1})^{1/2}} \tag{4.81}$$

Since the denominator in (4.81) is approximately $\sqrt{}$(generalized mass), it can be observed that this normalization procedure satisfies the relation

$$\boldsymbol{x}_{k+1}^{\mathrm{T}}\boldsymbol{M}\boldsymbol{x}_{k+1}=1, \tag{4.82}$$

which is referred to as 'mass orthonormality'. The frequencies are obtained approximately from Rayleigh's quotient (see §3.4.3). Before the next cycle of iteration the new matrix of trial vectors, $\boldsymbol{X}_{k+1}$, must be orthogonalized. A frequently used vector orthogonalization procedure in finite element analysis is the Gram–Schmidt method. This is equivalent to the 'mode sweeping' process of Equation (3.119) and is achieved by

$$\boldsymbol{x}^*=\boldsymbol{x}-\sum_{i=1}^{m}\alpha_i\boldsymbol{x}_i \tag{4.83a}$$

where

$$\alpha_i=\boldsymbol{x}_i^{\mathrm{T}}\boldsymbol{M}\boldsymbol{x}. \tag{4.83b}$$

In this procedure $\boldsymbol{x}_i$ are the lowest m trial vectors already orthogonalized, while $\boldsymbol{x}$ is the next lowest trial vector needing to be orthogonalized. Equation (4.83) may be obtained from (3.119) provided that the mass orthonormality rule, (4.82), applies and that the trial vectors are approximations to the true mode shapes.

A more sophisticated procedure was developed by Bathe (1971), which he termed 'subspace iteration'. A set of q trial vectors, $\boldsymbol{X}_k$, is substituted in the right-hand side of (4.80) and the set of modified vectors, $\bar{\boldsymbol{X}}_{k+1}$, is extracted from the equilibrium equation. These are used to project the structure stiffness and mass matrices, $\boldsymbol{K}$ and $\boldsymbol{M}$, onto a space of dimension q by means of

$$\boldsymbol{K}_{k+1}=\bar{\boldsymbol{X}}_{k+1}^{\mathrm{T}}\boldsymbol{K}\bar{\boldsymbol{X}}_{k+1} \tag{4.84a}$$

$$\boldsymbol{M}_{k+1}=\bar{\boldsymbol{X}}_{k+1}^{\mathrm{T}}\boldsymbol{M}\bar{\boldsymbol{X}}_{k+1}. \tag{4.84b}$$

There will be an eigensystem corresponding to this subspace given by

$$\boldsymbol{K}_{k+1}\boldsymbol{Q}_{k+1}=\boldsymbol{M}_{k+1}\boldsymbol{Q}_{k+1}\boldsymbol{\Lambda}_{k+1} \tag{4.85}$$

where $\boldsymbol{Q}_{k+1}$ is the matrix of eigenvectors of the subspace and $\boldsymbol{\Lambda}_{k+1}$ is the diagonal matrix of corresponding eigenvalues. Equation (4.85) can be solved by any eigenvalue method suitable for small problems. The improved approximation to the eigenvectors is found from

$$X_{k+1} = \bar{X}_{k+1} Q_{k+1}. \tag{4.86}$$

It should be apparent that the subspace mass and stiffness matrices, M_{k+1} and K_{k+1} given by (4.84), will approximate to diagonal matrices containing the generalized masses and stiffnesses. Therefore, Λ_{k+1} will converge onto the diagonal matrix containing the eigenvalues ω_i^2 and X_{k+1} will converge onto the modal matrix Φ.

Subspace iteration is undoubtedly the most widely used eigenvalue solution procedure for large problems, and has been implemented in many well known programs such as *SAP*IV (Bathe, Wilson and Peterson, 1974), *ADINA* (Bathe and Ramaswamy, 1980) and *ANSYS* (De Salvo and Swanson, 1985).

An alternative eigenvalue procedure was suggested by Lanczos (1950), but was not widely adopted initially because of certain deficiencies. However, in recent years the method has been improved upon considerably, and there is evidence that it is a serious rival to subspace iteration (Nour-Omid, Parlett and Taylor, 1983; Weingarten, Ramanathan and Chen, 1983; Sehmi 1986). First the standard eigenvalue equation

$$K\phi_i = \omega_i^2 M \phi_i \tag{4.87}$$

is transformed by making use of Cholesky factorization, $K = LL^{\mathrm{T}}$ (see §4.4.1), into the form

$$A\psi_i = (1/\omega_i^2)\psi_i \tag{4.88a}$$

where

$$A = L^{-1} M L^{-\mathrm{T}} \tag{4.88b}$$

and

$$\psi_i = L^{\mathrm{T}} \phi_i. \tag{4.88c}$$

Equation (4.88) is still a large eigenvalue problem. The basic step in the Lanczos algorithm is to transform A into a tridiagonal matrix T by using a set of orthonormal vectors $V = \{v_1, v_2, \ldots, v_N]$ such that

$$T = V^{\mathrm{T}} A V \tag{4.89a}$$

$$V^{\mathrm{T}} V = I. \tag{4.89b}$$

From (4.89) it is evident that

$$AV = VT = V \begin{bmatrix} \alpha_1 & \gamma_1 & & & \\ \gamma_1 & \alpha_2 & \gamma_2 & & \\ & \cdot & \cdot & \cdot & \\ & & \cdot & \cdot & \cdot \\ & & & \cdot & \cdot & \cdot \\ & & & & & \gamma_{N-1} \\ & & & & \gamma_{N-1} & \alpha_N \end{bmatrix} \tag{4.90}$$

to give the N equations

$$\boldsymbol{A}\boldsymbol{v}_i = \gamma_{i-1}\boldsymbol{v}_{i-1} + \alpha_i\boldsymbol{v}_i + \gamma_i\boldsymbol{v}_{i+1}; \quad i = 1, 2, \ldots, N. \tag{4.91}$$

This is a three-term recurrence formula from which, in conjunction with (4.89b), all the α_i, γ_i and $\boldsymbol{v}_i$ may be obtained once a suitable starting vector $\boldsymbol{v}_1$ has been chosen, e.g. a unit full vector.

Now, using the relationship

$$\boldsymbol{\psi}_i = \boldsymbol{V}\boldsymbol{z}_i, \tag{4.92}$$

the eigenvalue equation (4.88) can be transformed into the form

$$\boldsymbol{T}\boldsymbol{z}_i = (1/\omega_i^2)\boldsymbol{z}_i, \tag{4.93}$$

which is an eigenproblem with the same eigenvalues as (4.88). The Lanczos algorithm, as indicated in (4.90), may be truncated after q steps and therefore (4.93) is a reduced eigenvalue problem which may be solved by standard methods.

Versions of the Lanczos method are now available in some well known finite element programs including *NASTRAN* (1970) and *SAP* VII (1980), and it is likely to become increasingly popular as refinements are implemented.

4.4.3 Direct integration of equations of motion

The equations of motion of a structure subjected to arbitrary dynamic loading are

$$\boldsymbol{M}\ddot{\boldsymbol{u}} + \boldsymbol{C}\dot{\boldsymbol{u}} + \boldsymbol{K}\boldsymbol{u} = \boldsymbol{P}. \tag{4.94}$$

In Chapter 3, we saw that these could be solved by the method of modal superposition. However, there is a risk of stress detail being lost if the structure is very irregular in shape, or the loads are concentrated and contain shock-like fluctuations. Of course, more modes could be included to resolve the detail but at the expense of greater computation time. In these circumstances it is often preferable to solve the equations by direct integration. Indeed, there would be no alternative to direct integration of the equations of motion for a nonlinear structure.

Numerical integration of the equations of motion was introduced in Chapter 2. It is assumed that a step-by-step solution is sought so that displacement, velocity and acceleration are obtained at times 0, Δt, $2\Delta t, \ldots, t-\Delta t, t, t+\Delta t, \ldots$ with a regular interval of Δt. The important characteristics of an integration scheme are its *stability* and *accuracy*. There are three methods that are known to be unconditionally stable. These are the Newmark method (Newmark, 1959), the Wilson θ method (Wilson, Farhoomand and Bathe, 1973), and the Houbolt method (Houbolt, 1950). The first two are extensions of the linear acceleration method (see §2.9.3), and there is little to choose between them. The Houbolt method is less accurate, requiring about 50% more steps to achieve the same accuracy. We shall now study the Newmark method because it is probably the best known direct integration procedure.

Based on the linear acceleration method, the equations for forward integration of the velocities and displacements are

$$\dot{\boldsymbol{u}}_{t+\Delta t}=\dot{\boldsymbol{u}}_t+[(1-\delta)\ddot{\boldsymbol{u}}_t+\delta\ddot{\boldsymbol{u}}_{t+\Delta t}]\,\Delta t \tag{4.95}$$

$$\boldsymbol{u}_{t+\Delta t}=\boldsymbol{u}_t+\dot{\boldsymbol{u}}_t\Delta t+[(\tfrac{1}{2}-\beta)\ddot{\boldsymbol{u}}_t+\beta\ddot{\boldsymbol{u}}_{t+\Delta t}]\,\Delta t^2 \tag{4.96}$$

where δ and β are weighting factors. Careful study of these equations will reveal that the standard rules of kinematics are being used to predict the velocity and displacement at the end of an interval given the conditions at the start of the interval. The difference is that the acceleration is not constant and therefore weighting factors are used to obtain an average acceleration over the interval.

Using (4.96) the acceleration at the end of the interval is given by

$$\ddot{\boldsymbol{u}}_{t+\Delta t}=(1/\beta\Delta t^2)\,[\boldsymbol{u}_{t+\Delta t}-\boldsymbol{u}_t-\dot{\boldsymbol{u}}_t\Delta t-(\tfrac{1}{2}-\beta)\ddot{\boldsymbol{u}}_t\Delta t^2]. \tag{4.97}$$

Substituting in (4.95), we obtain the velocity

$$\dot{\boldsymbol{u}}_{t+\Delta t}=(\delta/\beta\Delta t)\,(\boldsymbol{u}_{t+\Delta t}-\boldsymbol{u}_t)+(1-\delta/\beta)\,\dot{\boldsymbol{u}}_t+(1-\delta/2\beta)\,\Delta t\,\ddot{\boldsymbol{u}}_t. \tag{4.98}$$

The equation of motion at the end of the interval is given by

$$\boldsymbol{M}\ddot{\boldsymbol{u}}_{t+\Delta t}+\boldsymbol{C}\dot{\boldsymbol{u}}_{t+\Delta t}+\boldsymbol{K}\boldsymbol{u}_{t+\Delta t}=\boldsymbol{P}_{t+\Delta t}. \tag{4.99}$$

Therefore, substituting (4.97) and (4.98) into (4.99) we obtain

$$\begin{aligned}\left[\boldsymbol{K}+\frac{1}{\beta\Delta t^2}\boldsymbol{M}+\frac{\delta}{\beta\Delta t}\boldsymbol{C}\right]\boldsymbol{u}_{t+\Delta t}&\\ =\boldsymbol{P}_{t+\Delta t}&+\left(\frac{1}{\beta\Delta t^2}\boldsymbol{M}+\frac{\delta}{\beta\Delta t}\boldsymbol{C}\right)\boldsymbol{u}_t\\ &+\left[\frac{1}{\beta\Delta t}\boldsymbol{M}-\left(1-\frac{\delta}{\beta}\right)\boldsymbol{C}\right]\dot{\boldsymbol{u}}_t\\ &+\left[\left(\frac{1}{2\beta}-1\right)\boldsymbol{M}-\left(1-\frac{\delta}{2\beta}\right)\Delta t\,\boldsymbol{C}\right]\ddot{\boldsymbol{u}}_t.\end{aligned} \tag{4.100}$$

Equation (4.100) is therefore used for forward integration of the displacements from time t to time $t+\Delta t$. Then Equations (4.97) and (4.98) are used to obtain the velocities and accelerations.

It should be apparent that (4.100) takes the form of an equilibrium equation, $\boldsymbol{Ku}=\boldsymbol{P}$, with the structural motion at time t modifying the load on the right-hand side. On the left-hand side the stiffness matrix is modified correspondingly. No special starting procedure is required. In the case of dynamic response of a linear system the modified stiffness need only be determined and factorized at the start of the solution. Then the right-hand side, which will vary with time, can be treated as a new load vector at each time step. Rapid evaluation of the full response history is therefore possible. If it is a nonlinear problem, the left-hand side will have to be evaluated at every time step and therefore a full equilibrium solution will also be required at each time step. Consequently, nonlinear dynamic response tends to be expensive in computer time.

A comprehensive study of the stability and accuracy of numerical integration procedures was carried out by Bathe and Wilson (1973). They recommended that the best results for the Newmark method were obtained with $\delta=\frac{1}{2}$ and $\beta=\frac{1}{4}$. The method is unconditionally stable with these values, and satisfactory accuracy can be achieved if $\Delta t \leqslant 1/20f$, where f is the highest frequency of interest. The effect of using greater step lengths is to filter out higher frequencies by elongation of their periods.

4.4.4 Reduction of stiffness and mass matrices

In the solution of most linear problems it is usually unnecessary to solve for all the system degrees of freedom. This is because structures tend to adopt smoother, less distorted shapes when vibrating. Therefore, it is often more efficient to reduce the number of degrees of freedom by a technique called 'dynamic matrix condensation'. The procedure was suggested by Guyan (1965) and is implemented in some commercial finite element programs (De Salvo and Swanson, 1985).

The matrix equation $\boldsymbol{F}=\boldsymbol{K}\boldsymbol{u}$ may be partitioned in the form

$$\begin{Bmatrix} \boldsymbol{F}_1 \\ \boldsymbol{F}_2 \end{Bmatrix} = \begin{bmatrix} \boldsymbol{A} & \boldsymbol{B} \\ \boldsymbol{B}^{\mathrm{T}} & \boldsymbol{C} \end{bmatrix} \begin{Bmatrix} \boldsymbol{u}_1 \\ \boldsymbol{u}_2 \end{Bmatrix} \tag{4.101}$$

where $\boldsymbol{u}_1$ are the chosen *master* degrees of freedom and $\boldsymbol{u}_2$ are the degrees of freedom to be eliminated. Clearly, there should be forces applied only at the master degrees of freedom and therefore $\boldsymbol{F}_2=0$, yielding

$$\boldsymbol{B}^{\mathrm{T}}\boldsymbol{u}_1+\boldsymbol{C}\boldsymbol{u}_2=0. \tag{4.102}$$

Substituting this in (4.101), it follows that

$$\boldsymbol{F}_1=\boldsymbol{A}\boldsymbol{u}_1-\boldsymbol{B}\boldsymbol{C}^{-1}\boldsymbol{B}^{\mathrm{T}}\boldsymbol{u}_1. \tag{4.103}$$

Therefore, the reduced stiffness matrix is

$$\boldsymbol{K}_1 = \boldsymbol{A} - \boldsymbol{B}\boldsymbol{C}^{-1}\boldsymbol{B}^{\mathrm{T}}. \tag{4.104}$$

It should be noted that this reduced stiffness matrix is exact since it contains all of the stiffness data. However, it will only yield displacements at the master degrees of freedom. Displacements at the *slave* degrees of freedom are determined from (4.102). This matrix condensation procedure is useful when analysing very large structures which can be broken down into *substructures* or *superelements* where master degrees of freedom occur only at substructure boundaries.

In dynamic analysis the mass matrix may be condensed also, by observing that the foregoing procedure amounts to the transformation

$$\boldsymbol{u} = \begin{Bmatrix} \boldsymbol{u}_1 \\ \boldsymbol{u}_2 \end{Bmatrix} = \begin{bmatrix} \boldsymbol{I} \\ -\boldsymbol{C}^{-1}\boldsymbol{B}^{\mathrm{T}} \end{bmatrix} \boldsymbol{u}_1 = \boldsymbol{T}\boldsymbol{u}_1 \tag{4.105}$$

where $\boldsymbol{I}$ is a unit diagonal matrix and $\boldsymbol{T}$ is the necessary transformation matrix. Therefore, if the structure kinetic energy is given by

$$\text{K.E.} = \tfrac{1}{2}\dot{\boldsymbol{u}}^{\mathrm{T}}\boldsymbol{M}\dot{\boldsymbol{u}}. \tag{4.106}$$

by substituting the transformation (4.105) it follows that

$$\text{K.E.} = \tfrac{1}{2}\dot{\boldsymbol{u}}_1^{\mathrm{T}}\boldsymbol{T}^{\mathrm{T}}\boldsymbol{M}\boldsymbol{T}\dot{\boldsymbol{u}}_1. \tag{4.107}$$

Hence if the structure mass matrix is partitioned in the same way as the stiffness matrix, so that

$$\boldsymbol{M} = \begin{bmatrix} \bar{\boldsymbol{A}} & \bar{\boldsymbol{B}} \\ \bar{\boldsymbol{B}}^{\mathrm{T}} & \bar{\boldsymbol{C}} \end{bmatrix}, \tag{4.108}$$

then the reduced mass matrix must be given by

$$\begin{aligned} \boldsymbol{M}_1 &= \boldsymbol{T}^{\mathrm{T}}\boldsymbol{M}\boldsymbol{T} \\ &= \bar{\boldsymbol{A}} - \bar{\boldsymbol{B}}\boldsymbol{C}^{-1}\boldsymbol{B}^{\mathrm{T}} - (\boldsymbol{C}^{-1}\boldsymbol{B}^{\mathrm{T}})^{\mathrm{T}}(\bar{\boldsymbol{B}}^{\mathrm{T}} - \bar{\boldsymbol{C}}\boldsymbol{C}^{-1}\boldsymbol{B}^{\mathrm{T}}). \end{aligned} \tag{4.109}$$

The master degrees of freedom should be chosen in such a way as to characterize the dynamic motion of the structure, particularly in the lowest modes of vibration. Translational degrees of freedom, distributed uniformly over a structure, are generally suitable, especially in regions of high mass. Rotational degrees of freedom, and those associated with stretching motion, usually correspond to higher modes and may be eliminated. There should be at least twice as many master degrees of freedom as there are required modes of the structure.

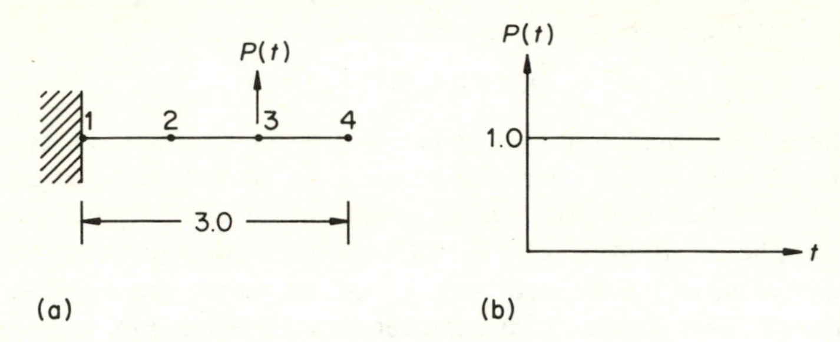

Fig. 4.18 Dynamic loading on a cantilever.

EXAMPLE 4.3

The uniform cantilever shown in Fig. 4.18 is subjected to a load which is applied suddenly at time $t=0$ and then remains constant at a value of $P(t)=1.0$. The cantilever is divided into three equal elements each of length 1.0 and the load is applied to node 3. The stiffness is given by $EI=1$ and the distributed mass by $m=420$. Reduce the order of the problem using Guyan reduction, with master degrees of freedom at v_2, v_3 and v_4. Determine the dynamic response for the first 20 seconds of motion using Newmark's method and ignoring damping.

Solution

The stiffness and mass matrices of each element, using (4.2) and (4.27), are given by

$$\boldsymbol{k}=\begin{bmatrix} 12 & 6 & -12 & 6 \\ 6 & 4 & -6 & 2 \\ -12 & -6 & 12 & -6 \\ 6 & 2 & -6 & 4 \end{bmatrix}, \quad \boldsymbol{m}=\begin{bmatrix} 156 & 22 & 54 & -13 \\ 22 & 4 & 13 & -3 \\ 54 & 13 & 156 & -22 \\ -13 & -3 & -22 & 4 \end{bmatrix}.$$

Noting that $\theta_1=0$ and $v_1=0$ at the support, the assembled stiffness and mass matrices of the complete structure will be given by

$$\boldsymbol{K}=\begin{bmatrix} 24 & & & & & \\ 0 & 8 & & & \text{symm.} & \\ -12 & -6 & 24 & & & \\ 6 & 2 & 0 & 8 & & \\ & & -12 & -6 & 12 & \\ & & 6 & 2 & -6 & 4 \end{bmatrix} \begin{matrix} v_2 \\ \theta_2 \\ v_3 \\ \theta_3 \\ v_4 \\ \theta_4 \end{matrix},$$

$$
\boldsymbol{M}=\begin{bmatrix} 312 & & & & & \\ 0 & 8 & & & \text{symm.} & \\ 54 & 13 & 312 & & & \\ -13 & -3 & 0 & 8 & & \\ & & 54 & 13 & 156 & \\ & & -13 & -3 & -22 & 4 \end{bmatrix} \begin{matrix} v_2 \\ \theta_2 \\ v_3 \\ \theta_3 \\ v_4 \\ \theta_4 \end{matrix} .
$$

It will be observed that banded matrices are obtained with the degrees of freedom arranged in the order of successive elements. Before employing Guyan reduction these matrices will be re-ordered, merely by exchanging rows and columns, so that the translational degrees of freedom are collected together. It can be verified quite easily that

$$
\boldsymbol{K}=\left[\begin{array}{ccc|ccc} 24 & -12 & 0 & 0 & 6 & 0 \\ -12 & 24 & -12 & -6 & 0 & 6 \\ 0 & -12 & 12 & 0 & -6 & -6 \\ \hline 0 & -6 & 0 & 8 & 2 & 0 \\ 6 & 0 & -6 & 2 & 8 & 2 \\ 0 & 6 & -6 & 0 & 2 & 4 \end{array}\right] \begin{matrix} v_2 \\ v_3 \\ v_4 \\ \\ \theta_2 \\ \theta_3 \\ \theta_4 \end{matrix} ,
$$

$$
\boldsymbol{M}=\left[\begin{array}{ccc|ccc} 312 & 54 & 0 & 0 & -13 & 0 \\ 54 & 312 & 54 & 13 & 0 & -13 \\ 0 & 54 & 156 & 0 & 13 & -22 \\ \hline 0 & 13 & 0 & 8 & -3 & 0 \\ -13 & 0 & 13 & -3 & 8 & -3 \\ 0 & -13 & -22 & 0 & -3 & 4 \end{array}\right] \begin{matrix} v_2 \\ v_3 \\ v_4 \\ \\ \theta_2 \\ \theta_3 \\ \theta_4 \end{matrix} .
$$

The matrices are partitioned as above so that

$$
\boldsymbol{K}=\begin{bmatrix} \boldsymbol{A} & \boldsymbol{B} \\ \boldsymbol{B}^{\mathrm{T}} & \boldsymbol{C} \end{bmatrix}, \quad \boldsymbol{M}=\begin{bmatrix} \bar{\boldsymbol{A}} & \bar{\boldsymbol{B}} \\ \bar{\boldsymbol{B}}^{\mathrm{T}} & \bar{\boldsymbol{C}} \end{bmatrix}
$$

in which the rotational degrees of freedom are to be eliminated. The inverse of $\boldsymbol{C}$ can be obtained by Gauss elimination, or any other method, and is given by

$$
\boldsymbol{C}^{-1}=\frac{1}{208}\begin{bmatrix} 28 & -8 & 4 \\ -8 & 32 & -16 \\ 4 & -16 & 60 \end{bmatrix} .
$$

Hence

$$C^{-1}B^{\mathrm{T}}=\frac{1}{208}\begin{bmatrix}-48 & -144 & 24\\ 192 & -48 & -96\\ -96 & 336 & -264\end{bmatrix}.$$

Then using (4.104) the reduced stiffness matrix is

$$K_1=A-BC^{-1}B^{\mathrm{T}}=\begin{bmatrix}24 & -12 & 0\\ -12 & 24 & -12\\ 0 & -12 & 12\end{bmatrix}-\begin{bmatrix}5.54 & -1.38 & -2.77\\ -1.38 & 13.85 & -8.31\\ -2.77 & -8.31 & 10.38\end{bmatrix}$$

$$=\begin{bmatrix}18.46 & -10.62 & 2.77\\ -10.62 & 10.15 & -3.69\\ 2.77 & -3.69 & 1.62\end{bmatrix}$$

Turning to the mass matrix, and omitting a few intervening calculations, it can be shown that

$$\bar{B}C^{-1}B^{\mathrm{T}}+(C^{-1}B^{\mathrm{T}})^{\mathrm{T}}(\bar{B}^{\mathrm{T}}-\bar{C}C^{-1}B^{\mathrm{T}})=\begin{bmatrix}35.93 & 12.44 & 27.20\\ 12.44 & -75.97 & -13.04\\ 27.20 & -13.04 & 38.79\end{bmatrix}.$$

Hence, using (4.109), the reduced mass matrix is given by

$$M_1=\begin{bmatrix}276.07 & 41.56 & -27.20\\ 41.56 & 387.97 & 40.96\\ -27.20 & 40.96 & 117.21\end{bmatrix}.$$

The calculations above may have appeared somewhat laborious, but it will be appreciated that the procedure is in a form that lends itself to computer implementation.

Next, the undamped equations of motion will be numerically integrated using Newmark's method. The highest frequency possible would be in the third mode which may be calculated from Equation (3.58), giving

$$f=\frac{(7.855)^2}{2\pi\times 3^2}\sqrt{\left(\frac{1}{420}\right)}=0.053.$$

It should be recalled, however, that there should normally be twice as many degrees of freedom as there are required modes of vibration. The corresponding period of vibration is 18.78 s and the recommended step length is therefore $\Delta t = 1$. Using $\delta = \frac{1}{2}$ and $\beta = \frac{1}{4}$, the equation of forward integration (4.100) becomes

$$[\boldsymbol{K} + 4\boldsymbol{M}]\boldsymbol{u}_{t+\Delta t} = \boldsymbol{P}_{t+\Delta t} + \boldsymbol{M}\{4\boldsymbol{u}_t + 4\dot{\boldsymbol{u}}_t + \ddot{\boldsymbol{u}}_t\},$$

which can be written as the following equilibrium equation:

$$\hat{\boldsymbol{K}}\boldsymbol{u}_{t+\Delta t} = \boldsymbol{R}_{t+\Delta t}$$

where

$$\hat{\boldsymbol{K}} = \boldsymbol{K}_1 + 4\boldsymbol{M}_1 = \begin{bmatrix} 1122.7 & 155.6 & -106.0 \\ 155.6 & 1562.0 & 160.2 \\ -106.0 & 160.2 & 470.5 \end{bmatrix}.$$

The inverse of $\hat{\boldsymbol{K}}$ is

$$\hat{\boldsymbol{K}}^{-1} = \frac{1}{1000}\begin{bmatrix} 0.9307 & -0.1184 & 0.25 \\ -0.1184 & 0.6784 & -0.2576 \\ 0.25 & -0.2576 & 2.2696 \end{bmatrix}.$$

The equations for updating velocities and accelerations are obtained from (4.97) and (4.98):

$$\ddot{\boldsymbol{u}}_{t+\Delta t} = 4(\boldsymbol{u}_{t+\Delta t} - \boldsymbol{u}_t - \dot{\boldsymbol{u}}_t) - \ddot{\boldsymbol{u}}_t$$

$$\dot{\boldsymbol{u}}_{t+\Delta t} = 2(\boldsymbol{u}_{t+\Delta t} - \boldsymbol{u}_t) - \dot{\boldsymbol{u}}_t.$$

The full calculations for the first 20 s of motion are set out in Table 4.1. It would be instructive to obtain the first two or three lines by manual calculation. However, after doing that it would be easier to write a simple program and run it on a personal computer. In addition to verifying the remainder of Table 4.1, it would be relatively straightforward to plot a graph of response against time, and then the behaviour over a longer period of time could be investigated.

4.5 DAMPING

4.5.1 Inclusion of damping in equations of motion

In the previous chapter (see §3.4), the equations of motion for a lumped parameter multi-degree-of-freedom system were derived. It was assumed that damping existed in the form of discrete viscous dashpots placed between lumped masses. The equation of motion was

$$\boldsymbol{M}\ddot{\boldsymbol{u}} + \boldsymbol{C}\dot{\boldsymbol{u}} + \boldsymbol{K}\boldsymbol{u} = \boldsymbol{P}. \tag{4.110}$$

Table 4.1 Dynamic response of cantilever by Newmark method

$t(s)$	$\boldsymbol{P}^{\mathrm{T}}$			$\boldsymbol{R}^{\mathrm{T}}$			$\boldsymbol{u}^{\mathrm{T}}/1000$			$\dot{\boldsymbol{u}}^{\mathrm{T}}/1000$			$\ddot{\boldsymbol{u}}^{\mathrm{T}}/1000$		
0	0	1	0	0	1	0	0	0	0	0	0	0	0	0	0
1	0	1	0	0.04	4.96	0.01	−0.12	0.68	−0.26	−0.24	1.36	−0.52	−0.47	2.71	−1.03
2	0	1	0	0.28	12.75	0.09	−0.55	3.36	−1.24	−0.62	4.00	−1.45	−0.29	2.58	−0.84
3	0	1	0	1.00	24.09	0.33	−1.23	8.59	−3.01	−0.74	6.46	−2.10	0.04	2.33	−0.46
4	0	1	0	2.60	38.62	0.86	−1.84	16.14	−5.21	−0.48	8.63	−2.31	0.50	2.00	0.04
5	0	1	0	5.50	55.91	1.84	−1.94	25.67	−7.35	0.27	10.43	−1.97	1.00	1.61	0.63
6	0	1	0	10.14	75.54	3.43	−1.04	36.81	−8.86	1.53	11.84	−1.04	1.50	1.21	1.24
7	0	1	0	16.84	97.09	5.80	1.35	49.16	−9.14	3.24	12.87	0.48	1.93	0.84	1.80
8	0	1	0	25.85	120.33	9.07	5.63	62.38	−7.65	5.32	13.56	2.51	2.22	0.55	2.26
9	0	1	0	37.22	144.72	13.36	12.09	76.16	−3.92	7.60	14.01	4.94	2.34	0.35	2.59
10	0	1	0	50.84	170.46	18.74	20.85	90.33	2.35	9.90	14.32	7.61	2.26	0.27	2.75
11	0	1	0	66.39	197.45	25.25	31.82	104.79	11.33	12.03	14.61	10.35	1.98	0.30	2.74
12	0	1	0	83.43	225.80	32.86	44.72	119.59	23.01	13.78	14.98	13.02	1.53	0.44	2.59
13	0	1	0	101.37	255.72	41.57	59.12	134.84	37.26	15.02	15.53	15.47	0.94	0.66	2.32
14	0	1	0	119.56	287.45	51.33	74.46	150.77	53.81	15.64	16.33	17.62	0.29	0.92	1.98
15	0	1	0	137.36	321.24	62.09	90.07	167.63	72.33	15.59	17.38	19.43	−0.38	1.19	1.64
16	0	1	0	154.17	357.28	73.82	105.33	185.67	92.50	14.92	18.69	20.92	−0.97	1.42	1.34
17	0	1	0	169.52	395.73	86.50	119.64	205.11	114.04	13.71	20.19	22.16	−1.45	1.58	1.15
18	0	1	0	183.08	436.60	100.17	132.54	226.11	136.77	12.10	21.80	23.29	−1.76	1.63	1.10
19	0	1	0	194.72	479.80	14.86	143.74	248.71	160.64	10.29	23.40	24.45	−1.87	1.57	1.21
20	0	1	0	204.46	525.11	130.67	153.13	272.85	185.77	8.48	24.89	25.80	−1.77	1.40	1.49

However, in practice it is not generally feasible to compute a damping matrix directly, since damping mechanisms cannot be quantified with sufficient precision. Damping data are usually available only in their overall effect on the structural performance. In particular, the critical damping ratio ξ, defined in Equation (2.15), is a quantity that may be measured experimentally. If suitable equipment is available, it is possible to measure the damping ratio ξ_m in any specific mode of vibration (Deinum, Dungar, Ellis *et al.*, 1982). With this information it is possible to obtain the transient response of a structure by stepwise integration of the uncoupled modal equations given by

$$\ddot{Y}_m + 2\xi_m\omega_m\dot{Y}_m + \omega_m^2 Y_m = Q_m(t)/M_m; \quad m = 1, 2, \ldots. \tag{3.132}$$

In the case of nonlinear transient dynamic analysis there is no option but to integrate the original coupled equations of motion, with their full matrices. A damping matrix is therefore required.

The most satisfactory procedure (Clough and Penzien, 1975) is to assume that the damping matrix is directly proportional to the mass and stiffness matrices in the form

$$\boldsymbol{C} = \alpha\boldsymbol{M} + \beta\boldsymbol{K} \tag{4.111}$$

where α and β are factors to be determined.

Since the mass and stiffness matrices satisfy the orthogonality condition, and therefore lead to uncoupled modal equations, it is evident that $\boldsymbol{C}$ will do the same. Therefore, it follows that there will be a generalized modal damping coefficient, as in Equation (3.130), given by

$$C_m = \boldsymbol{\phi}_m^{\mathrm{T}}\boldsymbol{C}\,\boldsymbol{\phi}_m = 2\xi_m\,\omega_m\,M_m, \tag{4.112}$$

where M_m is the generalized mass of the mth mode.

Pre- and post-multiplying Equation (4.111) by the mode shape we obtain

$$\begin{aligned}\boldsymbol{\phi}_m^{\mathrm{T}}\boldsymbol{C}\,\boldsymbol{\phi}_m &= \alpha\,\boldsymbol{\phi}_m^{\mathrm{T}}\boldsymbol{M}\,\boldsymbol{\phi}_m + \beta\,\boldsymbol{\phi}_m^{\mathrm{T}}\boldsymbol{K}\,\boldsymbol{\phi}_m \\ &= \alpha M_m + \beta K_m \end{aligned} \tag{4.113}$$

where K_m is the generalized stiffness of the mth mode. Therefore, using Equation (4.112) we find that

$$\xi_m = \alpha/2\omega_m + (\beta/2)\omega_m. \tag{4.114}$$

This equation can be used in conjunction with the experimental values of ξ_m to find the factors α and β and hence generate a full damping matrix from Equation (4.111). In practice some compromise is required in this procedure since two values of ξ_m are needed to define α and β. If damping ratios are known for the fundamental mode and for a high-frequency mode the resulting values of α and β will give a reasonable approximation of damping over that frequency range. It is more often the case that damping is only known in the fundamental mode and then the accepted procedure is to assume that $\alpha = 0$ and $\beta = 2\xi_1/\omega_1$. This will have the effect of greater damping in the higher modes. Clough and Penzien (1975) suggested an extension of Equation (4.111) in the form of a series expansion of the mass and stiffness matrices. This would enable a more accurate damping matrix to be formed provided that damping values were known in a sufficient number of modes.

4.5.2 Experimental values of damping

Damping in structures occurs because of energy losses during cycles of oscillation. Energy losses are mainly due to:

(a) internal hysteresis in materials arising from nonlinear stress–strain behaviour, intergranular friction and thermoelasticity; and
(b) friction in sliding or fretting of joints, supports, cladding or various other parts of the structure during relative motion.

Energy losses due to aerodynamic damping may be significant in the case of tall slender structures. Viscous forces are seldom relevant but, fortunately, theoretical viscous damping is usually a very good approximation to observed damping behaviour.

There have been numerous investigations of damping in many different types of structure, using a variety of measurement techniques including decay of free vibrations following initial disturbances, steady-state forced vibration, and spectral analysis of ambient vibrations. The wide variations in reported damping values for apparently similar structures may often be attributed to differences in measurement technique, though local foundation and support conditions are also important.

Recommended values of damping to be used in analytical or design studies are set out in Table 4.2. Damping is expressed as a percentage of critical

Table 4.2 Recommended values of damping

Type of structure	*Damping ξ (%)*
Material damping – steel	0.03–0.15
– concrete	0.15–1.0
– wire rope	0.4 –2.0
Bridges – all steel	0.2 –1.0
– composite construction	0.3 –1.6
– reinforced concrete or prestressed concrete	0.3 –1.6
Chimneys – steel	0.3 –0.8
– concrete	0.5 –1.0
Steel masts and towers	0.3 –2.9
Multi-storey buildings	0.7 –2.9
One- or two-storey houses	1.0 –5.0

damping, i.e. damping ratio ξ (%). Material damping refers to the least amount of damping observable due to internal energy losses alone. Steel possesses very little inherent damping capacity owing to its uniform metallurgical structure and freedom from internal cracks (Contractor and Thompson, 1940). The damping capacity of concrete, which is greater because of the presence of many internal defects, was reviewed by Cole and Spooner (1965). Internal damping in wire rope is due to friction on the contact patches between individual wires and it varies with the load range in a cycle compared with the mean load (Hobbs and Raoof, 1984).

Damping in complete structures is usually greater because of opportunities for absorption of energy due to friction in connections and at supports. However, in the case of structures such as bridges, these opportunities are often minimal owing to careful design of bearings and the lack of any attachments, except handrails and parapets. Steel bridges are particularly notorious for absence of damping capacity. Damping values can be as low as inherent material damping in some cases, although there are generally quite wide variations, chiefly due to relative motions at supports. The values in the table are based on recent data obtained by Eyre and Tilly (1977) and an excellent review by Tilly (1977). Oehler (1957) obtained generally higher values but this was probably due to his method of testing involving a test truck, there being interactions with its suspension.

Pritchard (1984) reviewed data on damping in steel chimneys. The lowest values relate to unlined chimneys on hard foundations in the absence of air spoilers or damping devices. Damping is increased by bolted flanges and refractory linings. Concrete chimneys generally possess more damping capacity, which is also affected by linings and multiple flues (Jeary and Winney, 1972; Milford, 1982). Damping in steel masts varies considerably with the type of construction, the lowest value being relevant to guyed masts of welded steel tube whereas the highest value was observed in a bolted lattice tower (Scruton and Flint, 1964).

The damping capacity of concrete multi-storey buildings is generally in the range of 0.7 to 1.3% if the vibrations are relatively small (Wyatt and Best, 1984; Williams, 1983). However, higher values are possible, especially at greater stresses which would be appropriate for earthquake loading. Under wind excitation, Lam and Lam (1979) observed a value of 2.9% in a ten-storey steel-framed structure with glass curtain walls. Cladding plays an important part in damping performance.

An extensive study of damping in one- and two-storey houses was carried out by Siskind, Stachura, Stagg and Kopp (1980), for the purposes of investigating vibration caused by blasting. Timber frame dwellings generally possessed greater damping than those constructed of brick or masonry.

An important consideration is the effect of frequency or mode shape on the damping capacity. It has been noted by several investigators that damping is often greater in the second or higher modes. Eyre and Tilly (1977) found that the damping in the second mode of highway bridges was often twice that in the fundamental, and usually greater still in higher modes. However, they noted some exceptions. Similar behaviour was observed in a concrete chimney by Jearey and Winney (1972) and in steel piles by Wooton, Warner, Sainsbury and Cooper (1972). However, Deinum, Dungar, Ellis *et al.* (1982) measured damping in the first six modes of a large concrete arch dam and found no correlation between damping capacity and frequency or mode shape. Kwok (1984) found no increase in the second-mode damping of the Sydney Tower. Therefore, a general rule may be unwarranted.

Choice of damping value should depend on the purpose of the analysis. An average value may be appropriate for a research investigation. On the other hand, it may be wise to adopt a low value in a design study to ensure that the results are conservative. There is a trend for modern forms of construction and materials to lack damping capacity.

4.6 FINITE ELEMENT MODELLING

Finite element analysis is an efficient method of determining the dynamic performance of structures for three reasons: (a) saving design time; (b) saving money in construction; and (c) increasing the safety of the structure. In the past it was often necessary to use advanced mathematical methods in the analysis of large structures. More accuracy generally required more elaborate techniques and therefore a large fraction of the designer's time could be devoted to mathematical analysis. Finite element methods free designers from the need to concentrate on mathematical calculations and allow them to spend more time on accurate representation of the intended structure and review of the calculated performance. By making use of programs with interactive graphical facilities (De Salvo and Swanson, 1985) it is possible to generate finite element models of complex structures with considerable ease, and to obtain the results in a convenient, readily assimilated form, thus saving valuable design time. Furthermore, more accurate analysis of structures is possible by the finite element method, leading to economies in materials and construction while also enhancing overall safety.

However, in order to use computer time and design time effectively, it is important to plan the analysis strategy carefully. It should be appreciated that the analysis is intended to provide a numerate answer to a specific question. Therefore it is important to identify the question clearly. Examples of typical questions arising in design might be as follows:

(a) Are the natural frequencies of a turbine support structure going to be well removed from the machine operating frequencies?
(b) Will the horizontal motion at the top floor of a 30-storey building, under storm wind loading, be disturbing to the occupants?

(c) Will the maximum stresses in an arch dam, when subjected to earthquake ground motion, be sufficient to rupture the concrete?

The ratio of costs to benefits of obtaining the answer is likely to have a significant influence on the method employed in the analysis. A simple two-dimensional frame analysis to obtain the natural frequencies of a turbine support structure could be carried out on a personal computer. Suitable low-cost programs are readily available. The accuracy must be sufficient to ensure only that resonance is avoided. The tall building problem is one of human response to vibration (see Chapter 11) and not a question of structural integrity. Root mean square horizontal acceleration at one point, namely the top of the building, is required. Finite element analysis may not be needed at all if a hand calculation, treating the building as an equivalent cantilever, showed that the motion is well within acceptable limits. On the other hand, the considerable expense of a fully three-dimensional analysis of an arch dam would probably be justified bearing in mind the consequences of a collapse.

4.6.1 Geometry

A finite element mesh that has been found satisfactory for static analysis would probably be adequate for dynamic appraisal, with certain provisos. One proviso is that structures tend to behave more three-dimensionally under dynamic loading. A two-dimensional model may neglect important out-of-plane modes of vibration which could be excited if their natural frequencies were in the range of the loading frequencies. A second proviso is that interaction between the structure and its boundaries tends to be more important under dynamic loading. The foundation material for an arch dam, for example, is generally more flexible than the material of the dam itself. Therefore, the finite element mesh should extend into the foundations and surrounding rock or soil. When designing the finite element model a wise procedure is to begin with a relatively coarse mesh, to obtain an approximate indication of the range of frequencies to be expected, and then to refine the model and examine the sensitivity of the dynamic performance to changes in the boundary conditions.

Greater deformation is likely to occur in regions of low stiffness, thus requiring a more refined mesh in these areas. A useful guide to the size of element required is to ensure that there are at least five elements between points of contra-flexure in the highest vibration mode of interest. If wave propagation is important, then it is recommended that the largest element dimension should be given by

$$a \leqslant \lambda/20 \tag{4.115a}$$

where λ is the wavelength,

$$\lambda = c/f, \tag{4.115b}$$

c being the wave velocity (see Chapter 3) and f being the frequency of excitation. In the case of impact, quite high frequencies are possible.

Foundation–structure interaction is a difficult problem, but very important in dynamic analysis, as we have seen already in Chapter 3. This is, first, because the material of the foundation is usually more flexible than that of the structure (Nath and Potamitis, 1982) and therefore displacements are likely to extend some considerable distance into the underlying strata. It is therefore advisable to extend the finite element mesh into the foundation material, as shown in Fig. 4.19. The ground behaves like a flexible support to the structure and neglect of this flexibility can lead to erroneous estimates of the

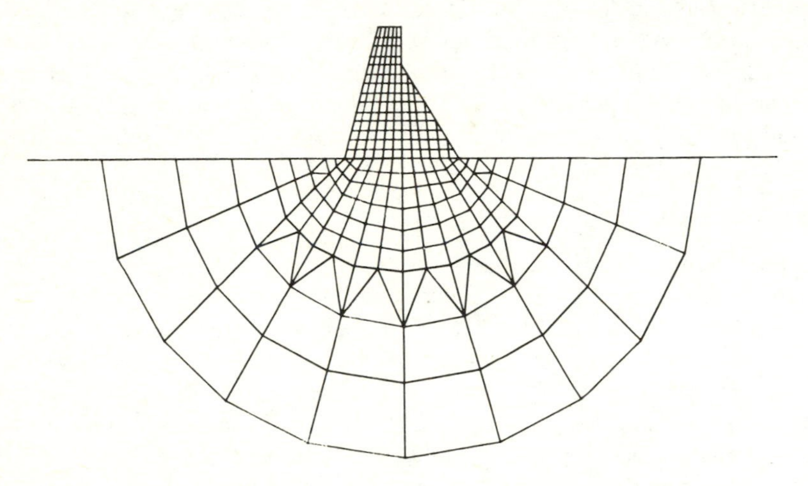

Fig. 4.19 Finite element mesh including foundations.

natural frequencies. Secondly, a soil or rock foundation behaves like a semi-infinite solid with wave motion occurring as a result of dynamic disturbance of the structure. Waves propagate away from the source of disturbance, absorbing dynamic energy in the process, and have the same effect as damping. This is known as 'radiation damping' (Warburton, 1978) and can have a significant influence on the dynamic performance. Unfortunately, the boundaries of a finite element model cause reflections of these waves and prevent proper radiation of the energy. To overcome this problem, a method of applying energy absorbing elements on the boundaries has been developed with some success (Lysmer and Kuhlemeyer, 1969; Seed, Lysmer and Hwang, 1975).

The mass distribution can be treated with somewhat less refinement. Satisfactory estimates of dynamic response can be obtained by a lumped mass approach with the distributed mass being concentrated at nodes. This procedure results in a diagonal mass matrix, which can offer savings in computation time. However, care should be taken to ensure that element nodes exist at the centres of gravity of members or other discrete parts of the structure. If this is not so, errors may be introduced. The simplest way of avoiding this problem is to use consistent mass matrices. Since most finite element programs include calculation of consistent mass matrices, at the expense of little extra computation time, this would be the recommended procedure.

Immersed structures, such as dams and offshore platforms, have the additional problem of hydrodynamic interaction. This arises from the acceleration of parts of the structure producing hydrodynamic pressures where they are in contact with water. A useful solution for a rigid vertical wall, with water on one side, was obtained by Westergaard (1933). His method allowed the calculation of a virtual mass which could be added to the structural mass. However, Westergaard's method makes no allowance for flexibility of the structure and some more refined methods have been employed in recent years (Zienkiewicz and Nath, 1963; Dungar, 1978; Altinisik, Karadeniz and Severn, 1981). It has been found that the presence of a water interface can reduce natural frequencies by as much as 40% (Selby and Severn, 1972). Haroun and Housner (1982) proposed a simple mechanical model for the dynamic response of liquid storage tanks. This took account of convective and hydrodynamic forces as well as impulsive forces due to flexure of the walls of the tank. These forces were then used to determine the maximum base shear. On the basis of experimental studies Maheri (1987) suggested some improvements to the Haroun and Housner model.

4.6.2 Accuracy of solution

Many dynamic problems are analysed by modal analysis, and therefore it is important to judge how many modes are required to obtain an accurate solution. Some useful information can be obtained by performing a spectral analysis of the loading which is, in effect, a study of the frequency content of its time variation. In Chapter 5 the response spectrum will be introduced. This is a graph showing how single-degree-of-freedom systems, with different natural frequencies, would respond to a given transient input. There is usually a cut-off frequency above which a structure would not respond dynamically. Clearly this would be a good guide to the number of modes that would be excited by the given input. Another factor is the distribution of the load on the structure. Evenly distributed loads tend to excite the lower modes predominantly. In higher modes, with more distorted shapes, some parts of the structures may be moving in the opposite direction to the loading, thereby having a cancelling effect. This is taken into account by the generalized force given by Equation (3.129). The generalized force tends to diminish for higher modes and is therefore a useful guide to the number of modes that need to be included in the analysis. If accelerations are required, more modes need to be included because acceleration is proportional to the square of frequency and, therefore, modal contributions diminish less rapidly than for displacements.

The choice of time increment is important in step-by-step analysis of transient responses. We have already noted that if the Newmark or Wilson θ methods are used for direct integration of the equations of motion, then the choice of step length should be governed by

$$\Delta t \leqslant 1/20f \tag{4.116}$$

where f is the highest natural frequency considered to be of interest. If wave propagation is important then

$$\Delta t \leqslant a/10c \tag{4.117}$$

where a is the size of a typical element and c is the wave velocity. A third consideration is to ensure that the time steps are short enough to describe the time variation of the loading accurately. A good guide is that there should be about five or six time steps along the shortest side of the excitation curve.

A note of caution should be introduced at this point concerning the calculation of stresses in a dynamic analysis. There is a tendency for stress detail to be lost when modal analysis is used for transient response problems. This is because stress concentrations are associated with high strain energy and severe local deformation. On the other hand, dynamic motion is associated with low strain energy and high kinetic energy, and therefore the most important modes are the lowest which generally have smooth deflected shapes. Hence, high stress peaks may be lost.

There are two methods of dealing with this problem. First, a sufficiently large number of modes could be used to obtain the required stress detail. However, the computation time for extraction of modes would increase and it may be just as efficient to use direct integration of the equations of motion. Secondly, a more efficient method is to use the 'mode acceleration' procedure (Hurty and Rubinstein, 1964). The essence of this approach is summed up in the equation

$$\begin{aligned}\text{Response} = {} & \text{Full static solution}\\ & + \text{Dynamic response of first } r \text{ nodes}\\ & - \text{Static portion of first } r \text{ modes.}\end{aligned} \tag{4.118}$$

In other words, the lost static response of the higher modes is extracted from the static response of the complete structure. Equation (4.118) may be expressed in full detail by referring back to Chapter 3 where it was seen that the equation of motion of the mth mode of a structure is given by

$$\ddot{Y}_m + 2\xi_m\omega_m\dot{Y}_m + \omega_m^2 Y_m = Q_m/M_m, \tag{3.132}$$

in which Y_m is the generalized coordinate, while Q_m and M_m are the generalized force and mass respectively. Using (3.122) the dynamic response using the first r modes is given by

$$\boldsymbol{u}_r = \boldsymbol{\Phi}_r \boldsymbol{Y}_r, \tag{4.119}$$

where $\boldsymbol{\Phi}_r$ is the modal matrix containing the first r mode shapes, and $\boldsymbol{Y}_r$ are the first r generalized coordinates at time t obtained from the solution of (3.132). Neglecting the time-dependent terms in (3.132), the static portion is given by

$$(Y_m)_{\text{st}} = Q_m / \omega_m^2 M_m. \tag{4.120}$$

Therefore, Equation (4.118) may be written as

$$\boldsymbol{u} = \boldsymbol{K}^{-1} \boldsymbol{F} + \boldsymbol{\Phi}_r (\boldsymbol{Y}_r - [\omega_r^2 \boldsymbol{M}_r]^{-1} \boldsymbol{Q}_r), \tag{4.121}$$

where $\boldsymbol{F}$ is the instantaneous load vector at time t, $[\omega_r^2 \boldsymbol{M}_r]$ is a diagonal matrix whose mth term is $\omega_m^2 M_m$, and $\boldsymbol{Q}_r$ are the first r generalized forces. In practice the stiffness matrix would not be inverted, the static displacements being obtained in the usual way by matrix decomposition. Finally, stresses may be recovered from the displacements given by Equation (4.121). Unfortunately, the above procedure is not necessarily a standard feature of commercial finite element programs although it has been demonstrated to be efficient for large structures (Anagnostopoulos, 1982).

5

Earthquake loading and analysis

5.1 INTRODUCTION

The importance of earthquake loading in structural design can be judged by the catastrophic consequences of major earthquakes. An earthquake in China in 1556 was reported to have cost 830 000 lives. In this century there have been many disastrous earthquakes, notably in San Francisco in 1906, in Tokyo in 1923 (leaving 99 000 people dead), and the recent 1985 earthquake in Mexico City. There are minor tremors somewhere in the world every day and there are a few earthquakes every year of sufficient strength to damage buildings. The careful study of earthquakes by man throughout history is revealed by historical records and artefacts, e.g. a Chinese bronze seismograph, made in 132 AD, was capable of detecting the existence of ground motion and its direction (ASCE, 1984).

The most important aspect of earthquake loading is the safety of buildings, in particular those which accommodate large numbers of people such as schools, flats, offices and hotels. Even more critical are hospital buildings which are urgently required in the aftermath of a seriously damaging earthquake. This was tragically illustrated during the 1971 San Fernando earthquake in California when four hospitals were severely damaged and put out of action. At one hospital ambulances were trapped under a collapsed shelter. Continued functioning of essential services and communications is vitally important and hence the need for survival of bridges, electrical installations, water supply facilities and telecommunications structures. The consequences of damage to large dams and nuclear power plants could affect the lives of large populations and therefore particular attention has to be paid to the earthquake resistance of these structures. Finally, there may be severe economic consequences of the loss of major industrial facilities. In the 1985 Chile earthquake, the refractory brick lining of a large industrial oven collapsed into its contents of molten copper, costing millions of dollars to replace. Large storage tanks at refineries and chemical plants are vulnerable structures and may contain hazardous materials.

In this chapter the nature of earthquakes will be reviewed, followed by a discussion of earthquake risk and assessment of design loading. Finally, there will be an introduction to methods of analysis of structures subjected to strong ground motion.

5.2 THE NATURE OF EARTHQUAKES

Very simply, earthquakes are vibrations of the earth's surface due to sudden movements of the earth's crust. In the geological theory of plate tectonics it is conjectured that the earth's crust consists of a number of solid rock plates floating like rafts on the molten rock interior of the earth. The plates, about twenty in total, drift on convection currents generated by hot spots deep within the earth. Movement is exceedingly slow, being on a geological time scale, but gives rise to enormous forces at the plate boundaries. Mountain ranges are thought to be evidence of flexural deformation where two plates are in collision. Examples are the Himalayas, being pushed up by the northward motion of India, and the Andes formed by the eastward motion of the South East Pacific plate. These examples of plate boundary bending have been likened to the 'rumples in rugs pushed together' (Matthews, 1973). Plates also slide past each other sideways rubbing edge to edge. This motion is not smooth or continuous but consists of a series of jumps, with periods of progressive stress build-up at the plate boundaries followed by sudden slips during which the energy is released (Reid, 1910). This is illustrated in Fig. 5.1(a) where a linear feature, e.g. a fence or a river, lies at right angles to a plate or fault boundary. With time, stress builds up resulting in the deformation shown in Fig. 5.1(b). When the shear stress exceeds the strength of the rock a rupture occurs, as in Fig. 5.1(c). In California there are some remark-

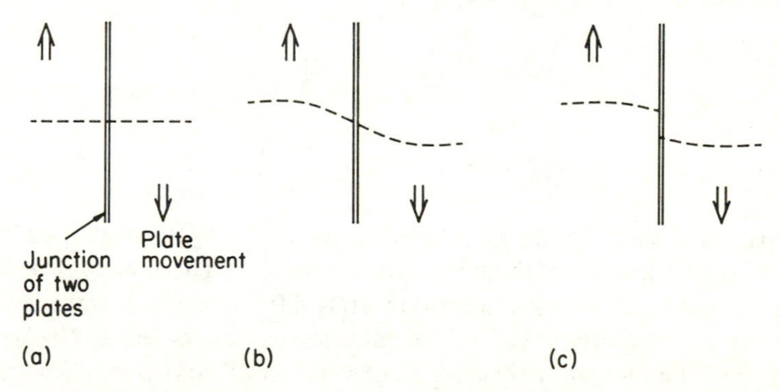

Fig. 5.1 Reid's theory of energy release at plate boundary.

able visible examples of progressive shearing displacement along the San Andreas fault. The bed of a river crossing the fault at right angles in the Carrizo plain has an offset of about 70 feet, much of which occurred during the Fort Tejon earthquake of 1857 when there was a 200 mile long rupture of the fault with a maximum displacement of about 30 feet (Canby, 1973).

The earth's crust, or mantle as it is called, is 30 to 100 miles thick and the origin of a geological fracture may be many miles below ground level, as in Fig. 5.2. The formation of new surfaces at this fracture must be accompanied by the release of a corresponding amount of energy. This energy is dissipated in the form of propagation of waves in the solid material making up the geological strata. The origin of the fracture is known as the hypocentre, or focus, of the earthquake.

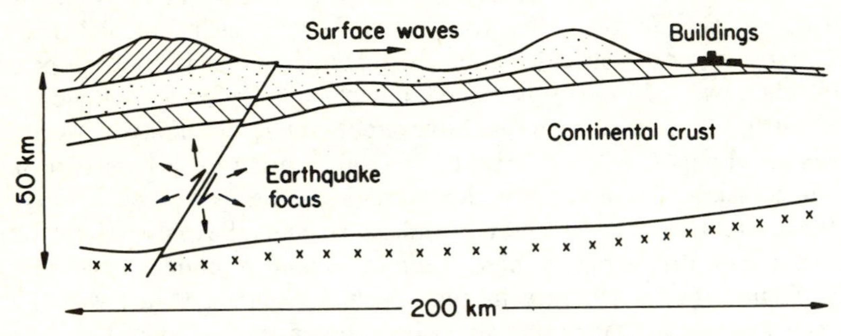

Fig. 5.2 Geological fault and propagation of seismic waves.

Two kinds of body wave are propagated through the bulk of the solid material. The first is the compressional or P wave, which is propagated as an expanding sphere of disturbance much like a sound wave. Secondly there is an S wave which is characterized by a shearing distortion without any volumetric change. The body-wave velocities in an infinite elastic solid were evaluated in §3.5.2 and were given by

$$\text{P wave velocity} = \alpha = \sqrt{\left(\frac{2(1-\nu)G}{(1-2\nu)\rho}\right)} \tag{5.1}$$

$$\text{S wave velocity} = \beta = \sqrt{\left(\frac{G}{\rho}\right)} \tag{5.2}$$

where G is the shear modulus, ρ the density, and ν is Poisson's ratio. Thus, when $\nu = 0.25$, the ratio of these velocities $\alpha/\beta = \sqrt{3}$. The S wave propagation velocity in the earth's crust is approximately 4.0 km/s. The P wave is the first indication of an earthquake and, in fact, seismologists locate the focus by noting the difference in arrival time between the P and S waves at various recording stations around the world. The S − P time delay is a measure of the distance of the focus from the station and by drawing circles centred on the various stations the focus may be located by intersection.

The point at which the body waves reach the surface immediately above the focus is called the epicentre of the earthquake. Here we have an interesting change in the physics of the wave motion. The existence of the free surface gives rise to two kinds of surface waves. The first are called Love waves, and consist of horizontal motion of the surface transverse to the direction of propagation. The second are called Rayleigh waves, in which surface particles move in vertical retrograde elliptical orbits. The waves propagate outward from the epicentre like the ripples on a pond, as illustrated in Fig. 5.3.

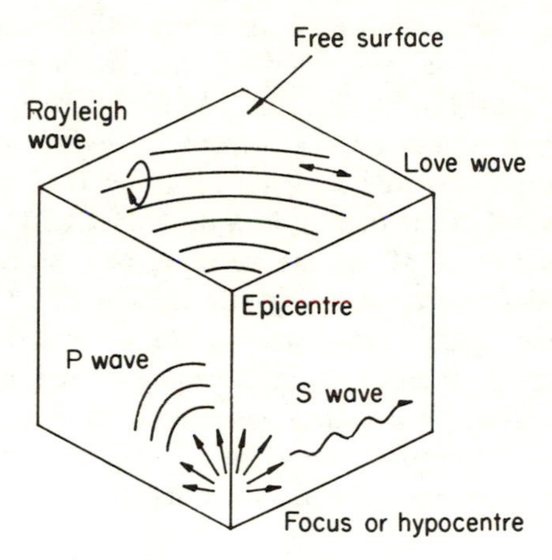

Fig. 5.3 Seismic waves.

The mechanics of wave motion in solid media are very complex, especially when layers are present, as in the case of geological strata. However, analysis of certain simplified models of the earth's crust has yielded useful information (Ewing, Jardetzky and Press, 1957). These were reviewed briefly in §3.5. The most important fact is that body-wave amplitudes decay at the rate of r^{-2}, where r is radial distance from the focus, whereas surface-wave amplitudes decay at the rate of $r^{-1/2}$. Therefore, the latter decay much less rapidly and contribute the most to earthquake damage. This fact also explains why earthquakes may be felt hundreds of miles from their epicentres.

Major earthquakes do not occur as isolated shocks since ruptures at a fault line, or plate boundary, tend to be uneven. There is often an initial fracture, accompanied by a release of energy, preceding the main shock by some hours or possibly even days. This is called a foreshock. The 1985 Chile earthquake was preceded by a magnitude 6.4 foreshock which frightened people out of their houses and was a major factor in reducing the number of casualties when the main shock of magnitude 7.8 came a little while later. There are also aftershocks which are caused by further progress of the fracture. Some very great aftershocks often occur and tend to add to existing tragedies as weakened buildings are further destroyed amidst rescue work.

5.2.1 Measures of magnitude and intensity

There are two useful definitions of the size of earthquakes which are sometimes confused, these being the magnitude and the intensity. The *magnitude* of an earthquake is related to the amount of energy released by the geological rupture causing it, and is therefore a measure of the absolute size of the earthquake, without reference to distance from the epicentre. A workable definition of magnitude was proposed by C. F. Richter (1958). He defined magnitude as the logarithm to the base 10 of the largest displacement of a standard seismograph situated 100 km from the focus. A correction can be made for seismographs situated at distances other than 100 km. Because of the logarithmic nature of the definition a difference of 1.0 in the magnitude represents a difference of 10 in the seismograph amplitude. Magnitude observations by different recording stations usually differ quite widely, often by as much as one degree. However, within a very short space of time following an earthquake there is international agreement on an official figure. The largest earthquake ever recorded was magnitude 8.9 on the Richter scale.

An approximate relationship between Richter magnitude, M, and the energy released by an earthquake, E, is given by

$$\log_{10} E = 4.8 + 1.5M \tag{5.3}$$

where E is measured in joules (Allen, St Amand, Richter and Nordquist, 1965). Thus the ratio of energies released by two earthquakes differing by 1.0 in magnitude is equal to 31.6. The ratio is 1000 for earthquakes differing by 2.0 in magnitude. Comparisons have been made between natural forces and nuclear weapons. The energy released by a 1 megaton hydrogen bomb is roughly equivalent to a magnitude 7.4 earthquake.

The *intensity* of an earthquake is a measure of the observed damage at a particular location. Thus the intensity will vary with distance from the epicentre and will depend on local ground conditions. The most widely accepted scale of intensity is that proposed by Mercalli and subsequently modified by Richter (1958). The Modified Mercalli Intensity scale (MMI) is based on a subjective assessment of the severity of an earthquake by interviewing eyewitnesses and making surveys of ground movement and structural damage. It is thus analogous to the Beaufort scale for wind. Observations are classified in twelve categories of intensity. An abbreviated version of the Modified Mercalli scale is given in Table 5.1.

Table 5.1 Modified Mercalli Intensity scale (abbreviated)

I.	Not felt.
II.	Felt by persons at rest, on upper floors, or favourably placed.
III.	Felt indoors. Hanging objects swing. Vibration like passing of light trucks. Duration estimated. May not be recognized as an earthquake.
IV.	Vibration like passing of heavy trucks or sensation of a jolt like a heavy ball striking the walls. Standing motor cars rock. Windows, dishes, doors rattle. Glasses clink. Wooden walls and frames creak.
V.	Felt outdoors. Sleepers wakened. Liquids disturbed, some spilled. Small unstable objects displaced or upset. Doors swing, close, open. Shutters, pictures move. Pendulum clocks stop.
VI.	Felt by all. Many frightened and run outdoors. Persons walk unsteadily. Windows, dishes, glassware broken. Knick-knacks, books, etc. off shelves. Pictures off walls. Furniture moved or overturned. Weak plaster and weak masonry cracked. Small bells ring (church, school). Trees, bushes shaken visibly.
VII.	Difficult to stand. Noticed by drivers of motor cars. Furniture broken. Damage to weak masonry. Weak chimneys broken at roof line. Fall of plaster, loose bricks, stones, tiles, cornices, unbraced parapets and architectural ornaments. Waves on ponds. Small slides and caving in along sand or gravel banks. Large bells ring.
VIII.	Steering of motor cars affected. Damage to ordinary quality masonry; partial collapse. Some damage to very good quality masonry; none to reinforced masonry. Fall of chimneys, monuments, towers, elevated tanks. Framed houses moved on foundations if not bolted down; loose panel walls thrown out. Branches broken from trees.
IX.	General panic. Weak masonry destroyed; good quality masonry seriously damaged. General damage to foundations. Frames racked. Conspicuous cracks in ground.
X.	Most masonry and frame structures destroyed with their foundations. Some well-built wooden structures and bridges destroyed. Serious damage to dams, dikes, embankments. Large landslides. Water thrown on banks of canals, rivers, lakes etc. Rails bent slightly.
XI.	Rails bent greatly. Underground pipelines completely out of service.
XII.	Damage nearly total. Large rock masses displaced. Lines of sight and level distorted. Objects thrown into the air.

Comparisons between magnitude and intensity are fraught with difficulty. First, intensity varies with distance from the epicentre. Secondly, a large earthquake may occur away from inhabited areas and therefore cause little apparent damage. Ground conditions and quality of building construction can have a considerable effect on subjective assessments of damage. However, a relationship between magnitude M, focal distance R (km) and Modified Mercalli Intensity was suggested by Esteva (1968) after analysis of many earthquakes in Mexico. The intensity is given by

$$I = 1.45M - 5.7 \log R + 7.9. \tag{5.4}$$

Magnitude–intensity relationships are no longer favoured for engineering purposes. A major factor affecting their unreliability is focal depth. Shallow earthquakes (for example Agadir in Morocco in 1960, $M = 5.6$) tend to be disproportionately more damaging because of the concentration of energy release and lack of dissipation.

5.2.2 Ground motion

For the design of structures to resist earthquakes it is necessary to have some knowledge of ground motions. During earthquakes the ground motions are very complex, being translations in any general direction combined with rotations about arbitrary axes. Modern strong motion accelographs are designed to record three translational components of ground acceleration, switching themselves on automatically once an earthquake ground motion reaches a certain threshold level, usually about $0.005g$. Typically, they are capable of recording accelerations in excess of 1.0 g with a frequency range of about 0.06 to 25 Hz. Rotational motions were often ignored in the past but can be recorded by the most modern instruments. The first complete record of strong ground motion was obtained during the 1940 E1 Centro earthquake in California. Over a period of years increasing numbers of strong motion recorders have been installed in many parts of the world and have yielded much useful data. In 1965 legislation passed by the city of Los Angeles required three accelographs to be installed in all new tall buildings, one in the basement, one at mid-height and one at the top. Thus there were about 250 instruments in Southern California during the 1971 San Fernando earthquake yielding a wealth of valuable scientific information.

An example of a strong motion record is shown in Fig. 5.4(a). The horizontal acceleration, after correction for instrument characteristics, may be digitized and can then be integrated to obtain traces of velocity and displacement as shown in Figs. 5.4(b) and 5.4(c). Recordings of ground motion during major earthquakes, such as those obtained at San Fernando, have been used in the dynamic analysis of many important structures.

A well established relationship between Mercalli Intensity and horizontal ground acceleration was given by Richter (1958) as

$$\log a = -0.5 + 0.33I \tag{5.5}$$

where a is in units of cm/s^2. More recently, Trifunac and Brady (1975) obtained empirical relationships between Mercalli Intensity and peak acceleration, velocity or displacement in both vertical and horizontal directions.

A relationship between earthquake magnitude, focal distance and ground

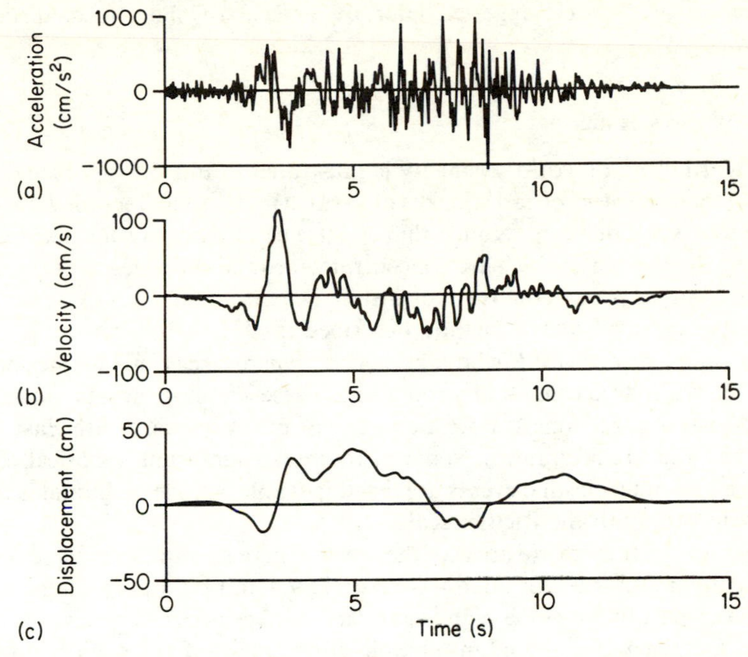

Fig. 5.4 Example of a strong motion record (San Fernando S16E, Pacoima Dam 1971).

motion was obtained by Esteva and Rosenblueth (1964), based on data from Mexican earthquakes. The peak horizontal ground motion is

$$Y = b_1 e^{b_2 M} R^{-b_3} \tag{5.6}$$

where b_1, b_2, b_3 are constants whose values are given in Table 5.2.

The focal distance R (km) is

$$R = \sqrt{(r^2 + h^2 + 20^2)} \tag{5.7}$$

where $r=$ epicentral distance (km), and $h=$depth to hypocentre (km). The

Table 5.2 Constants in Equation (5.6) (after Esteva and Rosenblueth, 1964)

Constants in (5.6)	*Y= acceleration (cm/s²)*	*Y= velocity (cm/s)*	*Y= displacement (cm)*
b_1	2000	16	7
b_2	0.8	1.0	1.2
b_3	2.0	1.7	1.6

correction of 20 was suggested later by Esteva for short focal distances (Esteva, 1967).

5.2.3 World seismicity

The distribution of world seismicity is illustrated in Fig. 5.5 in which circles denote the epicentres of earthquakes of various magnitudes recorded during a particular span of years. Notice that most earthquakes are located in well defined seismic zones which are concentrated around the edges of the tectonic plates. Many of them actually occur within the oceans, but their locations may still be determined by the method described in §5.2.1.

The most disastrous earthquakes occur where areas of high seismicity coincide with large centres of population. These are most notably in Japan, California, Central America, west coast of South America, South East Asia, Iran and the Mediterranean. Some of the most significant historical earthquakes and other seismic events are listed in Table 5.3, the magnitudes being given according to the Richter scale.

It is clear that there are areas of the world where strong earthquakes occur frequently, requiring that buildings be designed and built accordingly. Over the major part of the globe, significant earthquakes occur so rarely that their effects are generally ignored in building construction. Even so, it is possible for damaging earthquakes to occur almost anywhere. This was illustrated at Agadir, in Morocco, a region generally thought of as being aseismic, where a moderate earthquake in 1960 destroyed the town, killing two-thirds of its inhabitants. None of the buildings had been designed to resist earthquakes.

5.3 SEISMIC RISK

Specification of a design earthquake is a difficult problem because earthquakes are, by their very nature, notoriously unpredictable. Nevertheless, some means of assessing the safety of structures exposed to the risk of earthquakes is required. A rational procedure would be to determine the probability of exceeding a specified input ground motion and design the structure so that the probability of damage or collapse was appropriate for the class of structure being considered. Three classes of structure are suggested according to their uses as follows:

(A) *Critical structures.* Hospitals, police and fire stations; communications systems; electrical, water supply and other essential utilities; emergency control centres; large dams; nuclear power stations.

(B) *Important structures.* Hotels and office buildings; public auditoria, churches and grandstands; schools; large commercial and industrial facilities.

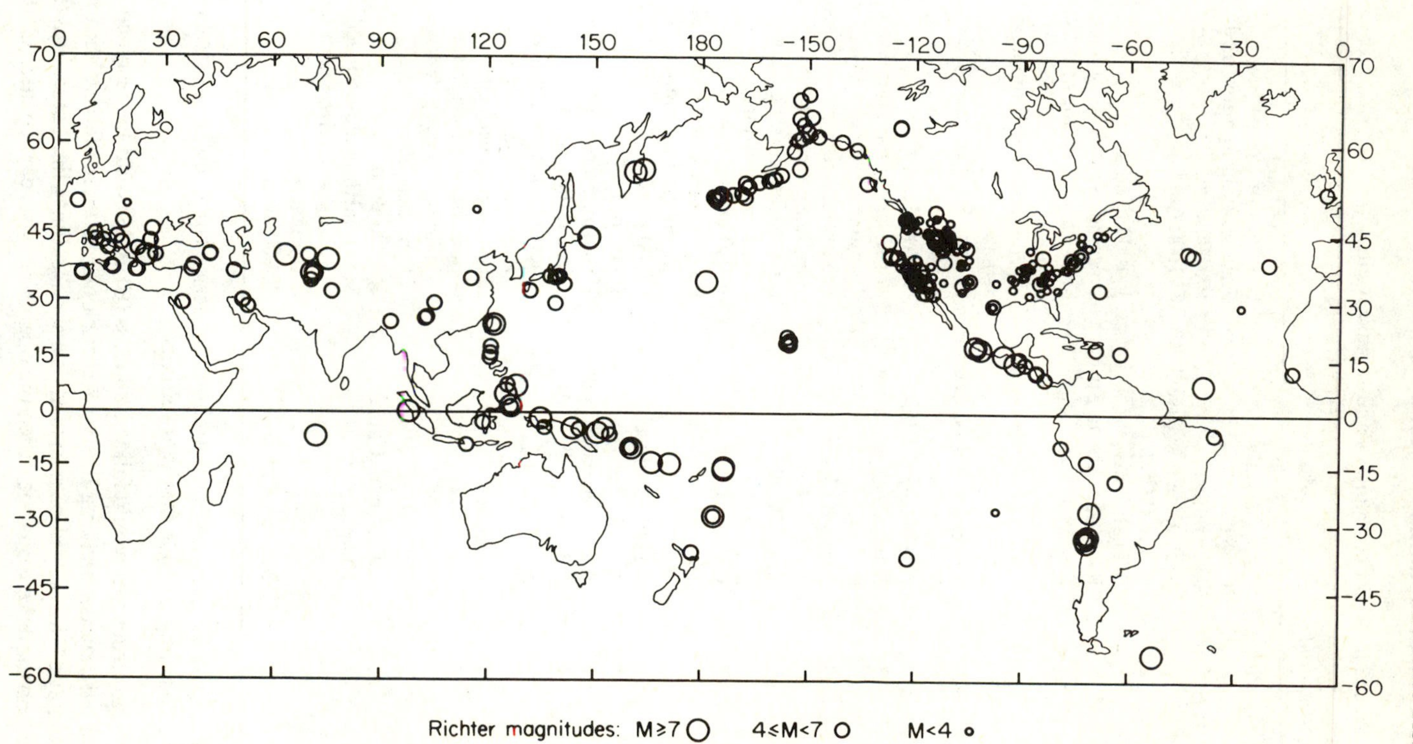

Fig. 5.5 World seismicity: locations of earthquake epicentres (1983–1986).

Table 5.3 Some important earthquakes and seismic events

Year	*Magnitude*	*Location*	*Remarks*
63AD		Pompeii, Italy	Eruption of Vesuvius
132		China	First seismograph invented
1201		Aegean Sea	100 000 dead
1556		Shensi, Yellow River Basin, China	830 000 dead
1737		Bengal, India	300 000 dead
1883		Krakatoa, Sunda Straits	Volcanic explosion, tsunami
1906	8.3	San Francisco, USA	First frame buildings subjected to severe earthquake; prompted Reid's rebound theory
1906	8.9	Colombia and N. Ecuador	Largest Richter magnitude
1908	7.5	Messina, Sicily	75 000 dead
1920	8.6	N. E. Kansu, China	180 000 dead
1923	8.3	Kanto, Tokyo, Japan	99 131 dead; first adoption of seismic coefficient of 0.1
1925	6.3	Santa Barbara	Prompted writing of Uniform Building Code, USA
1933	6.3	Long Beach	First useful strong ground motion recordings
1939	8.3	Chillan, Chile	30 000 dead
1940	7.1	El Centro, Calif., USA	First complete recordings of strong ground motions
1960	5.6	Agadir, Morocco	12 000 dead
1970	7.8	Ancash, Peru	40 000 dead
1971	6.6	San Fernando, Calif., USA	250 strong motion recorders in use; largest ever acceleration recorded at Pacoima dam ($1.25 + g$)
1975	7.3	Haicheng, China	Seismologists warned population to evacuate buildings – heavy damage, light casualties
1976	7.8	Tang Shan, China	250 000 dead
1985	7.8	Chile	Many earthquake-designed buildings survived with light damage
1985	8.1	Mexico City	20 000 dead

(C) *Less important structures.* Warehouses; agricultural buildings; advertising hoardings; single family dwellings.

Most large structures would fall within class (B) and should be designed to withstand earthquake shaking likely to occur once in their lifetime without serious damage. This would be regarded as a serviceability limit state and the associated earthquake would be called the operating basis earthquake (OBE).

A higher specification is the maximum credible earthquake (MCE) which would be expected to occur much less frequently, e.g. once in 200 or 500 years. The structure should be capable of withstanding the corresponding loading without overall collapse, even though a complete rebuild may be required. This would be regarded as an ultimate limit state. In the case of nuclear facilities a higher level of safety is demanded. The plant should be designed so that an exceedingly rare event, the safe shutdown earthquake (SSE), might occur without failure of safety-related equipment, loss of coolant, or melt-down of the reactor core. As the name implies, automatic controls should be activated which would shutdown the plant and allow the core temperature to be reduced.

The probabilities associated with the occurrence of major earthquakes may be assessed from historical data. Allen, St Amand, Richter and Nordquist (1965) established a relationship of the form

$$\log N = a - b\,M \tag{5.8}$$

where M = Richter magnitude, N = frequency of earthquakes exceeding magnitude M within a specified region, while a and b are constants.

An example obtained using Californian data is shown in Fig. 5.6, in which

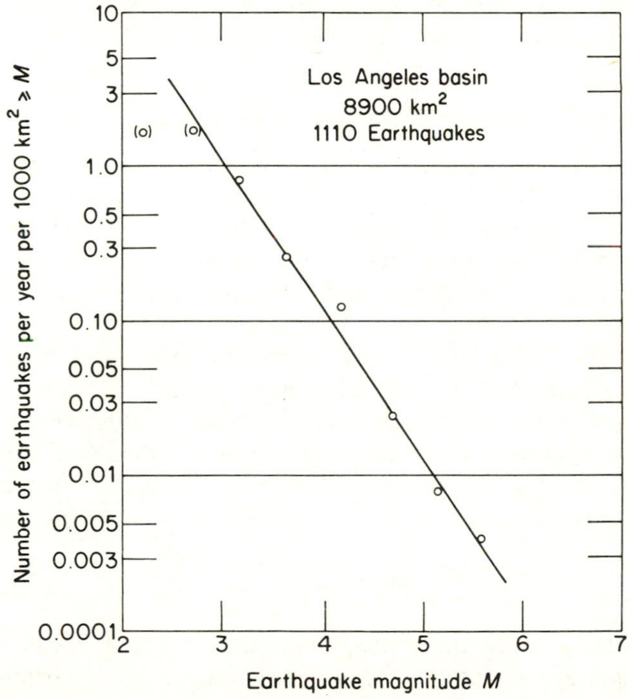

Fig. 5.6 Earthquake recurrence plot (after Allen, St Amand, Richter and Nordquist, 1965).

N is the average number of earthquakes exceeding magnitude M per year within an area of 1000 km^2. At small values of N its inverse is effectively the return period of an earthquake greater than the specified magnitude. In seismically active countries there are usually sufficient earthquake records to establish a curve of this kind. Even in the United Kingdom, not noted for its seismicity, Burton, McGonigle and Neilson (1981) were able to obtain a reliable curve for the Great Glen fault in Scotland. However, earthquake magnitude records have been in existence for not much more than 50 years and therefore only the OBE can be established with much confidence.

The next problem is that the exact location of the design earthquake is not known with certainty. If it were, it would be a simple matter to substitute the magnitude obtained from (5.8) into Esteva and Rosenblueth's formula, Equation (5.6), and obtain a ground motion at the known distance.

A rational procedure for calculating seismic risk at particular sites was developed by Cornell (1968). He considered three different descriptions of the earthquake origin, these being (a) origin at a specified point, (b) sources within an area, and (c) sources along a fault line.

Cornell assumed that the earthquake source was concentrated, i.e. it was a point source, but could be randomly located, except for type (a). We shall now review his method of assessing seismic risk.

5.3.1 Evaluation of seismic risk due to a point source

This is the simplest description in which it is assumed that the seismic region is so confined that it can be regarded as a point at which earthquakes occur randomly but with average frequencies given by Equation (5.8). This was the procedure adopted by Burton, McGonigle and Neilson (1981) for estimating the seismic risk at two sites in Scotland. The earthquake region was assumed to be within a confined area close to Inverness.

In the theory of earthquake risk assessment it is often assumed that earthquakes occur randomly as independent events. This is not strictly correct because of the known behaviour of geological faults. However, if this assumption is made, then earthquake events follow a Poisson distribution so that the probability of exactly n earthquakes in excess of magnitude M_0 occurring in any one year is given by

$$P(n)=\frac{N_0^n e^{-N_0}}{n!} \tag{5.9}$$

where N_0 is the average annual frequency of earthquakes greater than the smallest magnitude of interest, M_0. Given that such an earthquake has occurred, the probability density of its magnitude, $f(M)$, would be a diminishing function, as shown in Fig. 5.7. Therefore, the probability of its

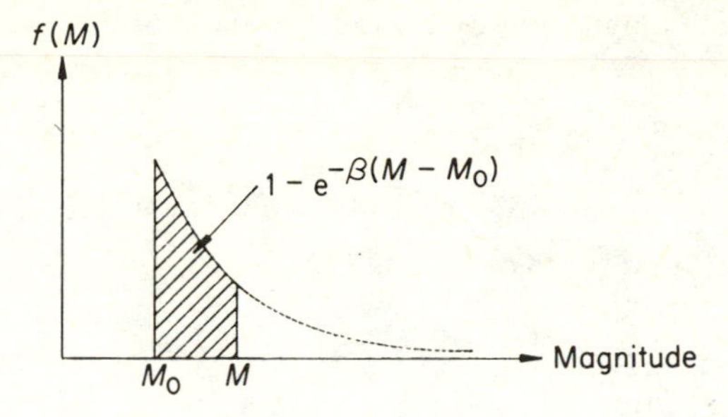

Fig. 5.7 Probability density function for earthquake magnitudes.

magnitude being between M_0 and M is

$$F(M)=\int_{M_0}^{M} f(M)\,\mathrm{d}M. \tag{5.10}$$

However, Epstein and Lomnitz (1966) suggested that a suitable model for large earthquakes would take the form

$$F(M)=1-\exp[-\beta(M-M_0)], \tag{5.11}$$

which is implied by Equation (5.8).

Substituting this in (5.10) it can be verified that

$$f(M)=\beta\exp[-\beta(M-M_0)] \tag{5.12}$$

and that the coefficient $\beta=1/(\bar{M}-M_0)$ where $\bar{M}$ is the average earthquake greater than M_0. The probability that there are exactly n earthquakes and that all are less than M may be obtained by the product law of conditional properties giving

$$P(n|\mathrm{mag}<M)=(N_0^n\mathrm{e}^{-N_0}/n!)\{1-\exp[-\beta(M-M_0)]\}^n, \tag{5.13}$$

where Equations (5.9) and (5.11) have been used. Thus the probability that no earthquakes in any one year will exceed M is given by the summation of such terms, yielding

$$\begin{aligned} P(M_{\max}<M) &= \sum_{n=0}^{\infty}(N_0^n\mathrm{e}^{-N_0}/n!)\{1-\exp[-\beta(M-M_0)]\}^n \\ &= \exp\{-N_0\exp[-\beta(M-M_0)]\} \end{aligned} \tag{5.14}$$

This then will be cumulative probability distribution for the annual maximum earthquake.

We require the probability that the annual maximum ground motion will exceed a specified value Y at a distance from the focus of R. Since ground motion is directly related to magnitude and focal distance by Equation (5.6) it is evident that this probability will be the same as the probability of the annual maximum magnitude exceeding the magnitude given by Equation (5.6). Hence

$$P(Y_{max} > Y) = P(M_{max} > M)$$
$$= 1 - \exp\{-N_0 \exp[-\beta(M - M_0)]\}. \quad (5.15)$$

But since, from Equation (5.6),

$$M = (1/b_2)\ln[(Y/b_1)R^{b_3}] \quad (5.16)$$

we can therefore write

$$P(Y_{max} > Y) = 1 - \exp[-N_0 C Y^{-\beta/b_2} R^{-\gamma}] \quad (5.17a)$$

where

$$C = e^{\beta M_0} b_1^{\beta/b_2} \quad (5.17b)$$

$$\gamma = \beta b_3/b_2. \quad (5.17c)$$

This can be further simplified by noting that in the case of large earthquake magnitudes the expression in brackets becomes very small. Therefore, Equation (5.17a) can be expressed approximately by

$$P(Y_{max} > Y) = N_0 C Y^{-\beta/b_2} R^{-\gamma}. \quad (5.18)$$

This equation gives the probability of ground motion exceeding a specific value in any one year. The inverse will be the return period of that ground motion. Therefore, if we know the intended lifetime of the structure we may determine the design ground motion from Equation (5.18). However, the ground motion would be obtained only in terms of peak acceleration, velocity or displacement. Information on frequency content would not be available. Actual strong motion records can be scaled by the peak value provided that these records have been obtained at locations having seismotectonic characteristics which are similar to the site in question.

5.3.2 Earthquake sources within an area

In this description it is assumed that earthquakes are equally likely to occur anywhere within a known seismically active region such as the rectangle *ABCD* shown in Fig. 5.8(a). The region may be subdivided into a number of annular areas as shown in Fig. 5.8(b), each annular area itself being made up of many arcs of area $\theta x\mathrm{d}x$. Notice that the origin of the earthquakes is at depth h below ground level. If the seismicity of the site is such that the annual frequency of the smallest earthquakes is N per unit area of the region then the contribution of one arc to the probability of exceeding a ground motion Y may be obtained from Equation (5.18) as follows:

$$\mathrm{d}P(Y_{max} > Y) = (N\theta x\mathrm{d}x)CY^{-\beta/b_2}R^{-\gamma}. \quad (5.19)$$

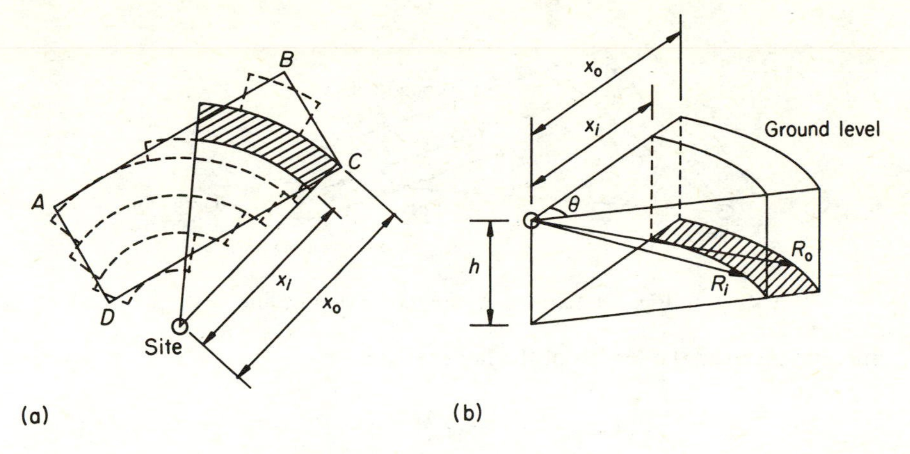

Fig. 5.8 Earthquake sources within an area.

Thus the total probability of exceeding the ground motion Y at the site can be obtained by integration of all the arcs making up the annular area. Hence, noting that $x^2+h^2=R^2$ and $x\mathrm{d}x=R\mathrm{d}R$, it follows that

$$P(Y_{\max}>Y)=NC\theta Y^{-\beta/b_2}\int_{R_i}^{R_0} R^{-(\gamma-1)}\mathrm{d}R$$

$$=NCGY^{-\beta/b_2}, \tag{5.20a}$$

where the geometry term G is given by

$$G=\frac{\theta}{(\gamma-2)R_i^{(\gamma-2)}}\left[1-\left(\frac{R_i}{R_0}\right)^{\gamma-2}\right]. \tag{5.20b}$$

The seismic risk associated with a region of general shape may be obtained by dividing it up into annular areas and summing the contributions calculated using Equation (5.20).

5.3.3 Sources on a fault line

A straight fault line, at a depth of h below the surface, is depicted in Fig. 5.9. If the average frequency of occurrence of earthquakes greater than the smallest of interest, M_0, is given by N_0 in any one year, then the frequency of occurrence per unit length of the fault is $N=N_0/L$. This assumes that earthquakes are equally likely to occur anywhere along the fault line. Therefore, using (5.18), the probability of the ground motion at the site exceeding Y owing to earthquakes occurring in the length dx is given by

$$\mathrm{d}P(Y_{\max}>Y)=N\mathrm{d}xCY^{-\beta/b_2}R^{-\gamma}. \tag{5.21}$$

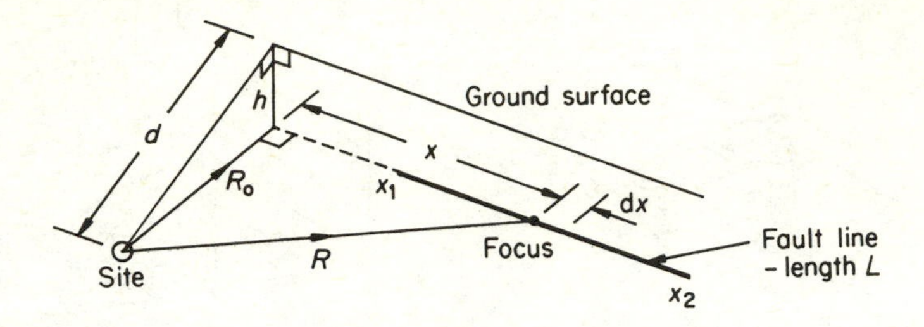

Fig. 5.9 Earthquake source on a fault line.

Integrating over the length of the fault we obtain

$$P(Y_{\max} > Y) = NCY^{-\beta/b_2} \int_{x_1}^{x_2} R^{-\gamma} dx. \tag{5.22}$$

But since $R^2 = x^2 + R_0^2$, we find that

$$P(Y_{\max} > Y) = NCGY^{-\beta/b_2} \tag{5.23}$$

where the geometry term is

$$G = \int_{x_1}^{x_2} (x^2 + R_0^2)^{-\gamma/2} dx. \tag{5.24}$$

This term may be evaluated by numerical integration.

5.3.4 Maps of seismic risk

The principles of seismic risk developed by Cornell (1968) were used by Algermissen and Perkins (1976) to produce a map of seismic risk in the USA. This required source zones and faults to be identified on a map, together with historical data on the frequency of earthquakes in the various zones, and attenuation laws relating ground motion to focal distances. A computer program was then used to obtain contours of different ground shaking intensities having equal probabilities of exceedance. The Algermissen–Perkins map is shown in Fig. 5.10. This map forms the basis of the seismic zoning map recommended for the design of buildings in the USA (ATC, 1978), and the same procedure has been adopted for seismic zoning in other parts of the world (Supplement to National Building Code of Canada, 1985). Cornell's method does not take account of the finite size of earthquake fractures, which can be quite large (Fort Tejon earthquake of 1857 resulted in a 200 mile long rupture). A line source model has been developed by Ang (1974) in an attempt to take account of this limitation.

Finally, it should be emphasized that seismic risk is not merely a matter of calculating probabilities. The input parameters themselves, e.g. location and

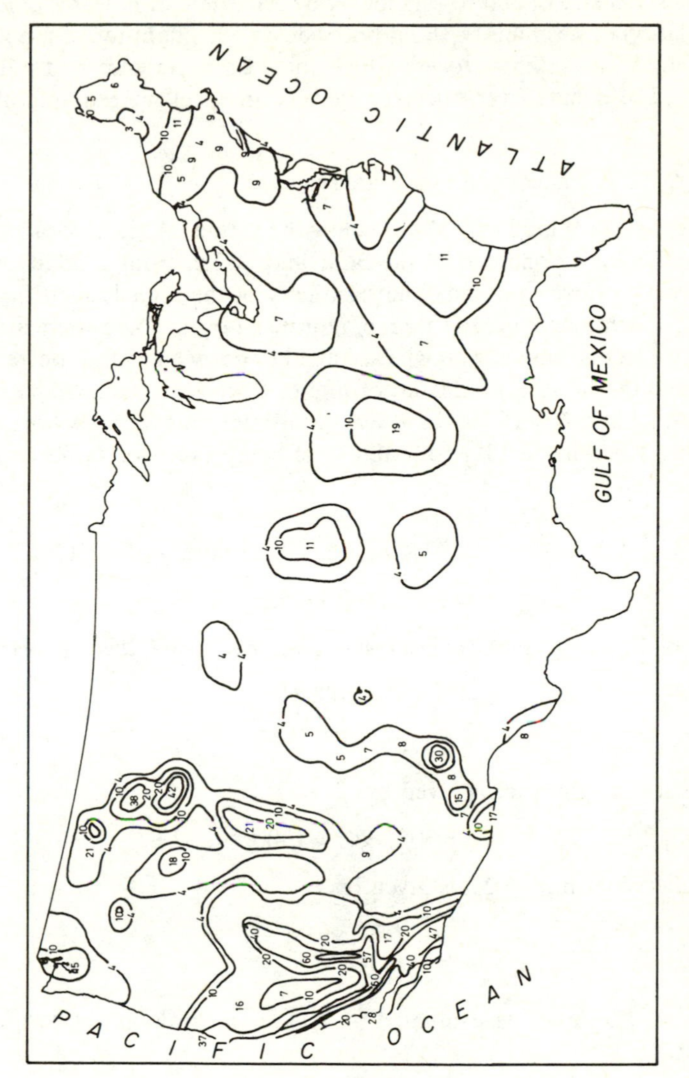

Fig. 5.10 Contours of peak horizontal ground acceleration (%*g*) with 10% probability of exceedance once in 50 years (after Algermissen and Perkins, 1976).

seismicity of faults, are very uncertain. Furthermore, the probabilities have to be considered in the light of what level of risk is acceptable to society at large, bearing in mind the cost implications of reducing that risk. Evidently, the acceptable level of risk depends on the class of the structure under consideration and the penalty of failure, whether it be economic or in terms of loss of life. Calculation of seismic risk should be seen as one quantitative input into the wider decision-making process which includes experience and political judgement. The debate over nuclear safety is an excellent example of this process.

EXAMPLE 5.1 Calculation of seismic risk

A fault line is 100 km long and 20 km below the surface. A site is located on a line perpendicular to one end of the fault and 30 km from a point on the surface directly above that end. The seismicity of the fault is such that, on average, 0.5 earthquakes greater than magnitude 4 occur somewhere along it per year. The attenuation law may be assumed to be given by the constants in Table 5.2 and the value of $\beta = 2.0$ (according to Cornell the value of β usually lies between 1.5 and 2.3). Using the above data, determine the peak horizontal acceleration which has a 10% probability of being exceeded in 50 years.

Solution

From Table 5.2, $b_1 = 2000$, $b_2 = 0.8$, $b_3 = 2.0$. Therefore, from (5.17c)

$$\gamma = 2.0 \times 2.0/0.8 = 5.0.$$

Since the smallest earthquake of interest is given by $M_0 = 4$, then using (5.17b)

$$C = e^{2.0 \times 4} \times 2000^{2.0/0.8}$$

$$= 5.332 \times 10^{11}.$$

The shortest focal distance is given by

$$R_o^2 = 30^2 + 20^2 = 1300.$$

If the geometry term in (5.24) is given by

$$G = \int_{x_1}^{x_2} f(x)\mathrm{d}x$$

the values of $f(x)$ may be evaluated at 10 km intervals as set out in the following table:

x	0	10	20	30	40	50	60	70	80	90	100
$f(x) \times 10^9$	16.41	13.64	8.39	4.41	2.64	1.12	0.59	0.33	0.19	0.12	0.07

Integrating using Simpson's rule, the geometry term is evaluated so that

$$G=\frac{10}{3}[16.41+0.07+4(13.64+4.41+1.12+0.33+0.12)+2(8.39+2.64+0.59+0.19)]=395.3\times10^{-9}.$$

The probability that the ground acceleration will exceed a specified value of Y is equal to 10% over a 50-year period. Therefore, in a one-year period

$$P(Y_{\max}>Y)=0.1/50=0.002.$$

The frequency of occurrence of earthquakes per unit length of the fault is

$$N=0.5/100=0.005.$$

Therefore using (5.23) we find that

$$0.002=0.005\times5.332\times10^{11}\times(395.3/10^{9})\times Y^{-2/0.8}.$$

Hence

$$Y=194.4\ \mathrm{cm/s^2}=0.198\,g.$$

This is the peak horizontal ground acceleration that would have a 10% probability of being exceeded once in 50 years.

5.4 RESPONSE SPECTRA

So far we have discussed earthquake ground motion in terms of its peak value. Obviously this is important, but so is its frequency content. The importance of the frequency content was well illustrated by the 1985 Mexico City earthquake (EEFIT, 1986). The main rupture occurred close to Lazaro Cardenas on the Pacific coast and the ground motion contained a broad band of frequencies, as is typical of many earthquakes. However, by the time the seismic waves had travelled the 400 km to Mexico City the higher frequencies had largely been attenuated. This effect, combined with the nature of the dried lake bed material on which the city is built, resulted in the ground shaking having a predominant period of 2 s. This coincided with the natural period of vibration of buildings in the medium height range (typically six to twenty storeys), many of which collapsed or suffered serious damage. Taller and shorter buildings were generally less severely affected. Hence, it is necessary to take account of the frequency content of earthquakes in the design of buildings.

The most widely adopted method for studying the frequency content of earthquakes is by the response spectrum. This can be defined as the variation of peak dynamic response of a single-degree-of-freedom system, for different values of its natural frequency, given a specified input transient motion. For example, consider the simple mass–spring system subjected to support motion as shown in Fig. 5.11. The equation of motion was given by (2.52) as

$$m\ddot{u}_r + c\dot{u}_r + ku_r = -m\ddot{u}_g(t) \tag{2.52}$$

where $u_r = u - u_g$ is the relative motion of the spring. Alternatively, this may be written

$$\ddot{u}_r + 2\xi\omega\dot{u}_r + \omega^2 u_r = -\ddot{u}_g(t). \tag{5.25}$$

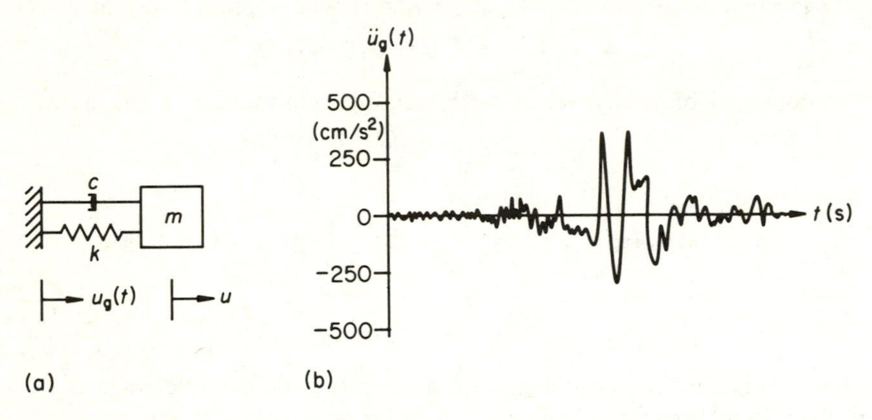

Fig. 5.11 Single-degree-of-freedom system subjected to earthquake ground motion.

We may now use an observed earthquake record, such as in Fig. 5.4, as input $\ddot{u}_g(t)$, and solve for the maximum response by evaluating Duhamel's integral. The procedure may then be repeated for different values of ω in order to obtain a graph of maximum response as a function of frequency. This graph is called a *response spectrum.* Response spectra for peak acceleration, peak relative velocity or peak displacement may be obtained for different values of the damping ratio, ξ. The relative velocity response spectrum is shown in Fig. 5.12, plotted as a function of the period ($T = 2\pi/\omega$).

An important relation exists between the peak relative displacement, $(u_r)_{max}$, and the peak acceleration of the mass, $\ddot{u}_{max}$. Noting that

$$\ddot{u}_r = \ddot{u} - \ddot{u}_g(t), \tag{5.26}$$

Equation (5.25) may be written

$$\ddot{u} + 2\xi\omega\dot{u}_r + \omega^2 u_r = 0. \tag{5.27}$$

The peak acceleration of the mass will occur simultaneously with the peak relative displacement, at which time the relative velocity will be zero. Therefore,

$$\ddot{u}_{max} + \omega^2 (u_r)_{max} = 0,$$

and hence

$$\ddot{u}_{max} = -\omega^2 (u_r)_{max}. \tag{5.28}$$

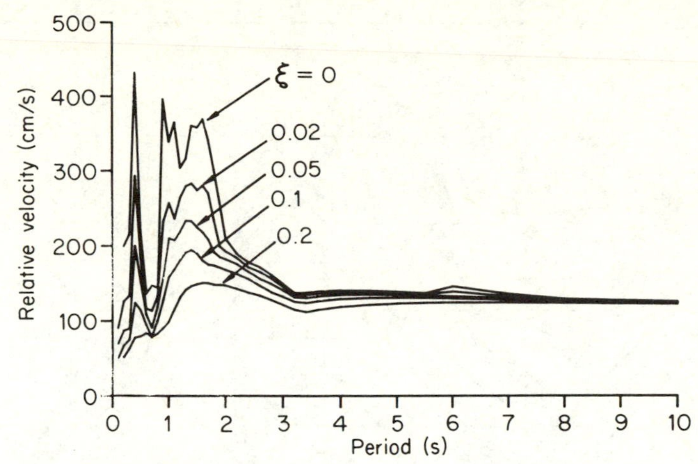

Fig. 5.12 Relative velocity response spectrum (San Fernando S16E, Pacoima Dam 1971).

The sign of the acceleration is unimportant and may be ignored. The relation is usually written as

$$S_a = \omega^2 S_d \tag{5.29}$$

where S_a is the spectral acceleration (of the mass), and S_d is the spectral displacement (relative). A third term, S_v, known as the spectral velocity, may be defined as

$$S_v = \omega S_d. \tag{5.30}$$

This is equal to the peak relative velocity in the case of zero damping only, and is therefore sometimes referred to as the spectral *pseudo-velocity*.

Because of the relationships (5.29) and (5.30) it is possible to plot all three parameters simultaneously on the three-way logarithmic scales shown in Fig. 5.13. It will be perceived that the curves provide the engineer with a powerful tool for earthquake appraisal. If the structure is assumed to be a single-degree-of-freedom system, and its period of vibration is known, then its maximum response may be read off the curves directly.

The curves shown were evaluated using the records from the 1971 San Fernando earthquake. This was a short duration but high intensity earthquake in a region where there were many recording instruments installed. These recordings have been used for many years by engineers all over the world for earthquake assessment of structures. They are considered to contain typical characteristics of earthquake ground motion.

5.4.1 Design response spectra

Other earthquakes yield different response spectra, resulting in different structural performance. In order to include the wide range of likely structural

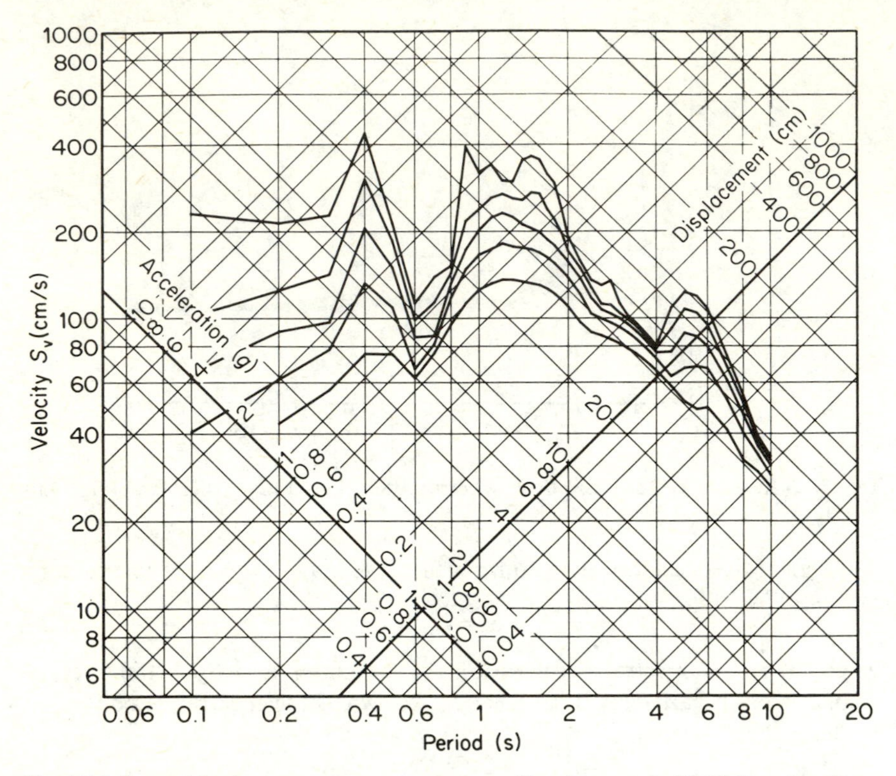

Fig. 5.13 Logarithmic response spectrum curves (San Fernando S16E, Pacoima Dam at $\xi = 0$, 0.02, 0.05, 0.1 and 0.2).

response that would be possible, design response spectra were developed by Newmark and Hall (1973). They averaged the response spectra of a large number of earthquake records, all scaled to a common peak ground acceleration, and, after some smoothing of the curves, produced the simplified normalized spectra shown in Fig. 5.14. The curves are all drawn for a maximum ground acceleration of 1.0 g, velocity of 122 cm/s, and displacement of 91 cm.

Fig. 5.14 represents ground motion that is much more severe than is generally considered reasonable for design. Realistic design spectra may be obtained by starting with predictions of peak ground motion, using the methods of §5.3, and plotting on a three-way logarithmic graph. The corresponding spectra are then constructed, for different values of damping, by using the Newmark–Hall amplification factors given in Table 5.4. Note that there is a transition region between periods of 0.02 and 0.167 s where the acceleration spectrum increases linearly from the ground motion to the fully amplified part of the spectrum.

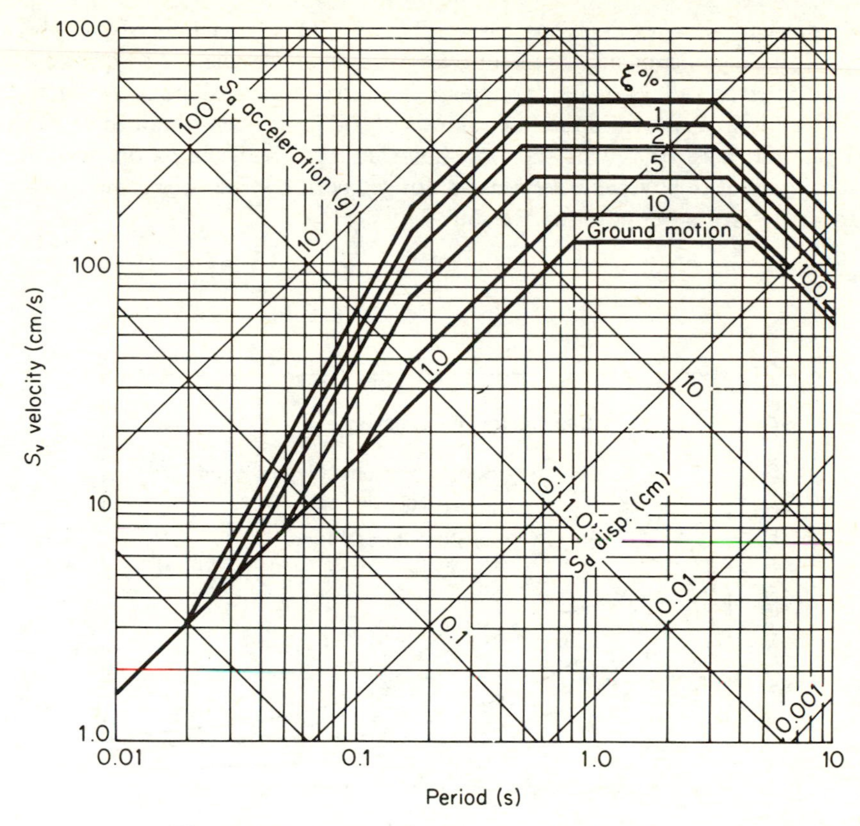

Fig. 5.14 Basic design spectra normalized to 1.0 g.

Table 5.4 Response spectrum amplification factors

ξ (%)	*Displacement*	*Velocity*	*Acceleration*
0	2.5	4.0	6.4
0.5	2.2	3.6	5.8
1	2.0	3.2	5.2
2	1.8	2.8	4.3
5	1.4	1.9	2.6
7	1.2	1.5	1.9
10	1.1	1.3	1.5
20	1.0	1.1	1.2

5.4.2 Site-dependent response spectra

The response spectra discussed so far have assumed that the earthquake waves have propagated over firm ground. However, local geological conditions can alter response spectra significantly. This is particularly so where there are significant depths of soft clay or alluvial deposits. Most cities are built in river valleys or near the coast where such conditions often exist.

Earthquake spectra on different kinds of soil have been obtained by Seed, Ugas and Lysmer (1976) and are shown in Fig. 5.15. These observations indicate that the dynamic response of longer-period structures is increased if the local ground conditions are soft.

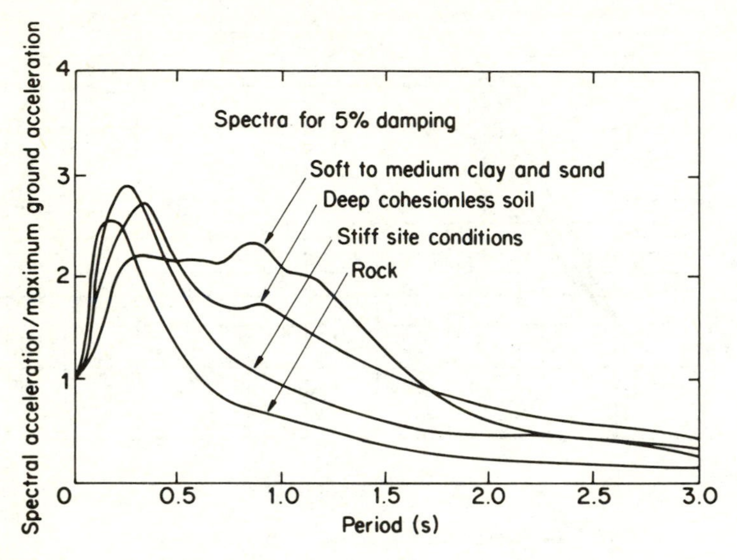

Fig. 5.15 Average acceleration spectra for different site conditions (after Seed, Ugas and Lysmer, 1976).

The dependence of response spectra on ground conditions is recognized in many earthquake-resistant design codes throughout the world (IAEE, 1980). Generally, site conditions may be described in three classes as follows:

(a) rock of any type with a shear wave velocity greater than about 750 m/s; shallow stable deposits of sands, gravels or stiff clays overlying rock;
(b) deep deposits (50–100 m) of cohesionless or stiff clay;
(c) soft to medium-stiff clays and sands with a depth of 10 m or more.

Balendra and Heidebrecht (1987) noted that surface ground motions may be amplified by as much as 2 or 2.5 compared with those of the underlying rock. They also drew attention to the importance of coincidence between the predominant period of the site and the period of vibration of the structure.

5.5 ANALYSIS

5.5.1 Equivalent lateral force procedure

The attractiveness of the equivalent lateral force procedure stems from the reduction of the dynamic problem to a static one. This is particularly useful in preliminary design when initial choices of member sizes have to be made. The method has the appearance of being over-simplistic but it has a sound basis in dynamic theory and has proved to result in buildings of adequate seismic resistance. It is the basis of most earthquake codes worldwide.

The basic concept of the method is that during strong ground motion the inertia of a building results in a horizontal shearing force at the base proportional to the weight of the building and to the imposed ground motion, thus:

$$V = C_s M g \tag{5.31}$$

where V is the total dynamic base shear, M is the total mass of building and its contents, C_s is the seismic coefficient, and g is the acceleration due to gravity. The seismic coefficient, C_s is equivalent to peak acceleration, as a fraction of g, which depends on the seismicity of the site, the local soil conditions, the natural period of vibration of the structure, and the ability of the structure to absorb dynamic energy by ductile deformation.

The seismic coefficient is expressed in many different ways in the aseismic design codes of various countries (IAEE, 1980). However, the following formula is a useful general version which serves to illustrate the most important features:

$$C_s = ASI/R, \tag{5.32}$$

where A is the site-dependent effective peak acceleration (as a fraction of g), S is the seismic response factor, R is a factor related to ductility of structure, and I is an importance factor.

The effective peak acceleration, A, is related to the seismicity of the site. Earthquake codes generally provide a seismic zoning map from which the appropriate value of A may be obtained. Since the spectral velocity is almost constant over the range of periods of most multi-storey buildings, the value of A is, in fact, related to seismic risk maps of peak velocity.

The seismic response factor, S, takes account of the period of vibration of the structure in its fundamental mode, and also the local soil conditions. A typical example is shown in Fig. 5.16. This is based on recommendations by the Applied Technology Council (ATC, 1978), and is clearly a simplified version of Fig. 5.15. The shape of the curved part is usually proportional to $T^{-1/2}$ or $T^{-2/3}$.

The factor R is intended to take account of the ability of a structure to absorb energy by ductile deformation without risk of collapse. This factor

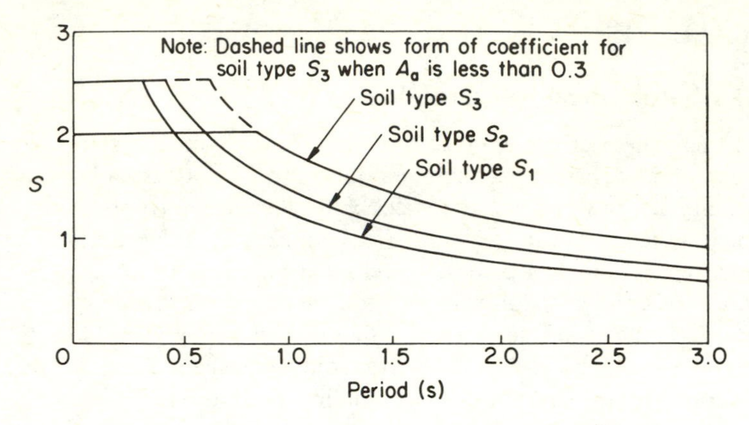

Fig. 5.16 Seismic response factor ($A = 1.0$) (from ATC, 1978).

may vary between 8 and 1.5 according to Table 5.5, which is based upon the Applied Technology Council report (ATC, 1978). It would be uneconomic to design buildings to remain within the elastic range when loaded by earthquakes of low probability since it is well known that structures have considerable reserves of strength beyond the point at which first yield occurs. Consider a typical inelastic load–deformation curve under static loading as shown in Fig. 5.17. This could be a curve of stress *versus* strain, load *versus* deflection, or moment *versus* curvature. A simplified curve, consisting of a straight-line segment up to the effective yield point followed by a level yield plateau, is superimposed on the actual curve. The ductility factor, μ, is defined as

$$\mu = \frac{\text{maximum deformation}}{\text{deformation at elastic limit}} = \frac{u_{\text{ult}}}{u_y}. \tag{5.33}$$

Notice that this also represents the ratio of the force that would have occurred under the specified ground motion, if the structure had behaved elastically, to the force that caused effective yield.

It is evident that the factor R is closely related to the formal ductility factor as defined in (5.33). However, the factor R also takes account of the ability of different forms of construction to develop alternative load paths when damaged by cyclic loading.

Many well known design codes deal with ductility differently. For example the Uniform Building Code of the USA (UBC, 1982) employs a seismic response factor that is effectively an elastic response spectrum scaled down by a suitable ductility factor. In addition there is a coefficient K which takes account of differences in type of construction. The Canadian Code (NBC, 1985) adopts a similar procedure to the Uniform Building Code.

It is apparent from Table 5.5 that moment-resisting frames can be designed to possess large ductility factors. On the other hand, elevated storage tanks

Table 5.5 Factor to take account of ductility

Type of structural system	*Seismic resisting system*	*R*
Building frame system: An essentially complete space frame providing support for vertical loads. Seismic resistance is provided by shear walls or braced frames.	Reinforced concrete shear walls	5.5
	Braced frames	5.0
	Unreinforced masonry shear walls	1.5
Moment-resisting frame: Similar to above except that seismic force resistance is provided by special moment-resisting frames capable of resisting the total prescribed forces.	Steel frames	8.0
	Reinforced concrete frames	7.0
Dual system: Seismic resistance is provided by a combination of special moment-resisting frames (at least 25% of prescribed seismic forces) and shear walls or braced frames.	Reinforced concrete shear walls	8.0
	Braced frames	6.0
Inverted pendulum structure: where the framing acts essentially as a vertical cantilever resisting seismic and vertical load, e.g. elevated storage tanks.	Special moment-resisting frame of steel or reinforced concrete	2.5

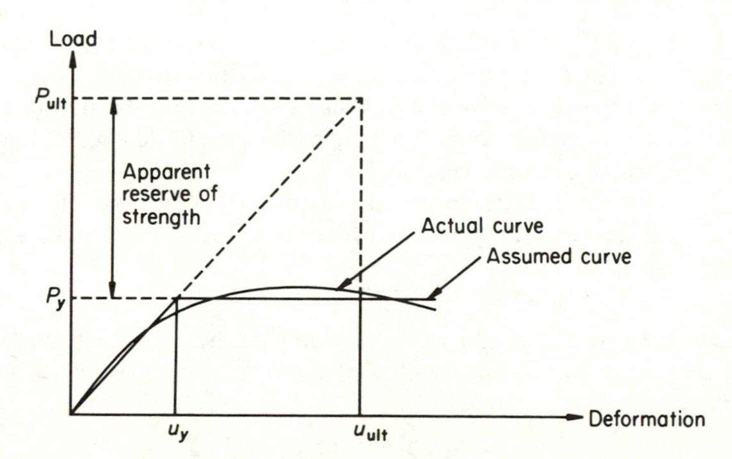

Fig. 5.17 Inelastic load–deformation behaviour.

are known to be vulnerable, and possess small reserves of strength, because collapse may be precipitated by the formation of a limited number of plastic hinges in the legs. Unreinforced masonry structures are known to perform badly in earthquakes and therefore they would be allocated a correspondingly low factor.

A note of caution should be added concerning moment-resisting frames. While it is true that they possess considerable reserves of ductility they are also susceptible to large deformations and inter-storey drifts which may not be acceptable.

The importance factor, I, is intended to take account of the consequences of collapse according to the class of the structure as discussed in §5.3. This would be equal to 1.0 for most structures but should be increased to 1.3 for critical facilities.

The lateral forces on the building are distributed over its height according to the formula

$$F_i = V\left(m_i h_i \Big/ \sum_{i=1}^{N} m_i h_i\right) \tag{5.34}$$

where F_i is the lateral force applied to the ith storey, h_i is the height of the ith storey, and m_i is the mass associated with the ith storey. Hence, the sum of the lateral forces is equal to the total base shear.

The above formulae are ideally suited to the preliminary design of a building since the only prerequisite is an estimate of the overall mass of the building and the distribution of mass over its height. This can be done by comparison with other similar buildings. The seismic forces would be combined with dead loads, and possibly some proportion of live load, when calculating the required sizes of the members.

5.5.2 Calculation of natural frequencies of tall buildings

It is evident from Fig. 5.16 that a knowledge of the period of vibration of the building in its fundamental mode is required. Fortunately, the natural frequencies of tall buildings are closely related to their heights and there are a number of useful formulae which may be used prior to detailed design and more sophisticated dynamic analysis.

Ellis (1980) obtained data on natural frequencies of 163 rectangular plan buildings and showed that the best fit formula for the fundamental natural frequency was given by

$$f = 46/H \tag{5.35}$$

where H is the overall height of the building in metres. The period, T, is merely the inverse of f. Ellis also showed that the formula

$$f = 10/N, \tag{5.36}$$

where N is the number of storeys, is a useful predictor.

Errors of the order of $\pm 50\%$ are not abnormal in the calculation of the natural frequencies of buildings. Furthermore, Ellis showed that sophisticated computer-based methods were not necessarily more reliable than the simple predictors, because of many uncertainties in the actual mass distribution, the stiffening effect of cladding, and the contribution of soil–structure interaction.

5.5.3 Multi-modal response spectrum analysis

It should have been evident that the equivalent lateral force procedure is a method based on the use of the response spectrum. It is ideally suited to simplified design codes and has been effective in the design of earthquake-resistant structures in many parts of the world. However, its main limitation is that the effect of the fundamental mode of vibration only is taken into account, according to the distribution of forces given by (5.34). Collapse in upper and intermediate storeys of rigid frame buildings in Mexico City in 1985 was evidence of the influence of higher modes in the actual dynamic response (EEFIT, 1986).

Multi-modal response spectrum analysis is an extension of the simple spectrum method and can be explained by referring to a lumped mass multi-degree-of-freedom system subjected to support motion, as shown in Fig. 5.18(a).

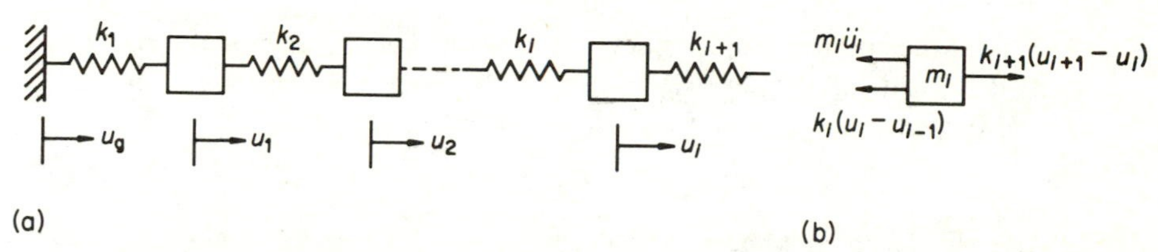

Fig. 5.18 Multi-degree-of-freedom system subjected to support motion.

By considering equilibrium of the ith mass, as in Fig. 5.18(b), its equation of motion may be written down as

$$m_i\ddot{u}_i + k_i(u_i - u_{i-1}) + k_{i+1}(u_i - u_{i+1}) = 0. \tag{5.37}$$

However, if the absolute displacement is replaced by the displacement relative to the support motion, the substitutions

$$(u_i)_r = u_i - u_g, \qquad (\ddot{u}_i)_r = \ddot{u}_i - \ddot{u}_g \tag{5.38}$$

can be introduced, giving

$$m_i(\ddot{u}_i)_r + k_i[(u_i)_r - (u_{i-1})_r] + k_{i+1}[(u_i)_r - (u_{i+1})_r] = -m_i\ddot{u}_g. \tag{5.39}$$

This is equivalent to the equation for a multi-degree system, but in terms of relative rather than absolute displacements, and with the ground motion term, $-m_i\ddot{u}_g$, replacing the applied force.

Hence the system can be analysed by the usual methods of modal analysis which yield uncoupled second-order equations for each mode of vibration. It is easily verified that damping can be included and therefore modal equations may be written,

$$\ddot{Y}_n + 2\xi_n\omega_n\dot{Y}_n + \omega^2 Y_n = -\ddot{u}_g\gamma_n; \quad n = 1, 2, \ldots, N, \tag{5.40}$$

where the generalized coordinates, Y_n, are related to the displacements relative to the support motion. γ_n is defined as the *mode participation factor* and is given by

$$\gamma_n = \frac{\sum_{i=1}^{N} m_i\phi_{in}}{\sum_{i=1}^{N} m_i\phi_{in}^2} \tag{5.41}$$

where

$$\boldsymbol{\phi}_n^{\mathrm{T}} = \{\phi_1\ \phi_2\ \ldots\ \phi_i\ \ldots\ \phi_N\}_n$$

is the *n*th mode shape. Equation (5.40) is the equation of motion of a single-degree-of-freedom system and therefore the maximum response may be obtained from the response spectrum. The maximum modal response is given by

$$(Y_n)_{\max} = \gamma_n d_n, \tag{5.42}$$

where d_n is the spectral displacement corresponding to the modal frequency, ω_n, and is read directly off the graph of the response spectrum. Thus the relative displacement of the *i*th mass in the *n*th mode is

$$[(u_{in})_r]_{\max} = \phi_{in}(Y_n)_{\max}. \tag{5.43}$$

Relative displacements between storeys, or *inter-storey drifts*, may be determined using

$$(\Delta_{in})_{\max} = [(u_{in})_r]_{\max} - [(u_{i-1,n})_r]_{\max}. \tag{5.44}$$

Dynamic forces on the structure may be obtained by calculating the horizontal base shear in each mode from

$$V_n = \gamma_n a_n \sum_{i=1}^{N} m_i\phi_{in}, \tag{5.45}$$

where a_n is the spectral acceleration corresponding to ω_n. Note that the generalized mass for the *n*th mode is used to multiply the acceleration response. Hence the lateral force at a particular storey level is

$$F_{in} = V_n \frac{m_i\phi_{in}}{\sum_{i=1}^{N} m_i\phi_{in}}. \tag{5.46}$$

Notice the similarity between this formula and Equation (5.34) of the equivalent lateral force procedure. Inter-storey drifts and moments may be calculated from the lateral forces by simple statics.

It is recommended that the required structural response, i.e. shear, moment, drift, be calculated for each mode separately before combining modal components. Since the maximum response in each mode would not necessarily occur at the same instant of time, it would be over-conservative to add the separate maximum modal responses. A well verified procedure is to combine the modal components by taking the square root of the sum of the squares of the modal maxima. In effect, this procedure is based on the assumption that the peak modal responses occur randomly with respect to time and that the square root of the sum of the squares (SRSS) represents the most probable maximum response (Goodman, Rosenblueth and Newmark, 1955). Total responses are calculated by the SRSS method using the formula

$$R \simeq \sqrt{\left(\sum_{n=1}^{N} R_n^2\right)}. \tag{5.47}$$

In the accurate application of the mode superposition procedure (see §3.4.4) it was explained that modal responses should be added, with due regard to sign, at every instant of time and, hence, the maxima are obtained. The multi-modal spectrum analysis procedure is a convenient and effective approximation which involves a minimum of computation.

Despite the widespread acceptance of the SRSS method, its use in the analysis of three-dimensional structures has come under heavy criticism. Wilson, Der Kiureghian and Bayo (1981) showed that, if there were important torsional contributions and closely spaced frequencies, the SRSS combination procedure could seriously underestimate or overestimate the total response. They strongly recommended the use of an alternative procedure known as the 'complete quadratic combination' (CQC) method. The combination equations are as follows:

$$R = \sqrt{\left(\sum_{n=1}^{N} \sum_{m=1}^{N} R_n \alpha_{nm} R_m\right)}, \tag{5.48}$$

where R_n and R_m are the responses in the nth and mth modes respectively, and

$$\alpha_{nm} = \frac{8(\xi_n \xi_m)^{1/2}(\xi_n + \rho\xi_m)\rho^{3/2}}{(1-\rho^2)^2 + 4\xi_n\xi_m\rho(1+\rho^2) + 4(\xi_n^2 + \xi_m^2)\rho^2} \tag{5.49}$$

where $\rho = \omega_m/\omega_n$.

Note that Equation (5.48) is of complete quadratic form with all cross-modal terms included. The expression for the cross-modal coefficient, α_{nm}, was obtained by the application of random vibration theory. Wilson, Der Kiureghian and Bayo (1981) showed that the CQC method gave very good agreement with full mode superposition, even for three-dimensional structures, and maintained that the extra computation was minimal.

A final comment on multi-modal spectrum analysis is that, strictly speaking, it does not apply to nonlinear structures because of the need to use the principle of superposition. However, it is usually assumed that inelastic response spectra can be derived from the elastic spectra, taking account of appropriate ductility factors, and that modal analysis provides reasonable approximations to the seismic forces and displacements (ATC, 1974).

The Canadian code (NBC, 1985) recommends that irregular structures should be analysed by multi-modal response spectrum analysis using an elastic response spectrum. A suitable spectrum could be obtained by the Newmark–Hall procedure which involves scaling a normalized spectrum, as in Fig. 5.14, by the design ground motion. Ductility is taken account of by scaling all the modal contributions by the ratio of the equivalent lateral force base shear, V, (which includes ductility) to the total multi-modal base shear, V_{m}, obtained by SRSS combination.

5.5.4 Direct numerical integration of the equations of motion

Spectrum analysis and the equivalent lateral force procedure should be used with caution since they involve a number of important assumptions. First, it is assumed that the structure can be independently analysed in its two principal orthogonal planes and the responses combined. Some allowance can be made for a limited amount of torsional response. Secondly, nonlinear behaviour should be well distributed through the building in order that modal analysis is a reasonable approximation. In order for structures to conform with these assumptions they should be regular in plan form, with centres of stiffness and mass being reasonably coincident, and there should not be any sudden changes of stiffness from one storey to another.

Structures with irregular plan forms may be particularly susceptible to torsion. Torsional modes are often very close to the frequencies of the lower translational modes and should therefore be considered very carefully. There have been a number of examples of structural failure due to torsion in major earthquakes (Selna and Tso, 1980). Structures which are irregular in the vertical direction may experience uneven distribution of plasticity and would therefore not conform to the requirements of modal analysis.

Thus, there are many cases when recourse to step-by-step integration of the equations of motion of a complex building structure would be required. There are a number of nonlinear finite element codes available for this purpose, e.g. *ANSYS* (De Salvo and Swanson, 1985) and *NONSAP* (Bathe, Wilson and Iding, 1974). These codes permit ground motions at the supports to be prescribed and provide for the elastic–plastic resistance of various structural materials to be defined. A detailed discussion of direct integration methods was given in §4.4.3.

Nonlinear time history analysis of complex three-dimensional structures is likely to make heavy demands on computer time and is consequently expensive. Furthermore, the reliability of nonlinear dynamic models is less well established than in the case of linear analysis and therefore results should be interpreted cautiously. Finally, it should be realized that structural response is likely to vary considerably according to the details of the earthquake record. One approach would be to carry out full analyses under a number of different recorded ground motions, and under simulated ground motions which conform to a prescribed response spectrum. However, the cost implications would be very great and only justifiable for structures of exceptional importance.

5.5.5 Multiple-support excitation of structures

It is usually assumed that all parts of the base support of a structure experience the same ground motion. This may be sufficiently accurate for tall buildings whose base dimensions are small relative to the wavelength of seismic motions. This wavelength could be typically 500 m given a Rayleigh wave velocity of 1000 m/s with an associated frequency of 2 Hz. Thus large dams, with lengths of 500 to 1000 m, and long multispan highway bridges could experience considerable discrepancies in ground motions at different support points over their lengths. Altinisik and Severn (1981) have studied this problem and have found that, in general, multiple-support or asynchronous excitation produces lower stresses than would be the case with synchronous excitation. However, they showed that pseudo-static stresses due to relative motion were not trivial and in some cases could result in increased total stresses.

An important example of multi-support excitation is in the dynamics of piping systems in industrial plant. This is particularly significant in the case of nuclear plant where pipe rupture could have dangerous consequences. Piping in a typical plant can extend over lengthy horizontal distances or may rise through many floors of a building, thus giving rise to differential motions at the various floor levels. This problem is the subject of intensive study in the nuclear industry (Subudhi, Bezler, Wang and Alforque, 1984) though there are indications that piping systems are more tolerant to earthquakes than theoretical calculations would suggest (Haas and Labes, 1985).

If the time histories of the support motions are known then these could be input as prescribed boundary conditions to a finite element model, with solution by direct integration. Unfortunately, in earthquake engineering the precise details of independent support motions are never known in advance, though one approach might be to impose a known earthquake motion on each support with an appropriate phase difference introduced. This is called asynchronous support motion. Altinisik and Severn (1981) developed a multi-mode spectrum solution which involved modified modal participation factors for each support excitation. The total modal response was added to the pseudo-static response.

EXAMPLE 5.2 Seismic analysis of a building frame

The ten-storey building shown in Fig. 5.19 is to be analysed by the equivalent lateral force procedure and also by the multi-modal response spectrum method. It may be assumed that the special moment-resisting steel frame has the following properties:

Mass at each storey level	$= 35\,000$ kg
Value of I for all beams	$= 23.99 \times 10^{-4}$ m^4
Value of I for columns from ground level to storey 5	$= 14.68 \times 10^{-4}$ m^4
Value of I for columns from storey 5 to storey 10	$= 2.99 \times 10^{-4}$ m^4
Value of E for steel	$= 200 \times 10^9$ N/m^2
Damping in structure	$= 5\%$ critical.

The effective peak horizontal ground acceleration is specified as $0.25\,g$ for design purposes and it may be assumed that the foundation soil conditions are soft.

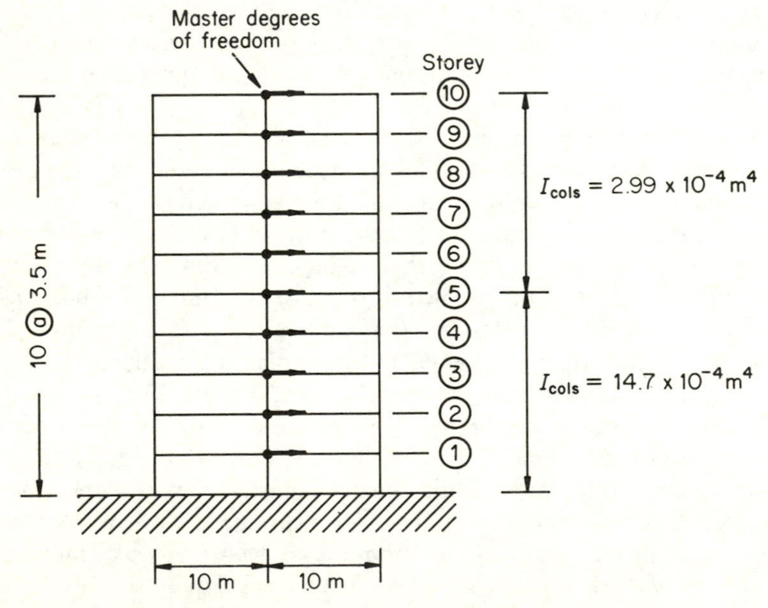

Fig. 5.19 Ten-storey rigid-jointed steel building frame.

Solution

Approximate values of the natural period of vibration may be estimated by either of Equations (5.35) or (5.36). Using (5.35) we find that

$$T = H/46 = 35/46 = 0.76 \text{ s}.$$

The total mass of the building is $M = 350\,000$ kg. The effective peak acceleration has been given as $A = 0.25$. The seismic response factor may be obtained from Fig. 5.16 for soil type 3 with $T = 0.76$ s, giving $S = 2.0$. The factor $R = 8$ for a special moment-resisting steel frame and the importance factor may be taken as $I = 1$. Therefore the seismic coefficient is given by (5.32) and is

$$C_s = 0.25 \times 2.0 \times 1.0/8 = 0.0625,$$

and hence the total design base shear is given by (5.31)

$$V = 0.0625 \times 350\,000 \times 9.81 = 214.6 \text{ kN}.$$

The lateral forces at each storey level are evaluated using (5.34) and are tabulated in the fifth column of Table 5.6. These forces may be used to design the structure. However, assuming that the member sizes finally chosen are given at the head of this example we can perform a dynamic analysis of the structure.

The dynamic analysis is not given here but may be carried out by the finite element method using beam elements and employing Guyan reduction with the master degrees of freedom as shown in Fig. 5.19. The mass normalized mode shapes ($\boldsymbol{\phi}^T \boldsymbol{M} \boldsymbol{\phi} = 1$) of the first three modes are given in Table 5.6. The mode participation factors may be evaluated from (5.41). Notice that m_i cancels out since the mass at each storey is constant. Also there will be a factor of 10^3 arising from the mass normalization.

The natural frequencies and periods of the first three modes are set out in Table 5.7. The spectral accelerations for each mode may then be obtained from the normalized response spectra for 5% damping (Newmark and Hall, 1973) shown in Fig. 5.14. These values are also given in Table 5.7. The modal base shears are determined using (5.45). These are greatly in excess of reasonable design values for two reasons. First, the spectral accelerations were obtained from the response spectrum normalized to 1.0 g. Clearly it should be factored down to the specified input ground motion. Secondly, no account has been taken of ductility or the energy absorbing properties of the type of construction.

The base shears will be corrected by the procedure recommended in the Canadian code (NBC, 1985). First, the most probable uncorrected base shear will be calculated from the modal base shears by the square root sum of squares (SRSS) method:

$$V_{\text{SRSS}} = \sqrt{(3422^2 + 1338^2 + 318^2)} = 3688 \text{ kN}.$$

Table 5.6 Mode shapes and seismic forces on building frame

Storey level	*Mode shapes* ($\times 10^{-3}$)			*Equivalent lateral force* (*kN*)	*Modal forces* (*kN*)			*SRSS storey shear* (*kN*)	*Equivalent lateral force shear* (*kN*)
	Mode 1	*Mode 2*	*Mode 3*		*Mode 1*	*Mode 2*	*Mode 3*		
10	2.66	−2.34	2.49	39.0	36.4	−27.7	14.3	47.9	39.0
9	2.54	−1.63	0.53	35.1	34.8	−19.3	3.0	87.0	74.1
8	2.33	−0.47	−1.80	31.2	31.9	−5.6	−10.3	116.0	105.3
7	2.01	0.82	−2.76	27.3	27.5	9.7	−15.9	137.8	132.6
6	1.62	1.88	−1.59	23.4	22.2	22.3	−9.1	155.2	156.0
5	1.17	2.37	0.76	19.5	16.0	28.1	4.4	169.5	175.5
4	0.93	2.27	1.58	15.6	12.7	26.9	9.1	184.8	191.1
3	0.68	1.89	1.86	11.7	9.3	22.4	10.7	199.2	202.9
2	0.41	1.26	1.49	7.8	5.6	14.9	8.6	209.6	210.7
1	0.16	0.51	0.67	3.9	2.2	6.0	3.8	214.1	214.6

Table 5.7 Multi-modal response spectrum analysis of building frame

Mode	*1*	*2*	*3*
Participation factor ($\times 10^3$)	0.508	0.228	0.111
Frequency (Hz)	0.95	2.59	4.36
Period (s)	1.05	0.39	0.23
Spectral acceleration (g)	1.35	2.6	2.6
Base shear (kN)	3422	1338	318
Corrected base shear (kN)	199.1	77.8	18.5

Then the modal base shears are corrected by a factor which is the ratio of the design base shear, V, to the most probable uncorrected base shear:

$$\text{correction factor} = (214.6/3688) = 0.0582.$$

Hence, the modal base shears may be corrected and are given in Table 5.7. Finally, the modal forces at each storey may be determined from (5.46) and are given in Table 5.6. It is recommended that a static analysis of each mode should be carried out separately, before combining responses such as stress or inter-storey drift.

For interest, the shears at each storey have been calculated by the SRSS combination method and are compared with those calculated by the equivalent lateral force procedure in the final two columns of Table 5.6. It will be noted that the latter method underestimates the shears near the top of the building. This is because the significant contribution of the higher modes in this region is ignored.

6

Earthquake-resistant design

6.1 GENERAL PRINCIPLES

6.1.1 Aims

Earthquakes are noted for their violent results. The Bible records one in which the foundations of a prison were so shaken that the doors flew open, giving the captives access to freedom (*Acts* 16:26). During the 1971 San Fernando earthquake it was reported that a parked 20-ton fire truck was physically thrown in the air and projected a distance of about 8 feet (Morril, 1971). If there are any weaknesses in the design of a structure, earthquakes will certainly find them.

Despite the generally catastrophic nature of severe earthquakes, there is evidence that buildings and other structures can be designed to perform relatively safely. During the 1985 Chile earthquake many modern structures that had been designed to recent earthquake regulations survived, while older buildings nearby suffered more serious damage. Experience during the 1985 Mexico earthquake was less typical because of the narrow frequency band of the shaking which affected medium-height buildings. However, it should be emphasized that some very tall buildings performed well during the unusually severe shaking.

After the 1971 San Fernando earthquake, teams of engineers studied damaged structures in great detail and accumulated a wealth of practical information on the behaviour of all kinds of different structures (Lew, Leyendecker and Dikkers, 1971). Case studies were reanalysed minutely and the resulting recommendations contributed significantly to the development of many current aseismic design codes. In recent years there has been a growing practice for teams of engineers and geologists to visit earthquake disaster areas in order to accumulate experience data. Examples of such groups are the Earthquake Engineering Field Investigation Team (EEFIT) of the UK, the Seismic Qualification Utilities Group (SQUG) of the USA and the Earthquake Engineering Research Institute of California. The resulting sharing of information throughout the international community will undoubtedly lead to more effective design standards.

The principal aims of earthquake-resistant design should be:

(a) *To prevent total collapse of structures.* In every major earthquake there have been examples of lives lost because of collapse of buildings that lacked continuity or lacked alternative load paths after failure of principal members. Brittle forms of construction are particularly notorious for sudden and total collapse. It should be possible to design buildings with sufficient ductility and redundancy so that people may be evacuated without loss of life even if gross distortions of buildings necessitate total reconstruction eventually.
(b) *To control damage to a repairable extent.* The economic consequences of an earthquake are usually very severe and therefore there is a strong incentive to implement methods of design that enable a structure to absorb the energy of an earthquake with minimal costs of subsequent repair. Furthermore, buildings and structures of strategic importance, such as hospitals, power generating facilities, telecommunications buildings, dams and bridges, should be capable of surviving a large earthquake and still keep functioning.

6.1.2 Structural configuration

To achieve safe and predictable performance of buildings during earthquakes, it is recommended that the following points of guidance be considered in the design:

(a) Structures should be geometrically simple. Uniformity and regularity in the structural configuration is highly desirable. Sudden changes in elevation or in stiffness should be avoided. This particularly applies to ground-floor columns which carry the greatest loads (see Fig. 6.1(a)). Unsymmetrical set-backs are undesirable (see Fig. 6.1(b)).
(b) Plan forms should be symmetrical as far as possible, including the distribution of shear walls or other stiffening elements (see Fig. 6.2). If the centre of mass of a building does not coincide with its centre of stiffness then torsional response would be induced and this often affects performance adversely. The vertical stiffening elements should not be concentrated too centrally, thereby reducing the torsional stiffness.
(c) Concentration of mass should be avoided, especially near the top of a building.
(d) There should be adequate clearance between adjacent high-rise blocks to ensure that large deflections during an earthquake do not lead to 'hammering'.

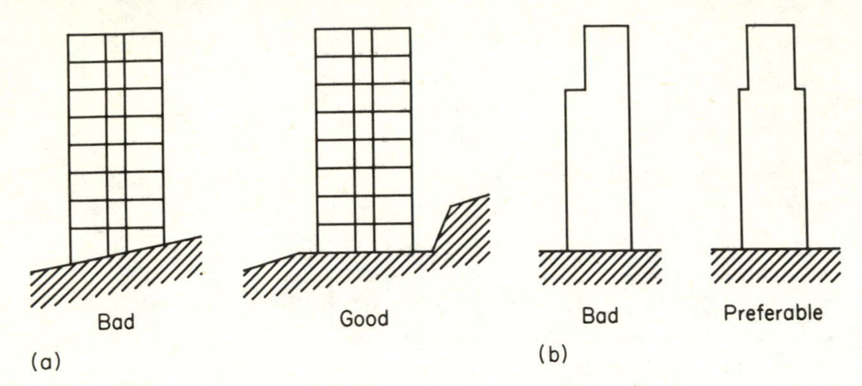

Fig. 6.1 Good and bad vertical configurations.

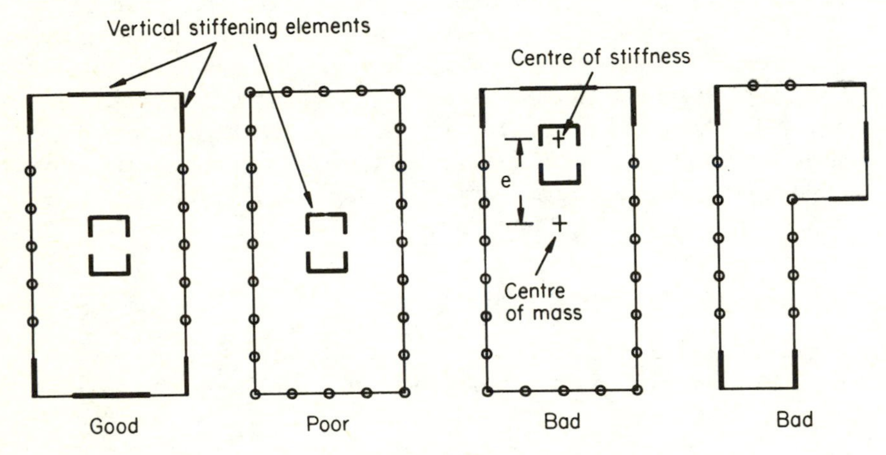

Fig. 6.2 Good and bad plan forms.

However, for architectural reasons it is not always possible to avoid a certain amount of irregularity, in which case more sophisticated dynamic analysis is advisable.

6.1.3 Ductility and redundancy

Collapse of framed structures in earthquakes often occurs by the formation of plastic hinges in columns in one of the lower storeys, as shown in Fig. 6.3(a). This concentration of ductility is inefficient. More energy could be absorbed if plastic deformation was distributed more uniformly throughout the building. This can be done if columns remain elastic with plastic hinges forming in the beams as shown in Fig. 6.3(b). Thus it is highly desirable to ensure that stable plastic hinges can be made to form in the beams while the columns remain elastic. This is the 'strong column–weak beam' principle of design. Coupled

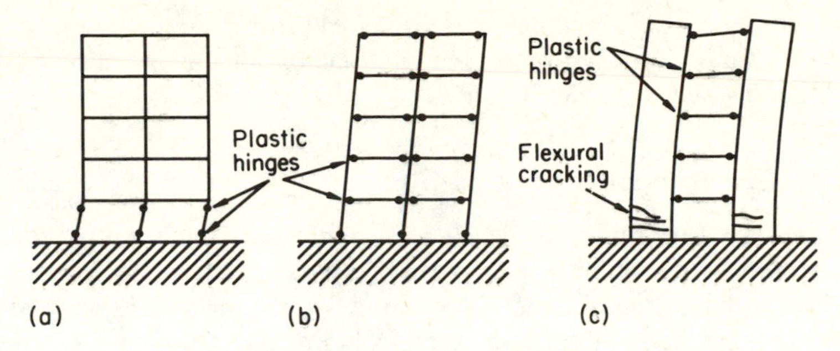

Fig, 6.3 Distribution of ductility.

shear wall structures can be designed to perform satisfactorily provided that the distribution of reinforcement is sufficient to ensure that flexural cracking and plastic hinge formation occurs as shown in Fig. 6.3(c).

It is desirable for structures to possess a degree of redundancy so that failed elements may shed their loads *via* alternative load paths. Simple statically determinate structures tend to be vulnerable in earthquakes. Single-column bridge piers have a bad record of earthquake performance and so have elevated water tanks. These are sometimes called 'inverted pendulum'-type structures (see Fig. 6.4). Multiple supports, with cross bracing or framing if possible, are recommended.

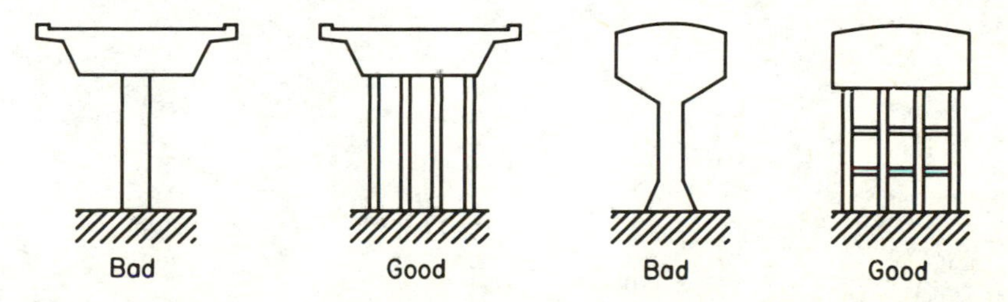

Fig. 6.4 Examples of inverted pendulum structures.

6.2 ASEISMIC DESIGN OF STRUCTURAL ELEMENTS

The main objective in the aseismic design of structural elements is to ensure that an adequate amount of ductility is achieved.

Now since plasticity in a tall structure is not going to be uniform, but tending to be concentrated towards the lower parts of the structure, the overall ductility factor will be a weighted average of the storey ductility factors. Thus, if an overall ductility factor of 3 to 5 is required, then storey ductility factors of between 3 and 10 may be required in the lower parts of the building. A similar argument applies when we determine the required ductility factors for individual members which will then probably have to range between 5 and 15 or even more. Ductility factors of this magnitude can be developed in well-designed members.

6.2.1 Concrete structures

Adequate ductility can be achieved with reinforced concrete provided that the disposition of the reinforcement satisfies certain criteria. During the 1971 San Fernando earthquake there were some significant lessons learned from the performance of reinforced concrete columns in certain of the structures (Lew, Leyendecker and Dikkers, 1971). Many columns were provided with inadequate transverse reinforcement in the form of shear hoops. Thus, under the severe shaking these columns burst and lost all their vertical strength. On the other hand, there were examples of spirally reinforced columns which contained vertical reinforcement and thus retained their strength despite cover concrete spalling off. Similar observations were made by EEFIT in Mexico City in 1985 (see Fig. 6.5).

The main principles underlying the design of concrete structures to resist earthquakes are:

(a) ultimate strength theory in design calculations for reinforced and prestressed concrete elements;

Fig. 6.5 Beam–column connection with inadequate reinforcement (reproduced with permission of the UK Earthquake Field Investigation Team).

(b) under-reinforced sections to ensure stable ductility;
(c) plastic hinges must form in beams while columns remain elastic;
(d) plastic hingeing should be in flexural and not shear mode, achieved by ensuring that shear capacity of a section exceeds the shear induced at full moment capacity; and
(e) core concrete in beam and column sections is confined by the presence of special transverse reinforcement.

By careful design of members and connections it is possible to develop very high ductility factors and it should be possible to absorb a certain amount of cyclic action at plastic hinges without catastrophic failure.

Recommendations for the detailing of reinforced concrete elements subjected to seismic loading are given in the Building Code of the American Concrete Institute, ACI Committee 318 (1983). The European Model Code for Seismic Design of Concrete Structures (CEB, 1984) adopts similar principles. The following is a brief review of recommendations for the detailing of reinforced concrete members.

Beams

Dimensional limitations are usually specified to ensure that beams behave as compact elements and to ensure that there is effective transfer of moments into the columns. For example:

(a) The ratio b/h should not be less than 0.25, where b is the breadth of the beam and h is its overall depth.
(b) The ratio l/h should not be less than 4 where l is the span of the beam.
(c) The beam should not be wider than the supporting column and should not frame into it eccentrically.

Calculation of the required amount of longitudinal reinforcement should follow the normal ultimate strength theory as explained fully by Park and Paulay (1975). Minimum amounts of reinforcement, depending on the concrete strength, are usually specified. Furthermore:

(a) There must always be reinforcement (a minimum of two bars) in both top and bottom of sections.
(b) The positive moment capacity at the face of columns must be at least 50% of the negative moment capacity.
(c) Within any potential plastic hinge region, the area of compression steel should be not less than 50% of the tension steel.

Transverse reinforcement should be provided to resist shear, with the amount calculated by the usual methods (Park and Paulay, 1975). The spacing of shear stirrups should not be greater than $d/2$, where d is the effective depth of the beam. In order to ensure adequate ductility and confinement of the concrete there should be additional special transverse reinforcement at the ends of beams or wherever plastic hinges could develop. The spacing should not exceed six times the diameter of the longitudinal bars. The latter recommendation is the result of experimental work by Park and Paulay (1975) who observed that cyclic loading can produce large cracks in the compression zone and possible buckling of the compression steel.

Splices in the longitudinal reinforcement should be kept well away from the column faces and should not be in regions of tension or reversed stress unless the concrete is adequately confined by stirrup ties.

The distribution of reinforcement in flexural members is illustrated in Fig. 6.6.

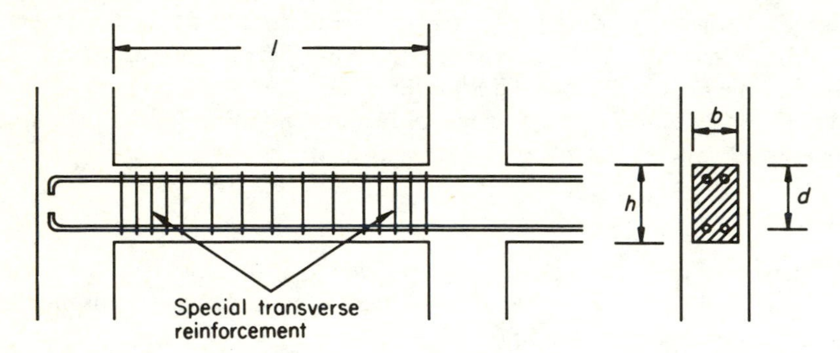

Fig. 6.6 Reinforcement in flexural members.

Columns

The ultimate capacity of columns should exceed the moment strengths of beams framing into them along each principal plane of the structure. This should ensure that plastic hinges would not form in the columns. However, this is only true if seismic loading occurred in only one principal axis of a structure. If seismic loading occurred in a more general direction the column stresses would be much greater and yielding could occur. Thus there should be adequate special transverse reinforcement to ensure confinement of the concrete. The ACI code gives formulae for the calculation of the amount of special transverse reinforcement to ensure confinement of the concrete core. Park and Paulay (1980) recommend a maximum spacing of six longitudinal bar diameters if cyclic reversed loading is a possibility.

Transverse reinforcement in columns should be provided in the form of positively anchored hoops and supplementary cross ties as shown in Fig. 6.7. Sheikh and Uzumeri (1982) suggested that columns with only four corner bars were inefficient and perhaps unsafe. They recommended the use of at least eight longitudinal bars. Spiral reinforcement is considered to be almost twice as efficient as rectangular hoops though it is less convenient for

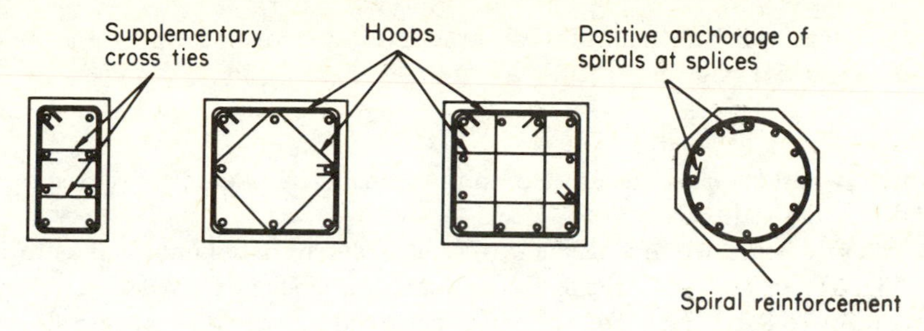

Fig. 6.7 Recommended arrangement of transverse reinforcement in columns.

rectangular column sections. However, the splices in the spiral should be positively anchored to prevent the spiral unwinding in the event of cover concrete spalling off. Lap splices of longitudinal reinforcement should be at mid height of columns.

Beam–column connections

There have been examples of failure in beam–column connections during earthquakes in the form of diagonal cracking as shown in Fig. 6.8(a). The behaviour of a connection is very complex, particularly under cyclic loading, when cracks can form at the column faces and cyclic shears can contribute to breakdown of the core concrete.

It is generally recommended that special confinement reinforcement should continue into the core of the connection and that there should be sufficient transverse reinforcement to carry the shear developed by the beams at ultimate capacity. The 'shear panel' analogy is considered to be an acceptable form of approximate analysis. This is illustrated in Fig. 6.8(b). The shear on the connection core is the sum of the forces exerted by the

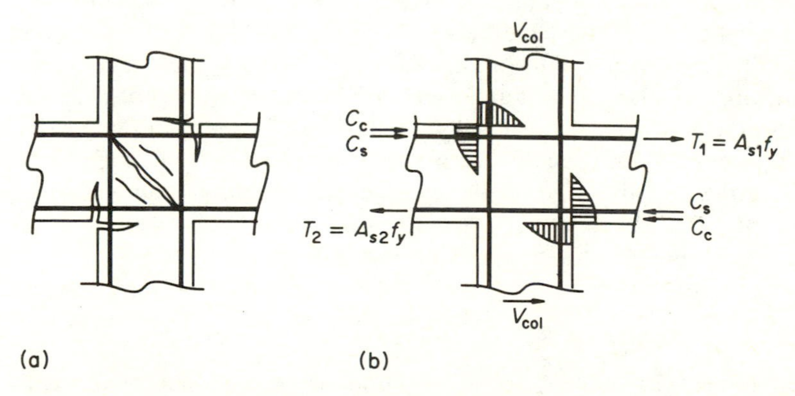

Fig. 6.8 Shear panel analogy for beam–column connection.

longitudinal reinforcement of the beams less the shear in the column above. Thus the shear is given by

$$V = A_{s1} f_y + A_{s2} f_y - V_{col} \tag{6.1}$$

where A_{s1} and A_{s2} are the areas of longitudinal steel in the beams as shown, and f_y is the yield stress of the steel.

The recommended arrangement of reinforcement in columns is illustrated in Fig. 6.9. In the case of external columns an efficient form of anchorage, according to Park and Paulay (1980), is achieved by providing a beam stub so that the longitudinal reinforcement may be anchored outside the column core.

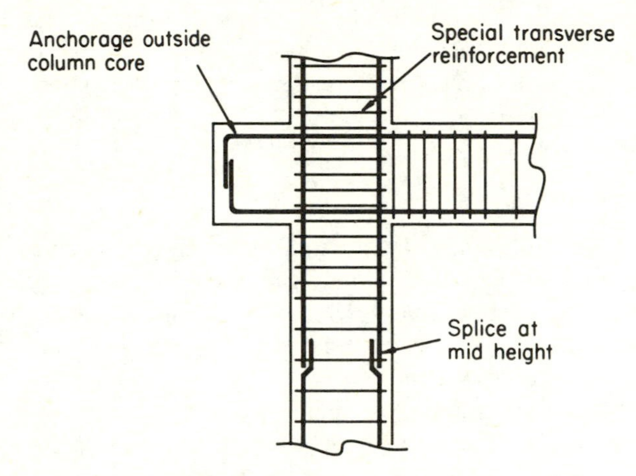

Fig. 6.9 Reinforcement in columns.

Reinforced concrete shear walls

Reinforced concrete shear walls have generally proved to perform well in earthquakes (Lew, Leyendecker and Dikkers, 1971). Examples of shear cracking failure can usually be attributed to lack of adequate shear reinforcement. Shear walls should be designed on the same theoretical basis as flexural members though it has been shown (Blakeley, Cooney and Megget, 1975) that better curvature ductility factors are obtained if flexural reinforcement is concentrated at the edges of the wall with the minimum amount over the main part of the web. Other precautions are that shear walls should not be too thin such as to risk compression instability.

6.2.2 Steel structures

Steel framed structures have a very good record of safe performance in earthquakes over a long period of years. Tall steel buildings have behaved safely in many major earthquakes, including San Francisco, 1906; Mexico City, 1957; San Fernando, 1971; Chile, 1985; Mexico City, 1985. The main reasons for this excellent record are the inherent flexibility and ductility of steel structures.

A specific problem that has occurred in braced frames is yielding in tension and buckling in compression of diagonal members (Haroun and Shepherd, 1986). Stretching of up to 15% was observed in San Fernando (Lew, Leyendecker and Dikkers, 1971), leaving bracing members totally slack after shaking had ceased. Many codes permit greatly increased stresses in steel under earthquake loading. This practice is under review. Failures of connections in braced structures have also been reported but could have been avoided if the connections had developed the full strength of the members.

Steel ductile moment-resisting frames should be designed on the basis of strong columns and weak beams to ensure plastic hinge formation in the beams. It is essential that connections are designed to develop the full plastic moment of the beams where they are connected to columns. Modern forms of high-strength bolted connection are quite capable of achieving this condition, though local reinforcement to flanges may be required to force plastic hinges to form outside the joint. Local buckling of a beam section at plastic hinges must be prevented by the addition of stiffening plates or limitations on the thickness of web and flanges.

A novel form of eccentric bracing was adopted for the 47-storey Embarcadero Centre in San Francisco (Merovich, Nicoletti and Hartle, 1982), as shown in Fig. 6.10. Fully concentric braces tend to attract high forces, with the risk of buckling, thus leading to heavy members. On the other hand, with no bracing horizontal drift is a problem. Eccentric braces may be used as a compromise with plastic hinges forming as shown.

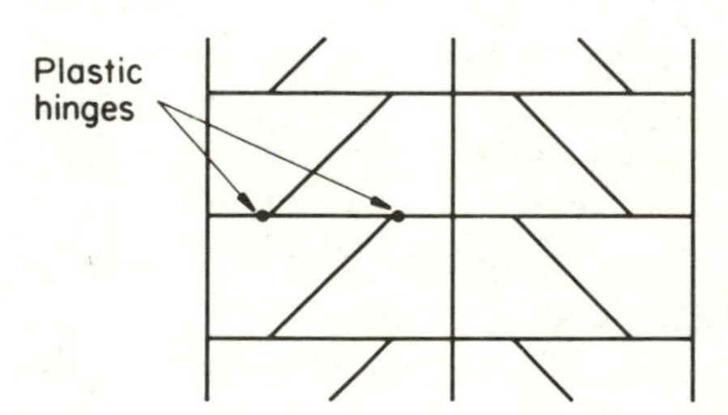

Fig. 6.10 Use of eccentric bracing.

6.2.3 Bridges

There have never been any serious problems with the behaviour of the main girders of highway bridges during earthquakes. When properly designed for highway loading they are capable of sustaining a substantial amount of transverse, vertical or longitudinal shaking. However, bridges frequently collapse during earthquakes because of failure of the piers or supports (Lew, Leyendecker and Dikkers, 1971).

Transverse inertia forces on the main girders, due to seismic motion, have to be carried in shear by the piers. The design of piers themselves should follow the normal rules for design in reinforced concrete or steel construction. Experience gained after the 1971 San Fernando earthquake revealed the need for multiple-column piers. Bridges with single-column supports proved to be extremely vulnerable since the lack of an alternative load path following the failure of that column led to total collapse. The redundancy and alternative load paths offered by multiple-column supports proved to be highly desirable. Wall-type piers also proved to behave well. Another advantage of multiple-column supports is that the overturning effect of the transverse load on the deck is resisted by a couple and the shearing load on the columns is reduced. Continuity between the piers and foundations is very important. This may seem an obvious remark but there were examples in San Fernando of columns being uprooted from the foundation piles owing to inadequate anchorage or continuity of the reinforcement.

The main reason for collapse of highway bridges in San Fernando was the inability of bearing pads to cope with excessive lateral ground movements. This led to bearings being literally pulled out from under the end supports of bridges, precipitating some catastrophic collapses. Large foundation movement is often caused by consolidation and compaction of loose strata during strong shaking. Both longitudinal and horizontal movements of bearings were observed. In many instances bearing areas were very small. Many collapses could have been avoided by increasing bearing areas. Longitudinal and transverse restraining devices could prevent girders falling off at the abutments but design would have to permit the required thermal movement.

It is obvious that there are limits to the improvements that are possible. In Chile in 1985 a bridge pier sank several feet a few days after the earthquake as a result of delayed liquefaction. Localized ground displacements of up to a metre or so have been known to occur. It would be unrealistic to attempt to design bearings to cope with such extreme movements.

Long-span suspension bridges have proved to be very robust in earthquakes. This is because of their inherent flexibility, which makes them capable of accommodating large ground motions. Abdel–Ghaffar and Rubin (1984) analysed the Golden Gate bridge under the effects of multiple-support excitation. They concluded that the main cable tension was increased by only 2% over that due to dead load. However, torsional stress at midspan reached 70% of yield stress under the worst conditions. Dumanoglu, Severn, Brownjohn and Taylor (1987) showed that the towers of a suspension bridge were the most critical components under dynamic loading. This was especially so when subjected to an earthquake wave travelling parallel to the deck with consequent multiple-support excitation.

6.2.4 Seismic performance of equipment in nuclear plant

Nuclear facilities must be designed to very rigorous margins of safety. Seismic damage to equipment is considered to be a significant contribution to the risk of a major failure and consequent contamination hazard (Bagchi and Noonan, 1985). A safe shutdown earthquake (SSE) is one during which there are no hazardous ruptures of critical containments or piping runs, and all safety-related control equipment continues to function normally, allowing a safe and orderly shutdown of the plant.

Examples of critical items of equipment are: motor control centres, switch-gear, motorized valves, pumps with their motors, emergency electrical supply equipment, and relays. Items of equipment such as these are normally required to be qualified by shake-table testing or analysis or a combination of both. Either of these options are extremely expensive. Analytical methods tend to be overconservative because of the difficulty of modelling detailed geometry and nonlinearities.

There is a shortage of information on the performance of such equipment in nuclear plant during actual earthquakes. However, similar and often identical equipment exists in conventional generating plant and various large industrial facilities. Nuclear plant operators in the USA formed a group called the Seismic Qualification Utilities Group (SQUG) and made strenuous efforts to obtain 'experience data' on the performance of critical items of equipment in various recent major earthquakes (EQE, 1984).

Data were obtained from 23 industrial facilities in five major earthquakes between 1971 and 1983. Data for earlier earthquakes was obtained by assessing detailed reports made at the time. Expert observers visited the sites of more recent earthquakes and were able to assess performance at first hand.

The results of the survey indicated that equipment in general was very rugged and failures were extremely rare. In fact the small number of failures were often caused by other items of plant falling or sliding because of inadequate anchorage.

An independent assessment of the work of industrial investigating teams was made by the Senior Seismic Review and Advisory Panel (SSRAP, 1985) who concluded that equipment installed in nuclear power plant is generally similar to, and at least as rugged as, equipment installed in conventional power plant. Most items of equipment had a good record of performance, with the exception of relays which tend to chatter during strong shaking. The need for adequate anchorage of equipment was emphasized.

A major project called the Active Components Test Program (ACT) has been carried out in Japan involving electric power companies, nuclear equipment manufacturers and universities (Komori, Ichihashi, Masuda *et al.*, 1985). The object was to develop simplified analytical models and evaluation criteria for safety-related classes of equipment. This involved testing many large items of equipment on shake-tables and subjecting them to continuous sinusoidal motion and artificial earthquakes with prescribed spectra. Maximum accelerations were typically very high, in excess of 0.9 *g* and as great as 10.0 *g*. Even under these severe conditions equipment generally functioned properly. Detailed finite element models were found to compare favourably with test results and, from observations of the most significant structural behaviour, simplified models were developed.

6.2.5 Pipework

Safety-related pipework in nuclear power plant is required to be designed to resist seismic loading (ASME, 1980). Examples are reactor coolant loop piping, main steam lines and feedwater pipes. Pipework usually consists of combinations of long horizontal and vertical runs connected with 90° bends and often including pumps, valves and other heavy fittings. Because of lengthy pipe runs the supports are required to accommodate thermal movements while providing adequate restraint against dynamic loading. Restraints are often provided in the form of mechanical snubbers, these being hydraulic piston-operated damping devices.

A wide range of analytical techniques has been employed when designing complex piping systems. Cautious assumptions involving low damping values and neglect of nonlinear effects have resulted in very stiff systems with large numbers of restraints. There has been much concern recently in the nuclear industry about potential overdesign of piping systems on the grounds of:

(a) the large number of seismic restraints causing congestion in the plants, together with high costs of installation and design; and
(b) actual performance in earthquakes proving to be contrary to design predictions.

SQUG sent a team of investigators to the site of the 1985 Chile earthquake and found many examples of very flexible piping runs that had never been designed seismically but had survived the 7.8 magnitude shock (Yanev and Smith, 1985). They discovered examples of seismically designed pipework, with snubbers, that were damaged while pipes without were undamaged.

Recent studies have underlined the excessive conservatism of current design methods. Haas and Labes (1985) calculated the response of piping systems in the El Centro steam plant by methods that would currently be employed in the design of nuclear stations. The calculations predicted high overstressing when considering actual recordings of the 1979 Imperial Valley earthquake. In fact the piping systems survived the earthquake without damage. It is now being recognized that piping systems possess larger reserves of strength, by virtue of ductility and greater damping, than is normally assumed in calculations (Ibanez and Ware, 1985).

On the other hand, the consequences of major failure of pipework in nuclear and some industrial plant are potentially very severe. Therefore, there is an urgent need for research to establish reliable methods of predicting the performance of pipework, so that soundly based design rules may be established.

6.2.6 Other types of structure

Masonry construction has a bad reputation in earthquake areas. This is largely because of total collapse of older unreinforced masonry structures leading to great loss of life. Masonry is an essentially brittle material. There were many examples of brittle shear failure of brick shear walls in the 1971 San Fernando earthquake (Lew, Leyendecker and Dikkers, 1971).

Nevertheless, by employing reinforced masonry construction principles it is possible to design substantial structures with satisfactory performance (Priestley, 1980). There are two principal forms of reinforced masonry construction. The first consists of two skins of blocks or brickwork with vertical and horizontal bars grouted into a central cavity of 50 to 100 mm width. The second consists of special bricks or concrete units possessing vertical and horizontal voids. The required amount of reinforcement is placed in the cavities, which are subsequently grouted up. The design of reinforced masonry shear walls follows the same principles that are involved in reinforced concrete design. The similarities between the materials are obvious. Ultimate strength theory should be employed. Vertical reinforcement carries the longitudinal bending forces, while the shear is carried by horizontal reinforcement.

Small buildings, such as individual family dwellings, have a good record of performance in earthquakes despite the lack of deliberate seismic design. Collapses that have occurred have often been due to large openings, such as garages, or abrupt changes in roof or floor elevation. By incorporating sufficient lateral ties at junctions and discontinuities it is possible to design adequately robust dwelling units (ATC, 1976).

Human life can also be endangered by the failure of what would seem to be quite minor fittings within buildings. For example, alarm was expressed concerning the widespread collapse of heavy lighting fittings, ventilating grilles and suspended ceilings in school buildings during the San Fernando earthquake. Fortunately, the main shock occurred early in the morning when the schools were empty, but risk of death or injury from heavy objects dropping from ceilings and walls was well appreciated. Shattering of glass in tightly fitting windows is a similar hazard which can be reduced significantly by appropriate attention to the architectural design of the framing.

7

The dynamic effects of wind loading on structures

7.1 INTRODUCTION

It is an everyday experience that the wind is not steady but fluctuates with time. This turbulence is characterized by sudden gusts superimposed upon a mean wind velocity. Difficulty in walking steadily in a storm wind is clear evidence of its gusty nature. Most structures are relatively stiff, so that their motions correspond directly to the wind velocity fluctuations, and hence a knowledge of the maximum gust speed is a sufficient basis for design. The corresponding pressure or drag force is then treated as a quasi-static loading.

Nevertheless, there are many forms of structure, particularly those that are tall or slender, that respond dynamically to the wind. The best known structural failure in the wind was the Tacoma Narrows Bridge which collapsed in 1940 in a wind speed of only 42 miles/hour. It had already earned the nickname of Galloping Gertie because of its large-amplitude vertical oscillations in quite light winds. But it eventually failed soon after it had developed a joint torsional and flexural mode of oscillation. There had already been a lengthy history of the susceptibility of suspension bridges to wind-induced oscillation, but this major failure prompted a concerted international research effort.

There are several different phenomena giving rise to dynamic response of structures in wind. These include buffeting, vortex shedding, galloping and flutter. Slender structures, whether they be towers or suspension bridges, are likely to be sensitive to dynamic response in line with the wind direction as a consequence of turbulence buffeting. Transverse or cross-wind response is more likely to arise from vortex shedding or galloping but may be excited by turbulence buffeting also. Flutter is a coupled motion, often being a combination of bending and torsion, and can result in instability.

An important problem associated with wind-induced motion of buildings is concerned with human response to vibration. This subject will be discussed in detail in Chapter 11. At this point it will suffice to note that humans are surprisingly sensitive to vibration, to the extent that motions may feel uncomfortable even if they correspond to relatively unimportant stresses. Chang (1973) estimated the wind-induced motions of some tall buildings in the USA, including the Empire State Building, and noted the comments by the occupants.

Another possible consequence of wind excitation is structural fatigue. This would not normally be a problem for buildings because the human response criterion would usually override it. However, slender masts or towers may experience many millions of cycles of stress within their lifetimes, and it is known that fatigue damage can accumulate under relatively small stresses (see Chapter 12).

This chapter is a brief introduction to the dynamic response of structures in wind. More detailed treatments by Simiu and Scanlan (1977) and by Lawson (1980) are strongly recommended.

7.2 THE DYNAMIC PROPERTIES OF THE WIND

7.2.1 The mean wind speed

At great heights above the surface of the earth, where frictional effects are negligible, air movements are driven by pressure gradients in the atmosphere, which in turn are the thermodynamic consequences of variable solar heating of the earth. This upper level wind speed is known as the *gradient wind velocity*. Closely spaced isobars on a weather map indicate high wind speeds. It should be noted that, owing to the curvature and rotation of the earth, a state of equilibrium is reached between the pressure gradient forces and the Coriolis forces acting on moving particles of air, with the result that the wind is almost perfectly directed along the isobars.

Closer to the surface the wind speed is affected by frictional drag of the air over the terrain. There is a *boundary layer* within which the wind speed varies from almost nil, at the surface, to the gradient wind speed at a height known as the *gradient height*. The thickness of this boundary layer, which may vary from 500 to 3000 m, depends on the type of terrain, as depicted in Fig. 7.1. It will be observed that the gradient height within a large city centre is much higher than it is over the sea where the surface roughness is less.

For structural design purposes a knowledge of the maximum wind speed at the appropriate height above the ground is required. In practice, it has been found useful to start with a reference wind speed based on statistical analysis of wind speed records obtained at meteorological stations throughout the country. The definition of the reference wind speed varies from one country to another. In the United Kingdom it is the extreme mean hourly wind speed

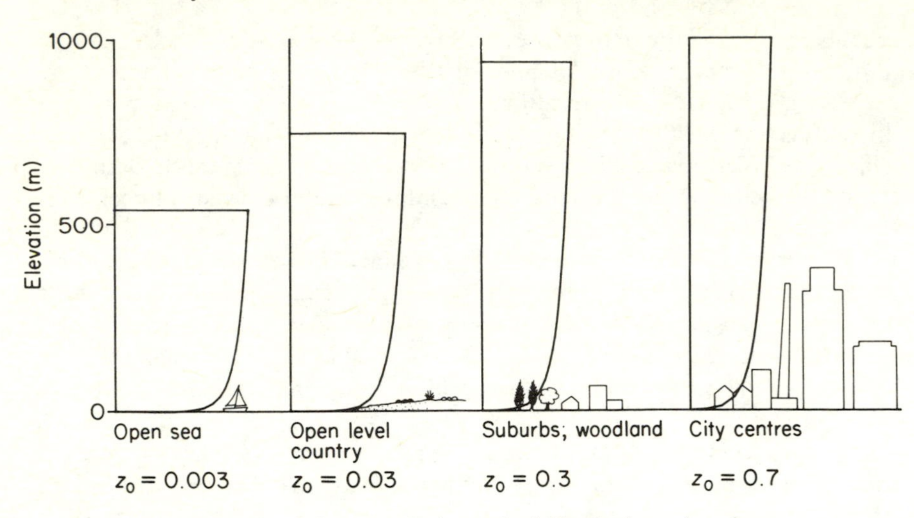

Fig. 7.1 Profiles of mean wind speed within the boundary layer.

with a 50 year return period at a height of 10 m above the ground assuming uniform open country terrain (ESDU, 1982). A similar definition is employed in Canada (NBC, 1985). Maps of reference wind speeds applying to various countries are usually available.

The reference wind speed may be modified by a factor to take account of the probability of the wind speed being exceeded in a period of N years. The probability of the wind speed, with a return period of 50 years, being exceeded in any one year is $1/50 = 0.02$. Therefore the probability of the same wind speed being exceeded in a period of 50 years is $1-(0.98)^{50} = 0.636$. However, it is often desirable to know the wind speeds having return periods of 5, 10 or 100 years but with the same probability of exceedance as the reference wind speed. An effective method of doing this is based on the Fisher–Tippett Type 1 extreme value distribution (Gumbel, 1958)

$$P(V) = 1 - \exp[-e^{-a(V-U)}], \tag{7.1}$$

which expresses the probability that a velocity V will be exceeded. U and $1/a$ are the mode and dispersion of the distribution and are derived by fitting available data. It is then possible to derive a factor K_N so that the desired wind speed may be evaluated from

$$V_N = V_{50} K_N \tag{7.2}$$

where V_N is the wind speed with the desired probability of exceedance in N years. Cook (1982) showed that V^2 (or the dynamic pressure $\frac{1}{2}\rho V^2$) is an even better variable in the Fisher–Tippett Type 1 distribution and obtained the factor

$$K_N = \left(\frac{5 + \ln N - \ln[-\ln(1 - P_N)]}{8.912}\right)^{1/2} \tag{7.3}$$

where P_N is the desired probability of exceedance.

Close to the surface of the earth the air flow closely approximates to two-dimensional turbulent flow of a fluid parallel to a rough wall such that the shear stress in the fluid is constant and is given by an equation similar to Newton's law of viscosity. This may be written

$$\tau_0 = (\mu + \eta)\,\mathrm{d}\bar{V}/\mathrm{d}z \tag{7.4}$$

where τ_0 is the constant shear stress in the fluid, μ is the viscosity of the fluid, η is the apparent viscosity due to turbulence, known as the eddy viscosity, and $\bar{V}$ is the mean speed of the fluid at distance z from the wall.

In effect this law states that the fluid shear stress is proportional to the velocity gradient close to the wall. However, the eddy viscosity η is not a constant property of the fluid but depends on the fluid motion and its density. It can be shown (Streeter and Wylie, 1979) that the solution to (7.4) is a velocity profile law of the form

$$\bar{V}/v_* = (1/k)\ln(z/z_0) \tag{7.5}$$

where v_* is the friction velocity and is defined by the surface shear stress τ_0 and the fluid density ρ, so that $v_* = \sqrt{(\tau_0/\rho)}$; k is the von Kármán's constant (von Kármán, 1934) $\simeq 0.4$; and z_0 is the roughness length, which is a constant for a surface of given roughness.

Equation (7.5) is known as the logarithmic law or the 'law of the wall'. Allowance should be made for the fact that the effective zero-plane is about one or two metres below the general height of buildings and trees that make up the roughness elements. Therefore it is recommended that z be replaced by $z–d$ in (7.5), where d is the height of the zero-plane above the ground. The parameter z_0 is obtained empirically from observed wind profiles. Typical values of z_0 and d are given in Table 7.1 (ESDU, 1985).

The friction velocity v_* is geographically determined and is related to the reference wind speed by Equation (7.5) with $z = 10$ m and $z_0 = 0.03$ m so that

$$\bar{V}(10)/v_* = 2.5\ln(10/0.03) = 14.52. \tag{7.6}$$

The logarithmic law assumes that equilibrium conditions exist. In practice, the wind should have blown over a fetch of 50 to 100 km of uniform terrain to ensure that the boundary layer is in equilibrium with the surface roughness. Unfortunately, such conditions seldom exist at a location of interest. Typically surface roughnesses might change from open level country to suburbs to city centres within a few kilometres. A method has been devised for evalu-

Table 7.1 Typical values of terrain parameters z_0 and d

	$z_0(m)$	$d(m)$
City centres, forests	0.7	15 to 25
Small towns, surburbs or large towns and cities, wooded country (many trees)	0.3	5 to 10
Outskirts of small towns, villages. Countryside with many hedges, some trees and some buildings	0.1	0 to 2
Open level country with few trees and hedges and isolated buildings; typical farmland	0.03	0
Fairly level grass plains with isolated trees. Very rough sea in extreme storms (once in 50 yrs)	0.01	0
Flat areas with short grass and no obstructions. Runway area of airports. Rough sea in annual extreme storms	0.003	0
Snow covered farmland. Flat desert or arid areas. Inland lakes in extreme storms	0.001	0

ating the effects of multiple step changes in terrain roughness upwind of the site in question (ESDU, 1982).

The logarithmic profile law is reasonably accurate within the lowest 30 m of the boundary layer. For heights up to 300 m Harris and Deaves (1980) have recommended an improved profile law. Historically a power law has been used but it is no longer recommended for engineering purposes. For the sake of simplicity the logarithmic law, with $(z–d)$ as the height parameter, will be employed in the remainder of this chapter.

7.2.2 Turbulence

Measurements of wind speed reveal that it is far from steady but fluctuates with time, as shown in Fig. 7.2. The existence of turbulence is demonstrated by the billowing motion of smoke rising from a chimney. A characteristic feature of this motion in a strong breeze is the formation of rotating 'whirlpools' of smoke. These are the result of eddies generated by the action of the wind blowing over obstacles such as tall trees or buildings. The generation of an eddy as the wind blows over a bluff object is shown in Fig. 7.3. The movement of wind around a sharp discontinuity results in a rotational vortex motion accompanied by a drop in pressure locally. The eddy is then carried off downwind and breaks down into smaller eddies in a cascade process. Further eddies are generated from the sides of the object resulting in a generally unsteady air stream.

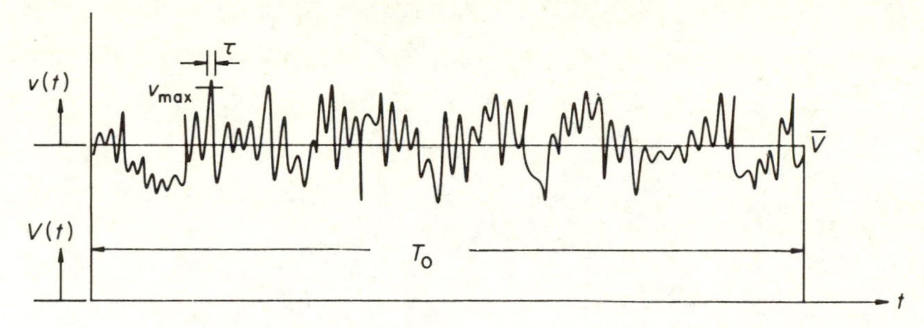

Fig. 7.2 Fluctuation of measured wind velocity at a point.

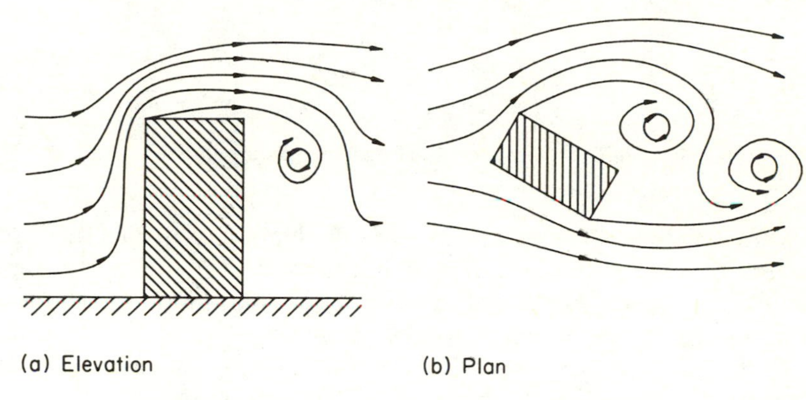

Fig. 7.3 Generation of eddies. (a) Elevation; (b) plan.

In general there will be fluctuating components of wind speed in all three orthogonal directions. The wind vector at a point may be regarded as the sum of the mean wind vector and a dynamic component

$$\boldsymbol{V}(z,t)=\bar{\boldsymbol{V}}(z)+\boldsymbol{v}(z,t). \tag{7.7}$$

This expression implies equilibrium conditions so that there is no variation with horizontal distance in the direction of the wind. If it is assumed that the mean wind vector is directed along the x axis of the coordinate system, then the magnitude of the dynamic wind vector is given by

$$v(z,t)=[v_x(z,t)^2+v_y(z,t)^2+v_z(z,t)^2]^{1/2}, \tag{7.8}$$

where v_x, v_y and v_z are the x, y and z gust components as shown in Fig. 7.4. The longitudinal component v_x is generally the greatest and is therefore the most important for the majority of structural applications. However, the cross-wind component v_y is often significant for sensitive structures such as tall chimneys and lamp standards, especially when combined with the effects of vortex shedding. The vertical component v_z is important in the case of

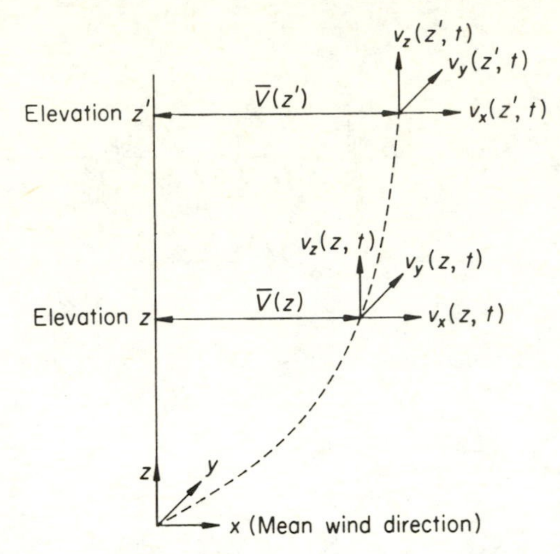

Fig. 7.4 Components of dynamic wind vectors.

long-span bridges which are usually more flexible vertically than they are in the transverse direction. The v_x, v_y and v_z components are almost independent of one another and therefore, in the discussion that follows, only the longitudinal gust component v_x will be considered.

A further consequence of turbulence is that dynamic loading on a structure depends on the size of the eddies. Large eddies, whose dimensions are comparable with the structure, give rise to well correlated pressures as they envelop the structure. On the other hand, small eddies result in pressures at various parts of the structure being practically uncorrelated. A simple procedure for dealing with this problem is to average the fluctuating wind speed over a brief interval representative of the time it takes for an eddy to pass a point on the structure. Thus a long averaging time would correspond to an eddy of large spatial extent. The following averaging times are employed in the United Kingdom wind loading code (BSI, 1972):

3 seconds: structures up to 20 m across
5 seconds: structures with a maximum dimension of 20 to 50 m
15 seconds: structures with a dimension larger than 50 m.

It is evident from Fig. 7.2 that the peak discrete gust speed, obtained by averaging over a time interval τ, is much greater than the mean velocity.

7.2.3 Analytical expressions for turbulence

Fluctuating wind velocity may be treated as a 'stationary random function'. It is *random* in the sense that the velocity cannot be predicted at any instant of time. However, it is *stationary* in the sense that certain statistical properties, which do not depend on the part of the record from which they were obtained, may be evaluated from a lengthy record. For example, the mean wind speed is given by

$$\bar{V}=\frac{1}{T_0}\int_0^{T_0} V(t)\,\mathrm{d}t \tag{7.9}$$

where T_0 is usually about one hour. An overall measure of the degree of turbulence is given by the mean square dynamic component of wind velocity or the variance

$$\sigma_x^2=\frac{1}{T_0}\int_0^{T_0} [v_x(t)]^2\,\mathrm{d}t. \tag{7.10}$$

Davenport (1961) suggested that the fluctuating velocity could be regarded as being compounded of a large number of harmonic components and thus could be represented by the sum of a Fourier series

$$v_x(t)=\sum_{j=1}^{\infty} v_j \sin(2\pi jt/T_0+\phi_j) \tag{7.11}$$

where j/T_0 is the frequency of the jth harmonic, v_j is its amplitude and ϕ_j is its phase angle.

Substituting (7.11) in (7.10) we obtain

$$\sigma_x^2=\sum_{j=1}^{\infty} v_j^2/2. \tag{7.12}$$

In the limit this may be expressed in the continuous form

$$\sigma_x^2=\int_0^{\infty} v_x^2(n)/2\,\mathrm{d}n \tag{7.13}$$

where n is the frequency (note that it is conventional to denote frequency by n in wind engineering). Notice that the term $v_x^2(n)/2$ is in a form that is associated with the energy of the wind. It is commonly referred to as the *spectral density* and, henceforth, will be denoted by $S_x(n)$. Therefore

$$\sigma_x^2=\int_0^{\infty} S_x(n)\,\mathrm{d}n. \tag{7.14}$$

It has been observed that if the energy of the wind is plotted against averaging time T_0, there is a minimum value, or *spectral gap*, at about one hour (Van der Hoven, 1957). Longer averaging periods are associated with diurnal and meteorological variations while shorter averaging times are associated with mechanical turbulence. This is the justification for the use of one hour as the averaging time for the reference wind speed.

Since the profile of the wind speed within the boundary layer is associated with turbulence it is not surprising that the r.m.s. velocities are also closely related to the friction velocity v_*. Approximate values of these (Davenport, 1983) near the ground are

$$\sigma_x = 2.5\,v_*$$
$$\sigma_y = 2.2\,v_* \qquad (7.15)$$
$$\sigma_z = 1.25\,v_*.$$

These values indicate the relative importance of the three components of the dynamic wind vector and are useful in evaluating the non-resonant or quasi-static response to turbulence buffeting.

Evaluation of the resonant response of a structure requires a knowledge of the frequency content of the turbulence which is represented by the spectral density $S_x(n)$. An expression for the spectral density was derived by Kolmogorov (1941) based upon the existence of an 'inertial subrange' through which energy is transferred from the largest eddies, which possess most of the energy, to the smallest eddies in which energy is dissipated by viscosity. The Kolmogorov formula is recognized as being substantially reliable in the high-frequency region of the spectrum (which is relevant for most structures) and is given in non-dimensional form by the expression

$$nS_x(z,n)/v_*^2 = 0.26\,f^{-2/3} \qquad (7.16)$$

where $f = nz/\bar{V}(z)$ is the reduced frequency. A useful expression for the full frequency range has been recommended by Simiu (1974):

$$nS_x(z,n)/v_*^2 = 200\,f/(1+50\,f)^{5/3}. \qquad (7.17)$$

This is compatible with (7.16). Expressions which take account of the scale of turbulence more accurately have been published by ESDU (1985). The form of Equation (7.17) is shown in Fig. 7.5. It will be observed that the natural

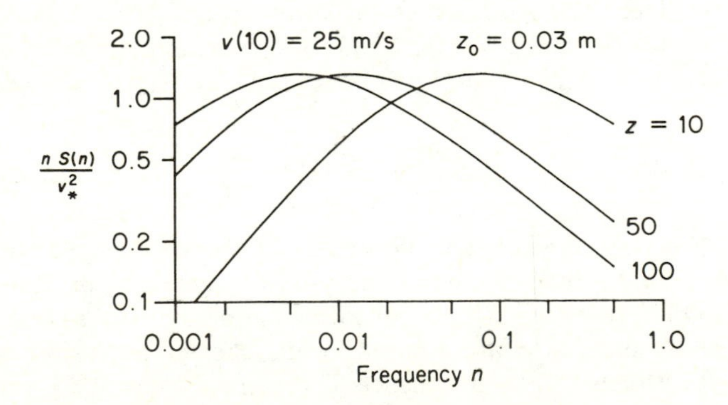

Fig. 7.5 Spectrum of horizontal gustiness.

frequencies of most tall structures will be at the high-frequency end of the curve.

We have already noted that if eddies are smaller than the overall dimensions of a structure then the spatial distribution of pressure will not be uniform. This is particularly so in the case of eddies in the range likely to cause dynamic response of large structures. In fact the pressures at two widely separated points on a structure may be almost uncorrelated and will often be acting in opposition, thus reducing the overall load on the structure. This cross-correlation phenomenon is clearly very important.

An expression for the *cross-correlation* of wind velocities may be obtained from wind velocity records at two points r and r' on a structure, as shown in Fig.7.6. During the evaluation of the dynamic response to turbulence we shall require cross-correlation terms of the form

$$C(r, r') = \overline{v(r, t)v(r', t)}, \tag{7.18}$$

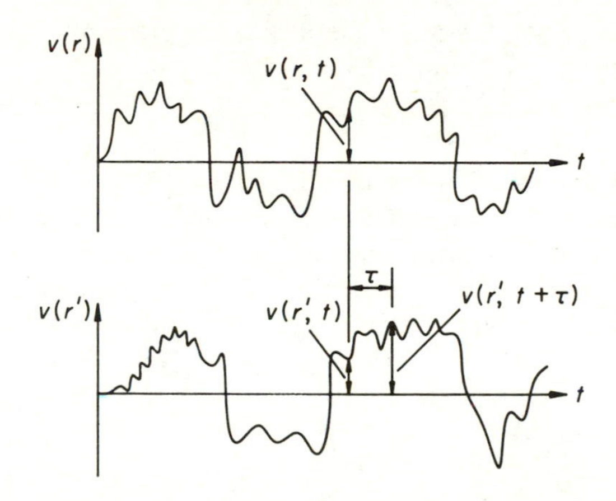

Fig. 7.6 Cross-correlation of wind velocities at two points r and r' on a structure.

where the bar denotes averaging over a fixed period of time. This is more useful in the non-dimensional form

$$R(r, r') = \overline{v(r, t)v(r', t)}/\sigma_x\sigma'_x \tag{7.19}$$

where σ_x and σ'_x are the r.m.s. velocities at the two points. In theory there are nine cross-correlation coefficients relating to the different components of fluctuating wind velocity. However, in structural applications it is the correlations between horizontal wind velocities at points separated either vertically (towers) or horizontally (bridges) that are of the most interest. From limited experimental data the following formula is recommended (ESDU, 1986;

Davenport, 1983)

$$R(\Delta r) = e^{-\Delta r/L} \tag{7.20}$$

where L is a lateral scale ($\simeq 60$ m).

A further important measure of cross-correlation is obtained if a time lag τ between observations at the two points is included. The resulting expression is called the *cross-covariance* and is given by

$$C(r, r'; \tau) = \overline{v(r, t) v(r', t+\tau)}. \tag{7.21}$$

By thinking of the velocities as being compounded of an infinite sum of harmonic components, as we did in the case of the spectral density function, it is possible to obtain

$$C(r, r'; \tau) = \int_0^{\infty} S_x(r, r'; n) \exp[\sqrt{(-1)} 2\pi n \tau] \, dn \tag{7.22}$$

where $S_x(r, r'; n)$ is the *cross-spectral density*. In fact this expression is one member of a Fourier transform pair. The relevant mathematics has been omitted for simplicity, the other member being given by

$$S_x(r, r'; n) = 2 \int_{-\infty}^{\infty} C(r, r'; \tau) \exp[\sqrt{(-1)}\, 2\pi n \tau] \, d\tau. \tag{7.23}$$

The significance of the $\sqrt{(-1)}$ is that, in general, the cross-spectral density is a complex quantity, having both in-phase and 90° out-of-phase components. However, in most structural applications the out-of-phase component is either zero or small enough to be neglected. The in-phase component may be expressed non-dimensionally as the *coherence* and is given by

$$\gamma^2(r, r'; n) = \frac{|S_x(r, r'; n)|^2}{S_x(n) S'_x(n)}. \tag{7.24}$$

A useful practical formula is given by

$$\gamma(\Delta r; n) = \exp(-C \Delta r n / \bar{V}_a) \tag{7.25}$$

where $\bar{V}_a$ is the mean wind velocity at a point midway between r and r'.

Experimental values of the coefficient C exhibit considerable scatter but reveal a distinct relationship with the separation and scale of turbulence. Expressions for evaluating C for both horizontal and vertical correlation are available in ESDU (1986). However, for simplicity, Davenport (1983) has suggested a value of C in the range 6–10 for most engineering applications.

7.3 RESPONSE TO TURBULENCE BUFFETING

The dynamic response of a structure subjected to random loading cannot be expressed in a deterministic form but instead has to be expressed in terms of probabilities. Just as we expressed wind in terms of its mean velocity, its variance and its frequency spectrum, we shall develop expressions for the response of a structure to the mean wind together with the variance of response to turbulence buffeting. Finally, using extreme-value statistics, estimates of the maximum likely total response will be made.

The following analysis will be developed for line-like structures. A line-like structure possesses a large length dimension compared with its breadth. A consequence of this is that it may be assumed that turbulent wind pressures are fully correlated over the face of the structure at any station along its length. This greatly simplifies the analysis because the correlation of wind pressures need only be considered along its length. Examples of line-like structures are cables, chimneys, slender towers and long-span bridge decks.

7.3.1 Response to the mean wind

Fig. 7.7 depicts a turbulent wind profile impinging on a tall slender structure. The mean intensity of loading at any point r will be given by the familiar expression

$$\bar{p}(z) = \tfrac{1}{2}\rho \bar{V}^2(z) C_D b \tag{7.26}$$

where ρ is the density of air $\simeq 1.2\ \text{kg/m}^3$; C_D is the drag coefficient; b is the width of the structure transverse to the wind; and $\bar{V}(z)$ is the mean wind speed at height z.

Therefore, by integrating this load intensity over the height of the structure, the load effect at some point x will be given by

$$\bar{E} = \int_0^H \tfrac{1}{2}\rho \bar{V}^2(z) C_D b \beta \, dz \tag{7.27}$$

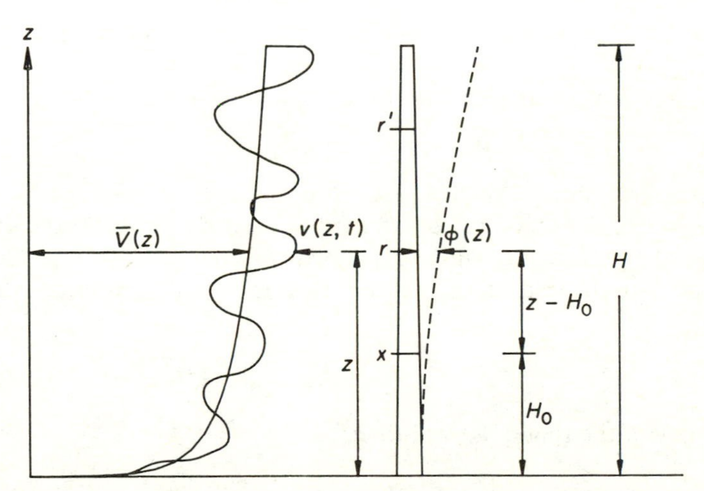

Fig. 7.7 Turbulent wind load on a vertical line-like structure.

where β is an influence function for the particular load effect. For example, if the load effect of interest is the bending moment at X then

$$\begin{aligned} \beta &= z - H_0 \quad z > H_0 \\ \beta &= 0 \qquad\quad z \leqslant H_0. \end{aligned} \tag{7.28}$$

Other load effects of interest are the deflection and the shear at a point, or the force in a member of a lattice structure.

7.3.2 Non-resonant or broad-band response to turbulence

We have already observed in Fig. 7.5 that the spectrum of horizontal gustiness covers a range of frequencies well below the natural frequencies of most structures. Only the lowest modes of vibration of very large structures are likely to be affected at the high-frequency end of the spectrum. Thus the response of most structures to turbulence can be treated as if the loading is quasi-static. This component of response is often called *broad-band* because it is evaluated over the full range of turbulence frequencies.

The distributed loading due to wind at height z is

$$F(z) = \bar{p}(z) + p(z,t) \tag{7.29}$$

where $\bar{p}(z)$ and $p(z,t)$ are the mean and fluctuating components respectively. In terms of velocities this becomes

$$F(z) = \tfrac{1}{2}\rho[\bar{V}(z) + v(z,t)]^2 C_D b. \tag{7.30}$$

Hence the fluctuating component will be given by

$$p(z,t) = \rho \bar{V}(z) v(z,t) C_D b \tag{7.31}$$

since the second-order term in $v^2(z,t)$ can be ignored in comparison with $\bar{V}^2(z)$.

The load effect on an element of the structure will be

$$dE(t) = \rho \bar{V}(z) v(z,t) C_D b \beta dz \tag{7.32}$$

where β is the influence coefficient defined in §7.3.1.

The effect of the randomly distributed fluctuating forces on the dynamic response of the structure will be explained with reference to Fig. 7.8. If the forces are considered as discrete forces then the total load on the structure will be

$$F = F_1 + F_2 + \ldots + F_i + \ldots + F_N = \sum_i^N F_i. \tag{7.33}$$

The square of the total load will therefore be

$$F^2 = F_1^2 + F_2^2 + \ldots + F_i^2 + \ldots + F_i F_j + \ldots + F_N^2 = \sum_i^N \sum_j^N F_i F_j. \tag{7.34}$$

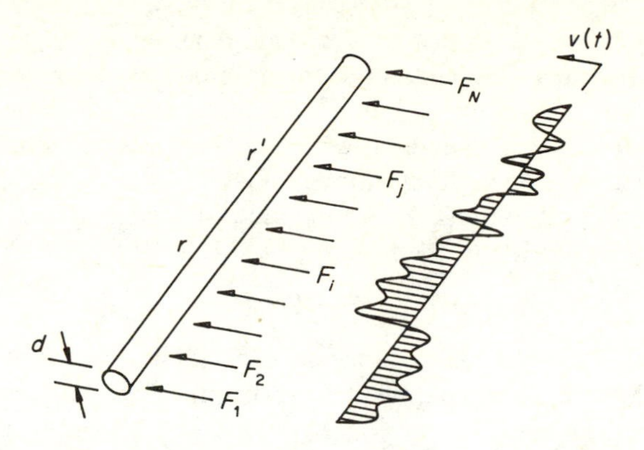

Fig. 7.8 Fluctuating wind pressures acting on a slender structure.

The mean square load will hence be obtained by averaging (7.34) over an appropriate period, e.g. one hour. Replacing F_i and F_j by distributed loading at r and r', as given by (7.31), and performing a double integral over the height of the structure we find that the mean square load effect is

$$\sigma_{\mathrm{B}}^2(E) = \rho^2 C_{\mathrm{D}}^2 b^2 \int_0^H \int_0^H \overline{[\bar{V}(z)v(z,t)\beta][\bar{V}(z')v(z',t)\beta']}\,\mathrm{d}z\mathrm{d}z'$$

$$= \rho^2 C_{\mathrm{D}}^2 b^2 \int_0^H \int_0^H \beta\beta'\sigma_x\sigma_x'\bar{V}(z)\bar{V}(z')R(z,z')\mathrm{d}z\,\mathrm{d}z' \qquad (7.35)$$

in which we have introduced the cross-correlation coefficient given by Equation (7.19).

It is reasonable to ignore the variation of σ_x^2 with height, using $\sigma_x = 2.5\,v_*$ as in (7.15), for example. Therefore, the expression becomes

$$\sigma_{\mathrm{B}}^2(E) = \rho^2 C_{\mathrm{D}}^2 b^2 \sigma_x^2 \int_0^H \int_0^H \beta\beta'\bar{V}(z)\bar{V}(z')R(\Delta z)\mathrm{d}z\mathrm{d}z', \qquad (7.36)$$

where the suffix B denotes the broad-band or background response.

The double integration has to be performed numerically in general. However, by recasting (7.36) in a non-dimensional form it has been possible to derive charts from which the response of a wide range of structures may be evaluated with ease (ESDU, 1976a; Wyatt, 1980).

7.3.3 The resonant or dynamic component of response

The dynamic response of elastic structures to random loads was discussed by Eringen (1953). However, Davenport (1962) showed that a much simpler solution could be achieved using modal analysis, by which means the equations for each degree of freedom may be uncoupled. Coupling of modes through damping and aeroelastic forces can generally be neglected for civil engineering structures, especially if the natural frequencies are well separated. The analysis that follows was originally developed by Davenport (1961) for single-degree-of-freedom structures and later extended to multi-degree-of-freedom line-like structures (Davenport, 1962).

The equation of motion of a single-degree-of-freedom system was derived in §2.2. It may be written in the form

$$m\ddot{Y} + c\dot{Y} + kY = Q(t), \tag{7.37}$$

where m, c and k are the mass, damping and stiffness constants, $Q(t)$ is the loading, and Y is the motion of the relevant degree of freedom. In the case of a periodic loading function,

$$Q(t) = q_0 \sin 2\pi nt, \tag{7.38}$$

the maximum steady-state dynamic response is given by

$$Y_{\text{max}} = (q_0/k)/[(1 - n^2/f^2)^2 + (2\xi n/f)^2]^{1/2} \tag{7.39}$$

where $f = \sqrt{(k/m)}$ is the natural frequency of the system and $\xi = c/[2\sqrt{(km)}]$ is the damping ratio. This result was essentially presented in Equations (2.25) and (2.27). The expression may be written in the form

$$Y_{\text{max}} = (q_0/k)H(n) \tag{7.40}$$

where $H(n)$ is called the *mechanical admittance* and is identical to the dynamic load factor (DLF) given in Equation (2.27). The form of this function is shown in Fig. 7.9 and should be studied carefully. It will be noted that over most of the frequency range the mechanical admittance is either 1.0 or zero. However, over a relatively small range of frequencies close to the natural frequency of the system it attains very high values if the damping is small. This property will be used later to simplify the analysis.

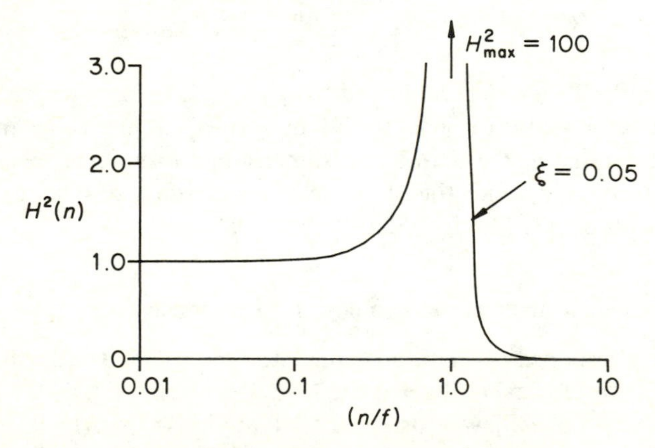

Fig. 7.9 Mechanical admittance function.

If the loading function $Q(t)$ is random we have to resort to methods which make use of the stationary properties of random functions, as we did in the case of wind velocities. We can assume that $Q(t)$ is compounded of an infinite sum of component harmonic loading functions. Following the same arguments that were employed in defining the spectral density of wind velocity, it can be shown that the mean square loading or variance is given by

$$\sigma^2(Q) = \int_0^\infty S_Q(n)\,\mathrm{d}n \tag{7.41}$$

where $S_Q(n)$ is the spectral density for the loading. Evidently the spectral density for the response is related to the loading by

$$S_y(n) = [H^2(n)/k^2]S_Q(n) \tag{7.42}$$

and hence the variance of the response is given by

$$\sigma^2(Y) = \int_0^\infty S_Y(n)\,\mathrm{d}n. \tag{7.43}$$

The square root of this is the root mean square (r.m.s.) response, which is a useful measure of the dynamic performance of tall buildings in relation to human sensitivity to vibration (see Chapter 11).

The same procedure may be applied to a multi-degree-of-freedom structure by writing the equations of motion in modal coordinates as follows:

$$M_m\ddot{Y}_m + C_m\dot{Y}_m + K_mY_m = Q_m; \quad m = 1, 2, 3, \ldots, N \tag{7.44}$$

where Y_m is a generalized coordinate or modal amplitude. M_m, C_m and K_m are the generalized mass, damping and stiffness of the mth mode while Q_m is the generalized force. This equation is identical to (3.131), which was derived for lumped parameter systems. It should also be compared with (3.82), which applied to beams with distributed mass and stiffness. The main point about (7.44) is that it consists of a set of uncoupled equations, each capable of being treated as representing a single degree of freedom.

First of all we need to focus on the generalized force, which is given by

$$Q_m(t) = \int_0^H \phi_m(z)p(z,t)\,\mathrm{d}z, \tag{7.45}$$

where $p(z, t)$ is the fluctuating component of the distributed wind loading, and $\phi_m(z)$ is the shape of the mth mode. This equation is analogous to (3.83). It is assumed that $\phi_m(z)$ is normalized so that it has a maximum value of 1.0.

The wind loading may be expressed in terms of the fluctuating component of wind velocity by making use of Equation (7.31). However, it has been found in practice that the presence of a structure tends to distort the turbulent flow. This applies particularly to the small high-frequency eddies. Davenport

(1961) suggested the introduction of a correction factor $\chi(n)$ to take account of these effects. This was termed the *aerodynamic admittance* function and when introduced into (7.45) together with (7.31) results in

$$Q_m(t) = \rho C_D b \chi(n) \int_0^H \phi_m(z) \bar{V}(z) v(z,t)\, dz. \tag{7.46}$$

The value of the aerodynamic admittance function has been studied in some detail by Bearman (1980), both experimentally and theoretically. He recommended the use of the following empirical formula suggested by Vickery (1965) which is in good agreement with experimental data:

$$\chi(n) = \frac{1}{1 + \left[\dfrac{2n\sqrt{A}}{\bar{V}(z)}\right]^{4/3}} \tag{7.47}$$

where A is the frontal area of the structure.

Clearly it is not possible to derive a deterministic form for the generalized force because of the random nature of the wind loading. Instead, a spectral method has to be employed as before. This involves obtaining the mean square load by averaging the square of (7.46) over a period of time, resulting in an expression of the form given by Equation (7.41). The integrand will then be the spectral density of the generalized force. Omitting the details of the procedure, the expression for the spectral density of the modal generalized force will be given by

$$S_{Q_m}(n) = \rho^2 C_D^2 b^2 \chi^2(n) \int_0^H \int_0^H \phi_m(z)\phi_m(z')\bar{V}(z)\bar{V}(z')S_x(z,z';n)\mathrm{d}z\,\mathrm{d}z' \tag{7.48}$$

where $S_x(z,z';n)$ is the cross-spectral density of wind velocity derived earlier in the chapter. This is approximately equal to

$$S_{Q_m}(n) = \rho^2 C_D^2 b^2 \chi^2(n) S_{x_0}(n) \int_0^H \int_0^H \phi_m(z)\phi_m(z')\bar{V}(z)\bar{V}(z')\gamma(z,z';n)\mathrm{d}z\,\mathrm{d}z' \tag{7.49}$$

where $\gamma(z,z';n) = \sqrt{(\text{coherence})}$ and $S_{x_0}(n)$ is the spectral density of horizontal wind velocity at a reference height. Since $S_x(n)$ varies slightly in form with height above ground, Equation (7.49) is applicable to horizontal structures such as long-span bridges. However, the error will not be great when applied to a vertical structure if a mean reference height is chosen.

A more comprehensive set of expressions for the spectra of generalized forces for vertical, horizontal and prismatic structures is given by ESDU (1976a) together with charts from which the integrals may be evaluated directly.

Having obtained the spectrum of generalized force it is now possible to determine the spectrum of generalized modal response from

$$S_{Y_m}(n) = [H_m^2(n)/K_m^2] S_{Q_m}(n) \tag{7.50}$$

where the mechanical admittance of the mth mode is given by

$$H_m^2(n) = \{[1-(n/n_m)^2]^2 + (2\xi_m n/n_m)^2\}^{-1} \tag{7.51}$$

in which n_m and ξ_m are the frequency and damping ratio of the mth mode.

The general form of Equation (7.50) is shown by the solid line in Fig. 7.10. The broad hump is governed by the shape of the spectrum for horizontal wind velocity which is only modified slightly by the aerodynamic admittance. However, the mechanical admittance has the effect of introducing a sharp spike in the vicinity of the natural frequency in that mode. We noted earlier that only the lowest natural frequencies of even very large structures are within the range of wind excitation. Therefore, it is usually sufficient to evaluate the response in the fundamental mode only, though the effect of the second mode of large structures may be slightly significant.

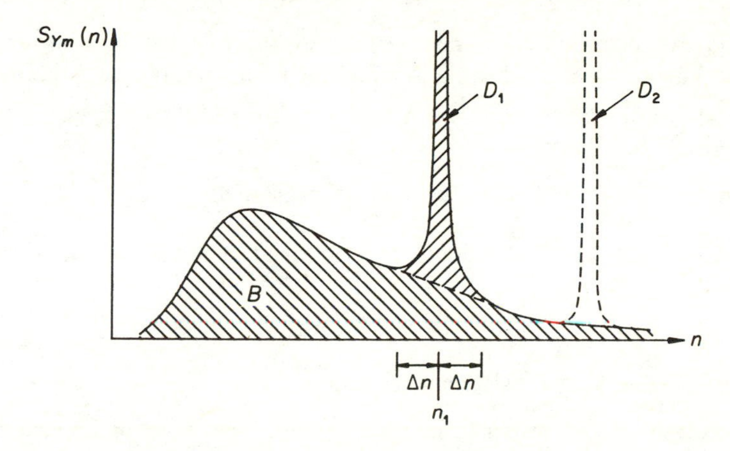

Fig. 7.10 Spectral density for modal response.

The next step is to obtain the variance of the modal response by integrating (7.50):

$$\sigma^2(Y_m) = \int_0^\infty S_{Y_m}(n)\,\mathrm{d}n. \tag{7.52}$$

This calculation is very much simplified if we treat the non-resonant and resonant responses separately. In effect we have already obtained the area indicated by B when we calculated the non-resonant or broad-band response in the previous section. A good approximation to (7.52) is obtained by integrating (7.50) over a narrow band in the vicinity of the natural frequency and adding to the broadband response. Therefore, designating the narrow-band response by $\sigma_D^2(Y_m)$ it can be shown that

$$\sigma_D^2(Y_m) = \int_{n_m - \Delta n}^{n_m + \Delta n} S_{Y_m}(n)\mathrm{d}n \simeq \frac{\pi n_m}{4\xi} \frac{S_{Q_m}(n_m)}{K_m^2}. \tag{7.53}$$

Note that the generalized stiffness K_m may be obtained from its relationship with the generalized mass M_m and the natural frequency. Therefore, making use of Equation (3.84) it is clear that

$$K_m = \omega_m^2 M_m = (2\pi n_m)^2 \int_0^H m_s \phi_m^2(z)\mathrm{d}z \tag{7.54}$$

where m_s is the mass per unit height of the structure and ω_m is its natural circular frequency in the mth mode. This integral for the first mode is shown hatched as area D_1 in Fig. 7.10. Area D_2 for the second mode may be obtained in a similar way.

Finally, modes may be combined using

$$\sigma_D^2(z) = \phi_1^2(z)\sigma_D^2(Y_1) + \phi_2^2(z)\sigma_D^2(Y_2) + \ldots \tag{7.55}$$

which gives the variance (and hence r.m.s. response) of deflection at height z.

Load effects such as shear and bending moment may be obtained by integrating the vibratory inertia forces in each mode over the height of the structure, by means of

$$\sigma_D(E_m) = \sigma_D(Y_m) \int_0^H m_s \omega_m^2 \phi_m(z) \beta \,\mathrm{d}z. \tag{7.56}$$

The total resonant load effect is then obtained by summing modal contributions

$$\sigma_D^2(E) = \sigma_D^2(E_1) + \sigma_D^2(E_2) + \ldots. \tag{7.57}$$

7.3.4 Evaluation of the maximum response

The maximum response of a structure subjected to random loading cannot be evaluated in a deterministic sense. However, the root mean square components of the dynamic response may be used to determine the probability of the response exceeding a certain magnitude. A practical procedure was suggested by Davenport (1961) by which he was able to derive a factor by which the r.m.s. component would be exceeded with a 50% probability. After further refinement Davenport (1964) recommended the following expression for this *peak factor*:

$$g = \sqrt{[2\ln(\nu T_0)]} + 0.577/\sqrt{[2\ln(\nu T_0)]} \tag{7.58}$$

where ν is the expected frequency that the fluctuating response crosses the zero axis with positive slope and T_0 is the period during which the peak response is assumed to occur. Usually $T_0 = 3600$ s.

In the narrow-band or resonant response region v is clearly equal to the natural frequency and therefore

$$g_D = \surd[2\ln(nT_0)] + 0.577/\surd[2\ln(nT_0)]. \tag{7.59}$$

In the broad-band region, Equation (7.58) has been evaluated (ESDU, 1976a), giving

$$g_B \simeq 3.5. \tag{7.60}$$

Using these factors the maximum total load effect may be obtained by summing the mean load effect together with the factored broad-band and narrow-band load effects as follows:

$$E_{max} = \bar{E} + \surd\{[g_B\sigma_B(E)]^2 + [g_D\sigma_D(E)]^2\}. \tag{7.61}$$

The reader is referred to the two Davenport references for a fuller discussion of the extreme-value analysis.

7.3.5 Aerodynamic damping

In addition to structural damping there is a further energy-absorbing mechanism arising from the relative velocity between the vibrating structure and the incident airflow. This is known as aerodynamic damping and its value is often comparable with structural damping in the case of slender flexible structures.

The mechanism of aerodynamic damping will now be explained with reference to Fig. 7.11 which illustrates a unit length of a line-like structure oscillating in the same direction as the mean wind. If the cyclic motion of the structure about its mean position is $Y(t) = Y_0 \sin\omega t$, as shown, then the aerodynamic force per unit length will be given by

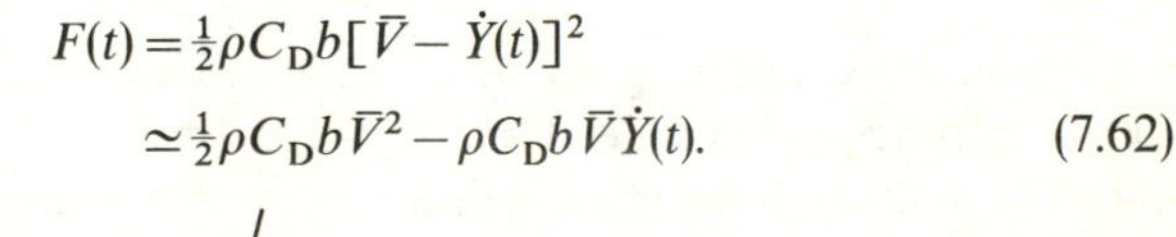

$$\begin{aligned} F(t) &= \tfrac{1}{2}\rho C_D b[\bar{V} - \dot{Y}(t)]^2 \\ &\simeq \tfrac{1}{2}\rho C_D b\bar{V}^2 - \rho C_D b\bar{V}\dot{Y}(t). \end{aligned} \tag{7.62}$$

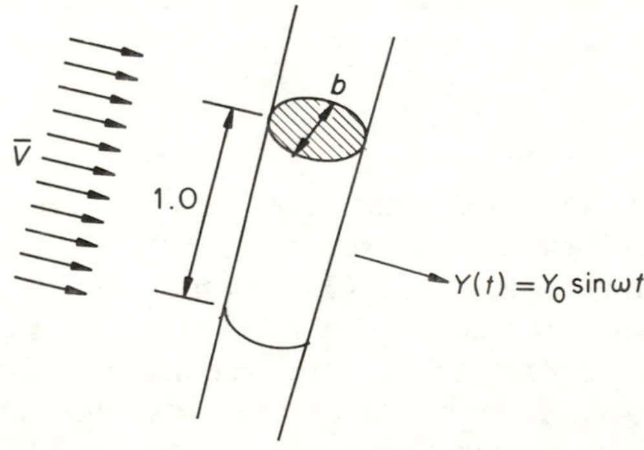

Fig. 7.11 Motion of a slender structure in the wind.

The work done by the force in full cycle of motion will be:

$$\text{Work done per cycle} = \int F\,\mathrm{d}Y = \int_0^{2\pi/\omega} F(t)(\mathrm{d}Y/\mathrm{d}t)\,.\,\mathrm{d}t$$

$$= Y_0\omega \int_0^{2\pi/\omega} [\tfrac{1}{2}\rho C_\mathrm{D} b \bar{V}^2 - \rho C_\mathrm{D} b \bar{V} Y_0 \omega \cos \omega t] \cos \omega t\,\mathrm{d}t$$

$$= -\rho C_\mathrm{D} b \bar{V} Y_0^2 \pi\omega. \tag{7.63}$$

The work absorbed by viscous damping is proportional to velocity and over a full cycle this will be:

$$\text{Work absorbed per cycle} = \int_0^{2\pi/\omega} 2\xi\sqrt{(km)}\,\dot{Y}(\mathrm{d}Y/\mathrm{d}t)\,.\,\mathrm{d}t$$

$$= 2\pi\omega\xi\sqrt{(km)}\,Y_0^2 \tag{7.64}$$

where $\xi = c/2\sqrt{(km)}$ is the damping ratio, k and m being the stiffness and mass of a single-degree-of-freedom system, per unit length.

Now it is evident that the work done by the aerodynamic force $F(t)$ may be interpreted as the result of negative damping, i.e. negative aerodynamic damping, which will be denoted by $-\xi_\mathrm{a}$. Therefore by making this substitution in (7.64) and equating to (7.63) we obtain

$$\xi_\mathrm{a} = \rho C_\mathrm{D} b \bar{V}/2\sqrt{(km)} = \rho C_\mathrm{D} b \bar{V}/2\omega m. \tag{7.65}$$

The total damping will be the sum of structural and aerodynamic damping

$$\xi = \xi_\mathrm{s} + \xi_\mathrm{a}. \tag{7.66}$$

A formula for evaluating the aerodynamic damping of complete structures, taking into account mode shape, is given by ESDU (1976a). There is some controversy over aerodynamic damping and a conservative approach would be to neglect it.

EXAMPLE 7.1

A tall multi-storey tower block is situated on the edge of a coastal city close to the shoreline. The dimensions of the structure are $H = 175$ m, $b = 40$ m and its fundamental natural frequency $n_1 = 0.2$ Hz with a mode shape that may be assumed to be linear. It is assumed that the structural and aerodynamic damping amount to $\xi = 0.05$. Estimate the maximum base shear in the once-in-50-year storm wind, given that the reference wind speed at 10 m height is 25 m/s. It may be assumed that the density of air ρ is 1.2 kg/m^3, the drag coefficient C_D is 1.3, and von Kármán's constant k is 0.4.

Solution

The friction velocity may be obtained from (7.6) giving

$$v_* = 25/14.52 = 1.722 \text{ m/s}.$$

Assuming that the extreme wind blows in from the sea, the roughness length is $z_0 = 0.003$ m, according to Table 7.1.

In the calculations that follow it will be assumed that the tower may be approximated by a line-like structure.

Response to the mean wind

The influence function for base shear is given by $\beta = 1$. Therefore, substituting (7.5) into (7.27), the mean load effect will be given by

$$\bar{E} = \tfrac{1}{2}\rho C_D b \frac{v_*^2}{k^2} \int_0^H \ln^2\left(\frac{z}{z_0}\right) dz.$$

The integrand is evaluated at five points over the height of the tower as follows:

z(m)	0	43.75	87.5	131.25	175
$\ln^2(z/z_0)$	0	91.9	105.7	114.2	120.4.

Using Simpson's rule the value of the integral is

$$\begin{aligned} \int &= (43.75/3)[0 + 120.4 + 4(91.9 + 114.2) + 2 \times 105.7] \\ &= 16\,863 \end{aligned}$$

Hence

$$\begin{aligned} \bar{E} &= \tfrac{1}{2} \times 1.2 \times 1.3 \times 40 \times (1.722/0.4)^2 \times 16\,863 \text{ N} \\ &= 9750 \text{ kN}. \end{aligned}$$

Non-resonant component

From Equation (7.36) the variance of the non-resonant or broad-band component is given by

$$\sigma_B^2(E) = \rho^2 C_D^2 b^2 \sigma_x^2 \frac{v_*^2}{k^2} \int_0^H \int_0^H \ln\left(\frac{z}{z_0}\right) \ln\left(\frac{z'}{z_0}\right) e^{-\Delta z/L}\, dz\, dz'$$

where (7.20) has been used for the cross-correlation coefficient, with $L = 60$ m.

In this case there is no alternative but to evaluate the double integral numerically. This may be done manually by calculating values of the integrand at a matrix of equally spaced points z and z', and performing a double application of Simpson's Rule. However, instead a short BASIC program has been written to perform the numerical computation as follows:

```
10 REM DOUBLE INTEGRATE BROAD BAND RESPONSE FUNCTION
20 REM VF = FRICTION VELOCITY
30 REM ZO = ROUGHNESS LENGTH
40 REM K = VON KARMAN CONSTANT
50 REM L = CROSS CORRELATION COEFFICIENT
60 REM H = HEIGHT OF TOWER
70 REM NF = NATURAL FREQUENCY
80 REM N = INTEGRATION STEPS
90 REM
100 READ VF,Z0,K,L,H,NF,N
110 DATA 1.722,.003,.4,60,175,.2,5
120 DZ = H/(N-1)                        'STEP LENGTH
130 ZI = 0 : ZJ = 0
140 SUM1 = 0
150 FOR I=1 TO N
160 ZI = (I-1)*DZ
170 SUM2 = 0
180 FOR J=1 TO N
190 ZJ = (J-1)*DZ
200 DELZ = (ZI -ZJ)^2
210 DELZ = SQR(DELZ)                    'SEPARATION
220 REM EVALUATE BROAD BAND FUNCTION
230 SP = EXP(-DELZ/L)
240 IF ZI<Z0 THEN ZI = Z0
250 IF ZJ<Z0 THEN ZJ = Z0
260 F = LOG(ZI/Z0)*LOG(ZJ/Z0)*SP
270 REM SIMPSON'S RULE
280 S2 = 4                              'J EVEN
290 IF 2*FIX(J/2) < J THEN S2 = 2       'J ODD
300 IF J=1 THEN S2 = 1                  'FIRST
310 IF J=N THEN S2 = 1                  'LAST
320 SUM2 = SUM2 + S2*F
330 NEXT J
340 I2 = SUM2*DZ/3                      'INNER INTEGRAL
350 REM SIMPSON'S RULE
360 S1 = 4                              'I EVEN
370 IF 2*FIX(I/2) < I THEN S1 = 2       'I ODD
380 IF I=1 THEN S1 = 1                  'FIRST
390 IF I=N THEN S1 = 1                  'LAST
400 SUM1 = SUM1 + S1*I2
410 NEXT I
420 I1 = SUM1*DZ/3                      'OUTER INTEGRAL
430 PRINT "DOUBLE INTEGRAL = ";I1
440 STOP
```

Using the program the value of the double integral is

$$\iint = 1.46 \times 10^6.$$

The value of the r.m.s. wind velocity, from Equation (7.15), is $\sigma_x = 2.5\, v_*$. Hence the r.m.s. broad-band response will be

$$\sigma_B(E) = 1.2 \times 1.3 \times 40 \times 2.5 \times 1.722 \times (1.722/0.4) \times \sqrt{(1.46 \times 10^6)}\text{N}$$

$$= 1397 \text{ kN}.$$

Resonant component

Assuming a linear form to the mode shape, $\phi = z/H$, the spectral density of the first-mode generalized force, Equation (7.49), is given by

$$S_{Q_1}(n) = \rho^2 C_D^2 b^2 \chi^2(n) S_{x_0}(n)(v_*^2/H^2k^2) \int_0^H \int_0^H zz' \ln(z/z_0) \ln(z'/z_0) e^{-Cn\Delta z/\bar{V}a} \, dz \, dz'$$

where (7.25) is used for $\sqrt{}$(coherence) with $c = 8$.

It will be necessary to evaluate this function at the first natural frequency $n_1 = 0.2$ Hz. Using (7.17) to obtain the spectrum of horizontal gustiness at mid-height we find that

$$\bar{V}(87.5) = (1.722/0.4) \ln (87.5/0.003) = 44.3 \text{ m/s}$$

$$f = 0.2 \times 87.5/44.3 = 0.395$$

$$nS_{x_0}/v_*^2 = (200 \times 0.395)/(1 + 50 \times 0.395)^{5/3} = 0.503.$$

Therefore

$$S_{x_0} = 7.46.$$

The mechanical admittance is obtained from (7.47):

$$\frac{2n\sqrt{A}}{V} = \frac{2 \times 0.2 \times \sqrt{(175 \times 40)}}{44.3} = 0.756;$$

$$\chi(0.2) = 1/(1 + 0.756^{4/3}) = 0.592.$$

The double integral must be evaluated numerically. An edited version of the program listed earlier may be employed for this purpose. The main differences in the integrand are the mode shapes and the cross-spectral density function where $\bar{V}a$ is the mean velocity at the average height of any two points. The value of the integral is

$$\iint = 11\,206 \times 10^6.$$

Hence the generalized force is

$$S_{Q_1} = [1.2^2 \times 1.3^2 \times 40^2 \times (0.592)^2 \times 7.46 \times 1.722^2/(175^2 \times 0.4^2)] \times 11\,206 \times 10^6$$

$$= 68\,978 \times 10^6.$$

From (7.53), (7.54) and (7.56), the r.m.s. resonant component of base shear is given by

$$\sigma_D(E_1) = \sqrt{\left(\frac{\pi n_1}{4\xi} S_{Q_1}\right)} \times \frac{m_s \omega_1^2 \int_0^H (z/H) \, dz}{m_s \omega_1^2 \int_0^H (z/H)^2 \, dz}$$

$$= \sqrt{\left(\frac{\pi \times 0.2 \times 68\,978 \times 10^6}{4 \times 0.05}\right)} \times \frac{3}{2} \text{ N}$$

$$= 698 \text{ kN}.$$

Finally, the resonant peak factor g_D is obtained from (7.59):

$$g_D = 3.79 \quad \text{and} \quad g_B = 3.5,$$

enabling us to obtain the maximum base shear from (7.61), which yields

$$\begin{aligned} E_{max} &= 9750 + \sqrt{[(3.5 \times 1397)^2 + (3.79 \times 698)^2]} \\ &= 15\,310\ \text{kN}. \end{aligned}$$

7.4 THE ACROSS-WIND RESPONSE OF SLENDER STRUCTURES

There are many examples of slender structures that are susceptible to dynamic motion perpendicular to the direction of the wind. Tall chimneys, street lighting standards, towers and cables frequently exhibit this form of oscillation which can be very significant especially if the structural damping is small. The principal forms of cross-wind excitation are turbulence buffeting, vortex shedding, galloping and flutter. Response to transverse components of turbulence may be estimated by the methods of the previous section, with certain modifications (Vickery and Clark, 1972; Milford, 1982).

7.4.1 Vortex shedding

The mechanism of vortex shedding is illustrated in Fig. 7.12. The air travels over the face of the body until it reaches the points of separation on each side where thin sheets of tiny vortices are generated. As the vortex sheets detach they interact with one another and roll up into discrete vortices which are shed alternately from the sides of the object. The asymmetric pressure distribution created by the vortices around the cross section results in an alternating transverse force as they are shed. If the structure is flexible, oscillation will occur transverse to the wind and the conditions for resonance would exist if the vortex shedding frequency coincided with the natural frequency of the structure. This situation could give rise to very large oscillations and possibly failure.

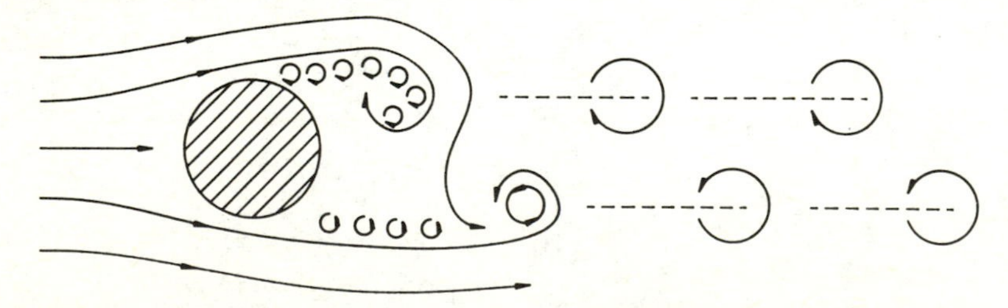

Fig. 7.12 Vortex formation in the wake of a bluff object.

Experiments by Roshko (1961) showed that the vortex-shedding frequency, n, could be related to a non-dimensional parameter called the Strouhal number, S, given by

$$S = nD/V, \tag{7.67}$$

where V is the wind speed and D is a typical dimension such as the diameter. The Strouhal number depends on the shape of the cross section and is practically independent of Reynolds number (or wind speed) for shapes with sharp edges such as squares and rectangles ($S \simeq 0.15$). For shapes with rounded profiles the Strouhal number varies with Reynolds number as the boundary layer develops from laminar to turbulent. Random vortex shedding tends to occur in the transition region of $4 \times 10^5 < R_e < 3 \times 10^6$, as shown in Fig. 7.13.

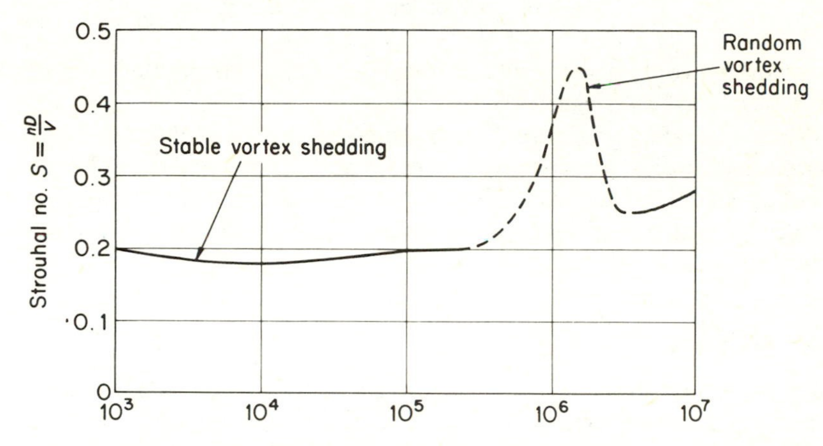

Fig. 7.13 Variation of Strouhal number with Reynolds number for a circular cylinder.

A further feature of vortex shedding is that at resonance the motion of the structure tends to result in the aerodynamic forces becoming better correlated over the length or height of the structure. This leads to the condition of 'lock on' in which the vortex shedding remains in resonance over a relatively wide band of wind speed. This is illustrated in Fig. 7.14, where the ratio of vortex shedding to the natural frequency (n/f) is plotted against the reduced velocity of the air flow. The amplitude of oscillation is also plotted against reduced velocity. It will be noted that the vortex-shedding frequency rises steadily with wind velocity until resonance is reached. Then large-amplitude motion results but the vortex-shedding frequency remains constant. Eventually the wind speed increases beyond a certain point and vortex shedding ceases together with the oscillation.

The cross-wind response of tall structures, such as chimneys, is modified considerably by turbulence. Turbulence in the air stream tends to inhibit the

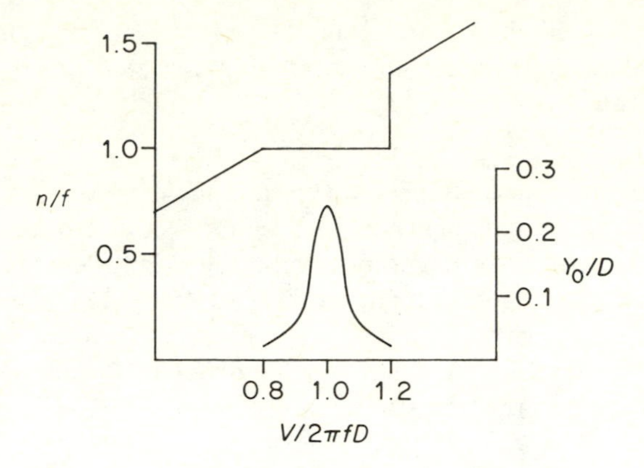

Fig. 7.14 Lock-on of vortex shedding.

formation of vortices and results in lower response. A method of analysis has been developed by Vickery (1978).

Vortex-shedding response of tall chimney stacks is difficult to avoid because the critical wind speed is usually well inside the range of practical wind speeds. The response can be alleviated by the use of aerodynamic spoilers such as shrouds or strakes near the top of the structure. These devices tend to break up the vortex formation.

7.4.2 Galloping

Galloping is an unstable phenomenon which arises from the aerodynamic forces generated on certain cross-sectional shapes as they displace transverse to the wind. The phenomenon may be understood by studying Fig. 7.15. Consider a square or rectangular section in a steady air stream of velocity V. If the body moves upwards with a velocity of v it is apparent that it experiences a resultant wind velocity on it with a downward angle of attack α. There will be a resulting lift force proportional to the lift coefficient C_L (conventionally denoted as positive when in the same direction as the angle of

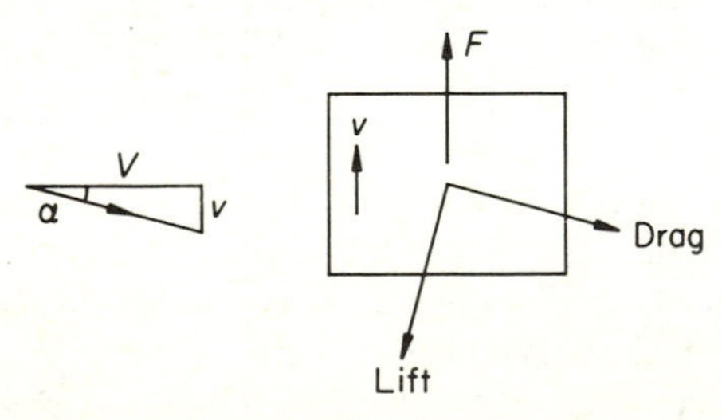

Fig. 7.15 Unstable lift force of a rectangular section.

attack). There will also be a drag force proportional to C_D. The net result will be a coefficient of cross-wind force given by

$$\begin{aligned} C_F &= -C_L(\alpha)\cos\alpha - C_D(\alpha)\sin\alpha \\ &\simeq -[C_L(\alpha)+\alpha C_D(\alpha)] \\ &\simeq -\alpha\left[\frac{d}{d\alpha}C_L(\alpha)+C_D(\alpha)\right]. \end{aligned} \tag{7.68}$$

If C_F is positive the product of F and v integrated over a cycle will represent positive extraction of energy from the wind and, hence, instability. This condition is known as Den Hartog's criterion (Den Hartog, 1947) and is given by

$$\frac{d}{d\alpha}C_L(\alpha)+C_D(\alpha)<0. \tag{7.69}$$

Lawson (1980) discussed some important qualifications regarding this simple criterion for the onset of galloping.

The lift coefficient term $dC_L(\alpha)/d\alpha$ is positive for a perfectly circular cross section but is often negative for some other shapes such as rectangles. Also, the accretion of ice on the leading edge of a cable can result in a unstable cross section.

7.4.3 Flutter

Classical flutter is an unstable coupled motion in any two degrees of freedom, such as combined bending and torsion. It is similar to galloping in that the aerodynamic forces depend on the motion. It differs in that there would be positive damping in either of the motions acting independently, but aerodynamic cross-coupling results in one of the motions giving rise to fluctuating forces affecting the other.

Flutter may occur in the decks of long-span suspension bridges if the torsional and bending frequencies are similar. Torsionally stiff cross sections are therefore required to separate the two modes. It has been found that the installation of turning vanes on the leading edges of aerodynamically shaped bridge decks helps to reduce vortex–flutter interaction (Tappin and Clark, 1985).

8

Foundations for industrial machinery

8.1 INTRODUCTION

There are many examples of heavy machines or industrial processes that give rise to large dynamic forces. These include reciprocating engines for power generation, large rotating machines with unbalanced masses, vibrating screens in the mining industry, and various kinds of hammer-forging processes. Vibrations may therefore be periodic, with constant amplitude over long periods of time, or in the form of sudden transients. In either case there is a risk of the vibrations being transmitted through the foundations and damaging the fabric of the enclosing or adjacent buildings. There is also the potential disturbing effect of the vibrations on people using the buildings.

8.1.1 Dynamic behaviour of large foundation blocks

Heavy industrial machines are normally mounted on large-mass concrete foundation blocks which distribute the load over the base area to the supporting soil or rock. On sites where the soil is insufficiently strong, such as soft clay or slit, a piled foundation would be used. In either case the foundation effectively provides an elastic restraint to the applied dynamic forces.

The dynamic behaviour of a large block supported on a soil foundation is an example of the problem of ground–structure interaction which was discussed in Chapter 3. The idealized model is depicted in Fig. 8.1 where a mass is shown resting on a semi-infinite elastic solid. It will be expected, therefore, that vibration of the block will result in wave motion in the foundation medium, requiring the analytical procedures of Rayleigh (1885) and Lamb (1904) which were discussed in §3.5. The theory was further developed by Reissner (1936) who obtained the natural frequency of a mass,

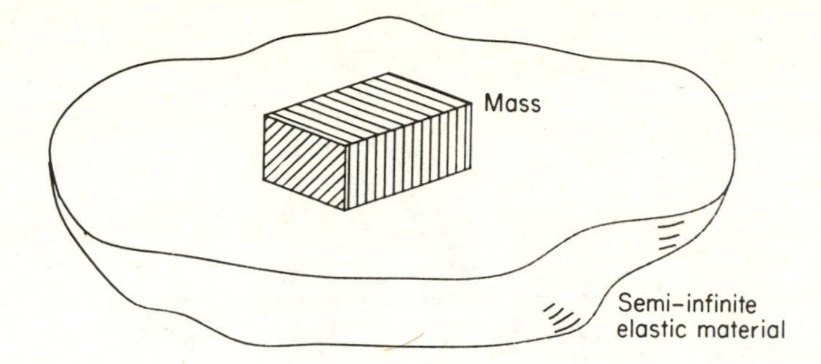

Fig. 8.1 Idealized model of a foundation block supported on soil or rock.

in the form of a rigid disk, vibrating on the surface of an elastic half-space. Arnold, Bycroft and Warburton (1955) and Bycroft (1956) found solutions to the same problem but for many practical cases of forced vibration.

Unfortunately, the mathematics required in the analysis of half-space problems is relatively advanced and not generally acceptable for use in design offices. Various attempts have been made to derive simplified formulae. From the elastic half-space theory it was readily appreciated that the soil foundation could be considered as providing an equivalent spring stiffness to restrain the motion of the mass. This would be accompanied by an equivalent soil mass thought of as acting in unison with the foundation mass. In addition, the waves propagating outwards from the disturbance carry away energy from the motion of the vibrating mass. This is equivalent to a damping mechanism which we have already referred to as *radiation damping*. Hsieh (1962) considered the six possible motions (three translational and three rotational) of a rigid block restrained by elastic stiffnesses and damping equivalent to those obtained from half-space theory by Arnold, Bycroft and Warburton (1955). His equations took account of the added mass of soil, radiation damping, and the coupling of the rocking and lateral degrees of freedom. However, his equations applied to a foundation block with an equivalent circular base area.

The method of analysis in most widespread use at the present time originates from the work of Barkan (1962) in the USSR. In his view the participating soil mass does not make a sufficient difference to the calculation of the natural frequency to warrant its inclusion in the equations, the error being unlikely to exceed 10%. Furthermore, provided that the machine frequency is sufficiently different from the natural frequency of the system, amplitudes of forced vibration may be calculated with reasonable accuracy by ignoring damping. The existence of damping would, in any case, reduce the amplitudes.

The simplified system is shown in Fig. 8.2 where the vertical soil stiffness is represented by elastic elements distributed about the base. Thus, provided

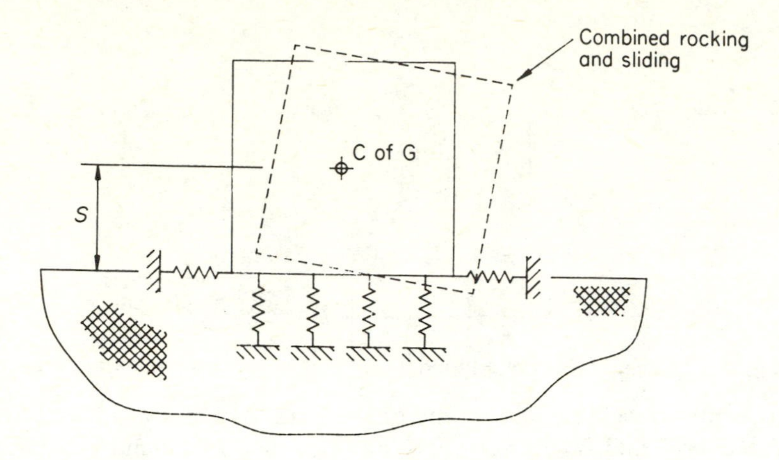

Fig. 8.2 Block on elastic supports.

that vertical components of dynamic loading act through the centre of gravity of the combined mass, vertical vibrations can be treated as a one-degree-of-freedom mass–spring system. The soil also provides horizontal shearing stiffness which is represented by the two horizontal springs in the diagram. However, horizontal sliding vibration will inevitably be accompanied by a rocking motion because of the elevation of the centre of gravity above the base. Thus, as will be shown later, the rocking and horizontal motion will constitute a coupled two-degree-of-freedom system. Similar interaction exists between pitching and horizontal motion in a plane perpendicular to the one shown. Finally, there is a yawing mode of vibration consisting of rotatory motion about a vertical axis through the centre of gravity. This occurs as a single-degree-of-freedom vibration.

8.2 DYNAMIC LOADING BY INDUSTRIAL MACHINERY

The dynamic loading from industrial machinery derives principally from the inertia effects of various moving parts. Every machine behaves differently and it is therefore the responsibility of the manufacturer to specify the dynamic loads for which the foundations must be designed. The following is a brief review of the nature of dynamic forces exerted by some common types of industrial machinery, which should help to clarify the recommended analysis procedures.

8.2.1 Reciprocating engines

Large multi-cylinder diesel engines are often used in electrical power generating stations, especially in remote places lacking indigenous fossil fuels. A typical arrangement of diesel engine and alternator mounted on a foundation block is shown in Fig. 8.3. A typical crank–piston linkage is shown in Fig. 8.4 where the masses of the crank, connecting rod and piston are m_1, m_2 and m_3 respectively. Engine speeds are normally quoted in revolutions per minute (r.p.m.), in which case the crankshaft rotation frequency is given by

$$\Omega = 2\pi N/60\,(\text{rad/s}) \tag{8.1}$$

where N is the engine speed. It is evident that there will be oscillatory inertia forces due to the motions of the masses m_1, m_2, m_3. The first two will have vertical and horizontal components while m_3 will have a vertical component only. The magnitude of these forces may be evaluated if the various masses and lengths are known. For example, it can be shown that the inertia force

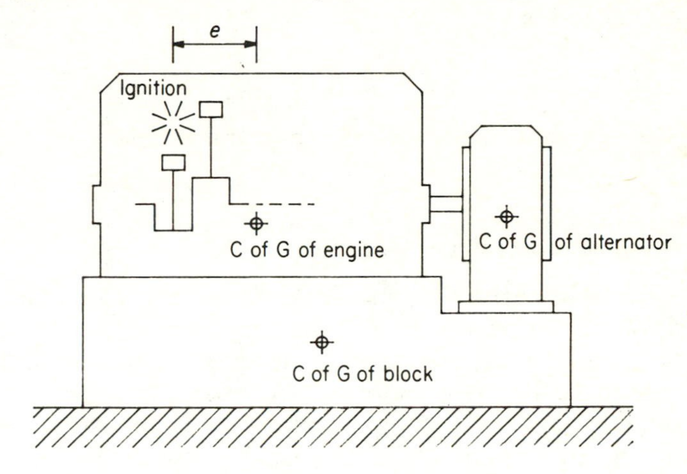

Fig. 8.3 Diesel engine and alternator.

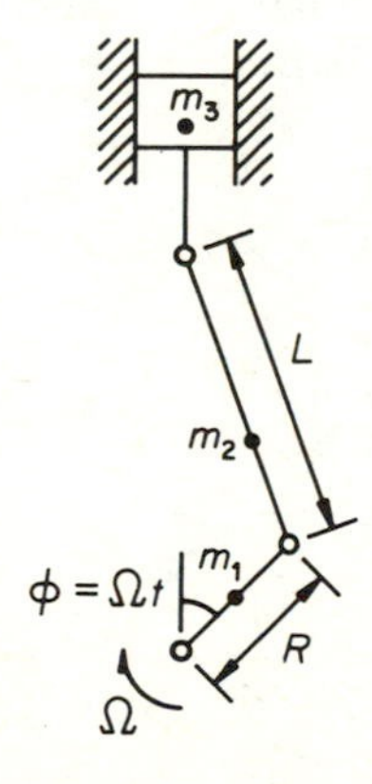

Fig. 8.4 Crank–piston linkage.

due to the piston will be

$$P(t)=m_3\Omega^2 R[\cos\Omega t+(R/L)\cos(R\Omega t/L)]. \tag{8.2}$$

This is not simple harmonic because of the second term. Therefore, in addition to the primary engine forces applied at engine speed, there will be higher harmonics applied at integral multiples of the engine speed.

In multi-cylinder engines it is possible to balance most of the inertia forces, depending on the number and arrangement of the cylinders. However, there will always be some residual unbalanced forces due to tolerances on weights and geometry. Furthermore, ignition of a cylinder will create a dynamic moment about the centre of gravity of the engine, as may be seen in Fig. 8.3. In a four-stroke multi-cylinder engine this will result in a pitching moment whose frequency will be

$$\Omega=\frac{2\pi}{60}N\times\frac{n}{4} \tag{8.4}$$

where n is the number of cylinders. The firing order is selected to minimize the magnitude of this moment.

8.2.2 Turbines and other rotating machines

Turbines, centrifugal pumps, fans and electrical generators are examples of high-speed rotating machinery. Even though considerable care is taken to balance the rotating parts, residual imbalances will always exist. Inertia forces are generated by the eccentricity of unbalanced masses about the centre of rotation as indicated in Fig. 8.5. Strictly speaking, these forces act radially and rotate with the shaft. But they will have oscillatory vertical and horizontal components which tend to excite the corresponding modes of vibration. This happens in everyday experience with a domestic spin dryer. It should also be noted that the speeds of turbines are many times greater than reciprocating engines of similar outputs and therefore the out-of-balance

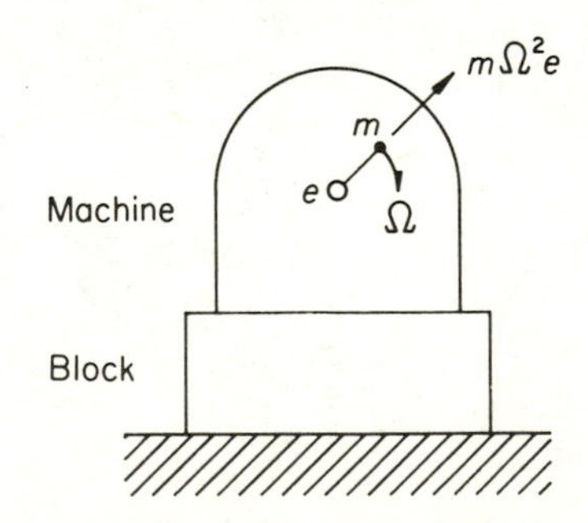

Fig. 8.5 Rotating unbalanced mass.

force will be significantly amplified because of the frequency being squared ($m\Omega^2 e$).

8.2.3 Transient torques in electrical machines

There are two important cases of transient dynamic loading that occur with driven electrical generators. The first is known as *short-circuit torque*. Consider an electrical alternator being driven by an engine as in Fig. 8.3. If a fault occurred, which had the effect of creating a short circuit in the output of the alternator, a very large current would be demanded (for a fraction of a second, perhaps). This would be experienced as a suddenly applied load or brake on the system. As a result a torque would be applied by the engine about the axis of the drive shaft. The second case is known as *faulty-synchronizing torque*. This occurs when an engine–generator system starts up and feeds power into the national grid. If the output of the generator is not synchronized with the a.c. waveform of the national grid a braking effect, or torque, is experienced by the generator until such time as it is in phase. This will have a similar effect to the short-circuit torque, though it is usually larger in magnitude.

In either case the engine–generator system will experience a suddenly applied external torque. The dynamic effect of a suddenly applied load is twice its static effect, as was noted in Chapter 2, §2.6. The holding-down bolts of the machinery need to be designed to resist this torque and the maximum stresses in the soil under the foundation block should also be checked.

8.2.4 Hammers

There are many industrial processes, typically impact forging, which require single or repeated blows with a hammer. Kinetic energy is given to the hammer head either by some external source of power such as steam, or more usually by gravity. Velocity is imparted to the anvil and workpiece by transfer of momentum. A schematic arrangement for a drop hammer is shown in Fig. 8.6. The mass of the hammer head is denoted by m_0, the anvil by m_2 and the foundation by m_1. Some kind of elastic layer, often hardwood, is interposed between the anvil and the foundation block. The block is either supported elastically by the foundation material or by specially designed springs to minimize the transmission of vibration to nearby buildings. Thus the system may be observed to have two degrees of freedom.

The velocity of the hammer head before impact is given by

$$v = \sqrt{(2g\, h_0)}. \tag{8.5}$$

In the case of pneumatic or steam-powered drop hammers Barkan (1962) found that, in practice, this should be reduced by an empirical factor of 0.65 to allow for friction and the resistance of exhaust air or steam.

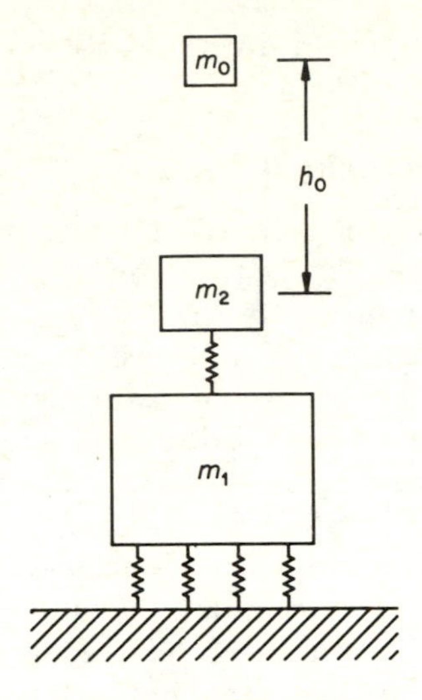

Fig. 8.6 Schematic arrangement for a drop hammer.

Following the method of Barkan (1962), conservation of momentum can be expressed by

$$m_0 v = m_0 v' + m_2 v_0 \tag{8.6}$$

where $(-v')$ is the rebound velocity of the hammer head and v_0 is the velocity imparted to the anvil. The relative velocity after impact depends on the elastic characteristics of the colliding bodies and is obtained from the expression

$$C_r = (v_0 - v')/v \tag{8.7}$$

where C_r is the coefficient of restitution. This constant varies between 0 (fully plastic) to 1 (perfectly elastic). Thus the velocity of the anvil after impact may be obtained from Equations (8.6) and (8.7) and is given by

$$v_0 = \frac{1 - C_r}{1 + \mu_0} v \tag{8.8}$$

where $\mu_0 = m_2/m_0$. This velocity may be used as an input to the equations of motion of a two-degree-of-freedom system. It is very likely that the hammer might strike the anvil eccentrically, thus imparting a rotational component. This case is also considered by Barkan (1962).

Novak and El Hifnawy (1983) have verified that the above procedure is satisfactory provided that the duration of the impact is much shorter than the natural period of the foundation. This may not be so if the foundation support is very stiff, e.g. piles. It would then be necessary to take account of the force–time function of the impulse.

8.2.5 Miscellaneous

There are many other industrial processes that give rise to significant vibrations and the reader is referred to Barkan (1962) or Major (1962) for further information. Furthermore, new manufacturing techniques will give rise to different conditions of dynamic loading. A few examples will be discussed briefly as follows:

(a) *Vibrating screens.* These are used extensively in the mining industry for washing and separation processes. Ore is passed over a horizontal screen which is supported at its four corners. The screen is vibrated vertically and horizontally by a motor-driven eccentric crank. Sinusoidal inertia forces are thus applied to the supporting frame structure and to the foundations.

(b) *Rolling mills.* As a bloom or billet enters the rollers, the resistance acts like a suddenly applied torque to the shaft of the driving motor. The resulting dynamic couple applied to the foundation block is analogous to the 'short-circuit' torque in electrical generators.

(c) *Grinding mills.* Cement grinding is carried out in huge horizontal rotating drums full of cast steel balls through which the cement clinker is passed. Dynamic forces are exerted by the centrifugal effects of the moving masses.

(d) *Rotary rock crushers.* A rotary rock crusher is depicted in Fig. 8.7. It consists of a large steel drum into which uncrushed rock is loaded. In the centre of the drum is a conical spindle which is positioned eccentrically with respect to its drive shaft. As it rotates, rock is crushed against the walls of the drum and is discharged at the bottom when small enough. The eccentric rotation of the crusher spindle gives rise to a periodic force as shown.

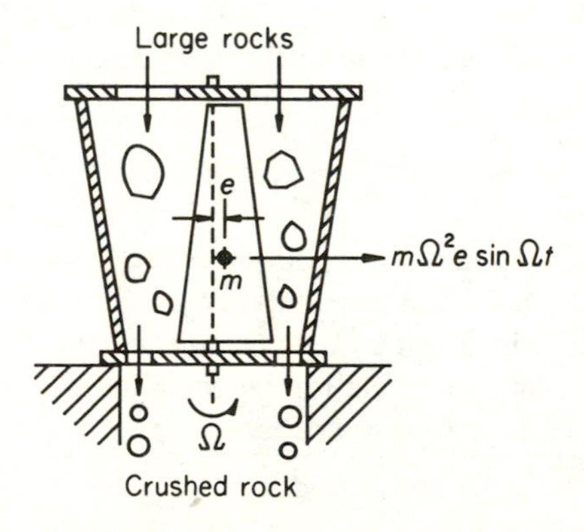

Fig. 8.7 Rotary rock crusher.

8.3 DYNAMIC PROPERTIES OF FOUNDATION SYSTEM

8.3.1 Stiffness of foundation material

We have already said that the recommended method of analysis, as developed by Barkan (1962), treats the supporting material as an equivalent uniform elastic stiffness. Soils are notoriously unreliable in their stiffness properties and exhibit nonlinear behaviour under load. However, the dynamic components of foundation stress are very small compared with the total vertical loading and therefore it is permissible to treat the material as elastic for vibration analysis.

Values of Young's modulus and Poisson's ratio, to be used in calculations, vary considerably from one material to another and geotechnical information is required for the specified site. It is recommended that the site investigation should provide information to a depth at least three times the mean plan dimension of the foundation. Dynamic tests for obtaining the appropriate elastic properties are to be preferred. A well-known method is by measurement of the velocity of propagation of surface waves. The surface wave is sufficiently close to that of a shear wave to permit the following simple expression to be used:

$$E = 2(1+\nu)\rho\beta^2 \tag{8.9}$$

where ν is the Poisson's ratio (taken to be somewhere between 0.3 and 0.5), ρ is the soil density, and β is the shear wave velocity taken to be the same as the measured velocity.

8.3.2 Coefficients for soil deformation

The classic problem of a static load applied to the surface of an elastic half-space was first solved by Boussinesq (1885) who obtained the following formula for the vertical displacement of the surface at a radius r from a point load of magnitude P:

$$w = \frac{(1-\nu^2)}{E}\frac{P}{\pi r}. \tag{8.10}$$

The surface displacements under a rectangular region loaded with a uniform pressure were determined by Schleicher (1926). The displacements varied from the centre to the edges, but by calculating the average displacement he was able to obtain the following relation between uniform surface pressure σ_z and deflection w of an effectively rigid base:

$$\sigma_z = C_z w. \tag{8.11}$$

C_z is the *coefficient of uniform compression*, or modulus of subgrade reaction, and is given by

$$C_z = \frac{E}{(1-\nu^2)\sqrt{A}} \times$$

$$\left[\pi\sqrt{\alpha}\left(\ln\frac{\sqrt{(1+\alpha^2)}+\alpha}{\sqrt{(1+\alpha^2)}-\alpha} + \alpha\ln\frac{\sqrt{(1+\alpha^2)}+1}{\sqrt{(1+\alpha^2)}-1} - \frac{2}{3}\frac{(1+\alpha^2)^{3/2}-(1+\alpha^3)}{\alpha}\right)^{-1}\right] \tag{8.12}$$

where α = ratio (B/D) of the sides of the rectangular loaded area and A = area of foundation (BD). The factor in square brackets, denoted by β, was evaluated by Schleicher for different values of B/D and is given in Table 8.1.

Table 8.1 Values of the factor β

B/D	*1*	*1.5*	*2*	*3*	*5*	*10*	*100*
β	1.06	1.07	1.09	1.13	1.22	1.41	2.71

The coefficient of non-uniform compression, C_ϕ, is relevant to rotational deformation of the soil about a horizontal axis as shown in Fig. 8.8. The relation between moment, M, and rotation, ϕ, about any horizontal axis is given by

$$M = C_\phi I_{x,y}\phi \tag{8.13}$$

where $I_{x,y}$ is the second moment of area of the base about the x or y axis. The coefficient of uniform shear C_x is defined by

$$\tau_x = C_x u \tag{8.14}$$

where τ_x and u are shear stress and corresponding displacement in the x direction. The same relation applies to shear in any horizontal direction.

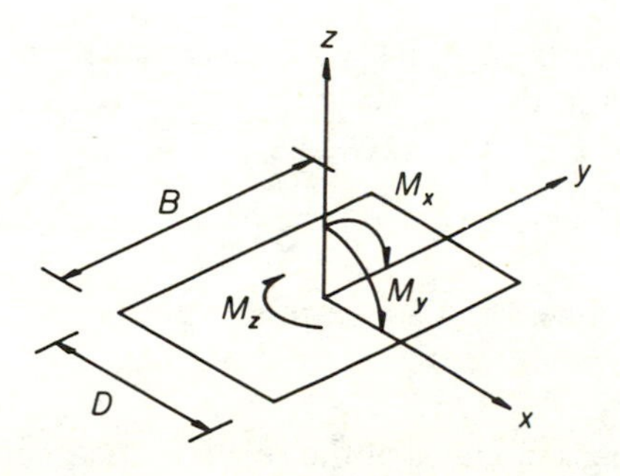

Fig. 8.8 Moments acting on base area of foundation.

The coefficient of non-uniform shear, C_ψ, is relevant to yawing rotation about a vertical axis and is defined by

$$M_z = C_\psi I_z \phi_z \tag{8.15}$$

where M_z and ϕ_z are the moment and corresponding rotation about z and I_z is the polar second moment of area of the base contact area.

Theoretical relationships exist between C_z, C_ϕ, C_x and C_ψ and may be obtained by procedures similar to Schleicher's. However, on the basis of large-scale experiments Barkan (1962) recommended the following:

$$C_\phi = 2.0\, C_z \tag{8.16a}$$

$$C_x = 0.5\, C_z \tag{8.16b}$$

$$C_\psi = 0.75\, C_z. \tag{8.16c}$$

Stiff clay and gravel are suitable materials for machine foundations. Rock is even better because the greater stiffness should make it possible to design a block with a natural frequency greater than the operating speed of the machine. Furthermore, the stiffness properties of rock are not affected by settlement and do not change with time, as is often the case with soils. However, cracks and fissures in a rock foundation can affect the stiffness and care should be taken to ensure that the material is well grouted.

8.3.3 Piled foundations

Soft soils such as peat and alluvial silt are not suitable for machine foundations which should then be carried on piles. Data for calculating the stiffness of friction piles and pile groups in vertical compression are given by Major (1962) and Srinivasulu and Vaidyanathan (1976). The stiffness of a single pile is given by

$$k_z = C_p A_s \tag{8.17}$$

where A_s is the surface area of the pile, and C_p is the coefficient of elastic resistance.

An end-bearing pile in very soft material may be treated as an elastic rod whose stiffness is given by

$$k_z = AE/L, \tag{8.18}$$

where A is the cross-sectional area of the pile, E is its Young's modulus and L its length.

The horizontal stiffness of a single pile is given by

$$k_x = 12\, EI/h^3 \tag{8.19}$$

where E is the Young's modulus of the pile material, I is the second moment of area of the pile cross section, and h is the length of pile that is free to move laterally between fixed ends as shown in Fig. 8.9. This length will be very difficult to estimate as it will depend on the horizontal resistance of the soil.

The stiffness of a pile group in vertical or horizontal deformation will be the single-pile stiffness multiplied by the number of piles in the group. A group reduction factor would be applied to friction piles.

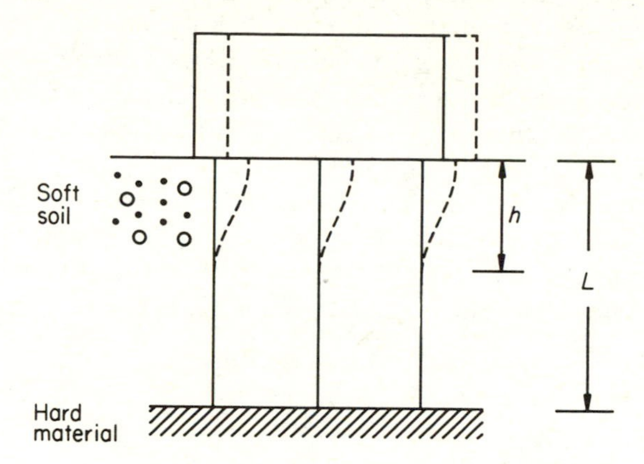

Fig. 8.9 Horizontal deformation of pile group.

Since piles provide effective point supports the rotational stiffness relations for pile groups are given by

$$M_x = k_z \Sigma\, y^2\, \phi_x \tag{8.20a}$$

$$M_y = k_z \Sigma\, x^2\, \phi_y \tag{8.20b}$$

$$M_z = k_x (\Sigma\, x^2 + \Sigma\, y^2) \phi_z\,, \tag{8.20c}$$

where x and y are the distances of individual piles from the y and x axes respectively, as shown in Fig. 8.10.

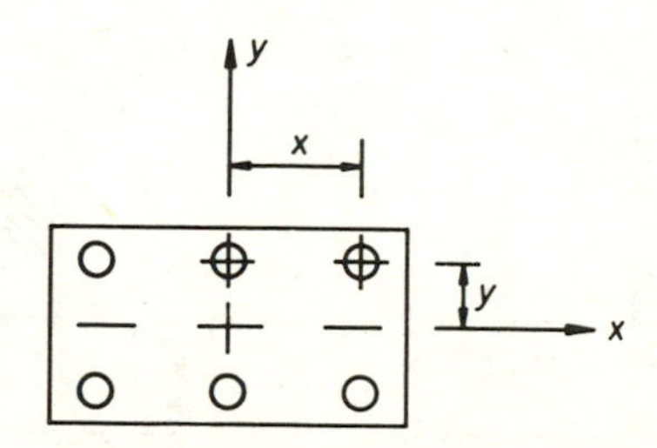

Fig. 8.10 Block on piled supports.

8.3.4 Mass moments of inertia

Calculation of translational modes of vibration will require the combined mass of machinery and foundation block. However, rocking, pitching or yawing modes will require rotational inertias in the calculations. These are referred to as the mass moments of inertia. It will be necessary to determine the centroid of the foundation block, together with the machine, and its mass moments of inertia about the centroid. Then, because the centroid of the machine and block will not be coincident with the centroid of the elastically resisting base surface, it will be necessary to calculate the inertia properties with respect to the centroid of this surface. The methods of performing these calculations will now be explained.

For convenience we shall refer to the coordinate system shown in Fig. 8.11. Assuming that the foundation block is made up of a number of brick-like volumes we can locate its centre of gravity by taking the first moments of the various volumes about the origin (see Fig. 8.12) using

$$x_{\mathrm{b}}=\Sigma(abcx)/\Sigma(abc) \tag{8.21}$$

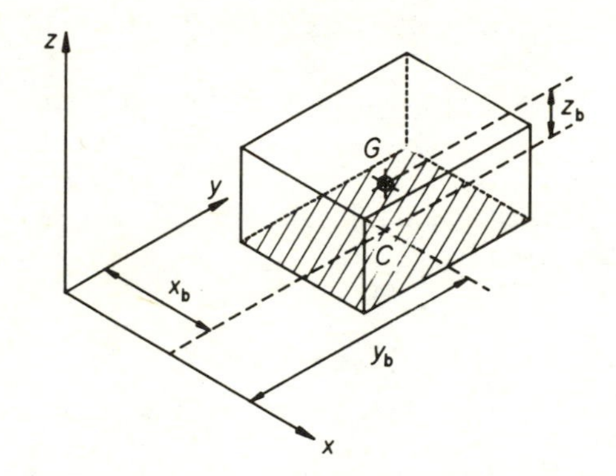

Fig. 8.11 Coordinate system.

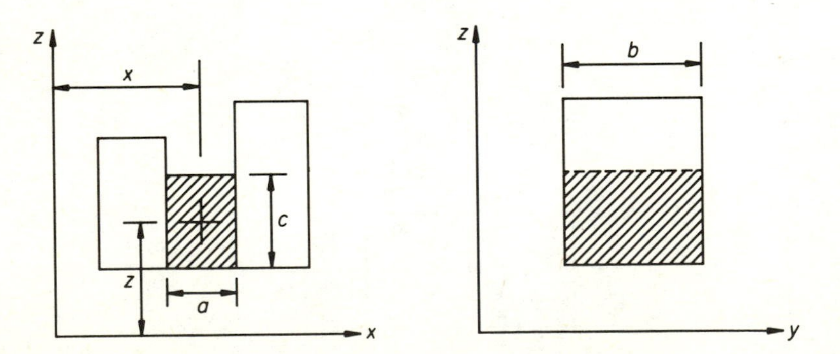

Fig. 8.12 Subdivision into volumes.

where the summations include all volumes comprising the block. The same procedure is followed to obtain y_b and z_b.

The mass moment of inertia of a brick-like volume about the centre of gravity of the block is given by

$$\Delta\theta_y = \rho abc\left[\frac{a^2+c^2}{12}+(x_b-x)^2+(z_b-z)^2\right] \tag{8.22}$$

for rotation about a line parallel to the y axis, where ρ is the density of the block. The first term within the brackets represents the mass moment of inertia of the volume about its own centre of gravity. The remaining terms utilize the parallel-axes theorem to give the mass moment of inertia about a different axis, namely one passing through the centre of gravity of the entire block. Terms such as Equation (8.22) are summed for all volumes comprising the block and a similar procedure is followed to obtain inertias about the remaining axes.

The masses, inertias and centres of gravity of the machinery will have been provided by the manufacturers. These are combined with those of the foundation block, using the same methods as above, so as to obtain the inertias and centres of gravity of the complete system. Finally, it is necessary to calculate the inertias about the base surface by means of the parallel-axes theorem.

EXAMPLE 8.1

These procedures will now be illustrated with the simple example shown in Fig. 8.13. This consists of a concrete foundation block for a diesel engine driving an electrical alternator. Because of symmetry the centres of gravity of block, engine and alternator will lie in the plane of the chosen YZ axes. This will simplify the calculations. The block is composed of three distinct volumes as indicated by the dashed lines.

The centre of gravity of the block is given by

$$x_b = 0$$

$$y_b = \frac{(2.5\times5.5\times1.5)\times2.75+(4\times5.5\times1.5)\times2.75+(4\times1.5\times2)\times6.25}{20.62+33.0+12.0}$$

$$= 3.39\text{ m}$$

$$z_b = \frac{20.62\times2.25+33.0\times0.75+12.0\times1.0}{65.62}$$

$$= 1.267\text{ m.}$$

Then, from Equation (8.22), the mass moment of inertia of the block about a

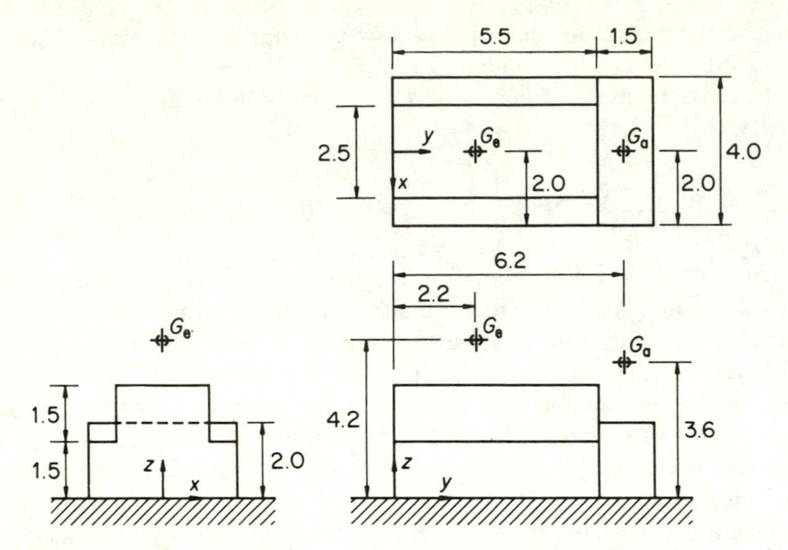

Fig. 8.13 Foundation block for engine–generator system (dimensions in m).

line through the centre of gravity parallel to the Y axis is

$$\theta_{yb}=2400\left\{20.62\left[\frac{2.5^2+1.5^2}{12}+(1.267-2.25)^2\right]\right.$$

$$+33.0\left[\frac{4^2+1.5^2}{12}+(1.267-0.75)^2\right]$$

$$\left.+12.0\left[\frac{4^2+2^2}{12}+(1.267-1.0)^2\right]\right\}$$

$$=274.57\times 10^3\ \mathrm{kg\,m^2}$$

where the density of concrete is 2400 kg/m^3.

The inertias about the X axis will not be calculated in this example (though they would be required in full design calculations), since the pitching mode of vibration is less important. The mass moment of inertia about a vertical axis through the centre of gravity is given by

$$\theta_{zb}=2400\left\{20.62\left[\frac{2.5^2+5.5^2}{12}+(3.39-2.75)^2\right]\right.$$

$$+33.0\left[\frac{4^2+5.5^2}{12}+(3.39-2.75)^2\right]$$

$$\left.+12\left[\frac{4^2+1.5^2}{12}+(3.39-6.25)^2\right]\right\}$$

$$=787.86\times 10^3\ \mathrm{kg\,m^2}.$$

The mass of the block is

$$M_b = 2400 \times 65.62 = 157\,488 \text{ kg}.$$

The masses and inertias of the engine and alternator are provided by the manufacturer as follows:

$$M_e = 45\,000 \text{ kg}$$

$$\theta_{ye} = 40 \times 10^3 \text{ kg m}^2$$

$$\theta_{ze} = 120 \times 10^3 \text{ kg m}^2$$

$$M_a = 20\,000 \text{ kg}$$

$$\theta_{ya} = 15 \times 10^3 \text{ kg m}^2$$

$$\theta_{za} = 10 \times 10^3 \text{ kg m}^2.$$

The centres of gravity of the engine and alternator are shown in Fig. 8.13. Therefore the centre of gravity of the complete system will be given by

$$\bar{x} = 0$$

$$\bar{y} = \frac{157\,488 \times 3.39 + 45\,000 \times 2.2 + 20\,000 \times 6.2}{222\,488}$$

$$= 3.40 \text{ m}$$

$$\bar{z} = \frac{157\,488 \times 1.267 + 45\,000 \times 4.2 + 20\,000 \times 3.6}{222\,488}$$

$$= 2.07 \text{ m}.$$

Therefore, the mass moment of inertia of the complete system, about a line through the centre of gravity of the complete system and parallel to the Y axis, is given by

$$\theta_{\bar{y}} = 274.57 \times 10^3 + 157\,488 \times (2.07 - 1.267)^2$$

$$+ 40 \times 10^3 + 45\,000 \times (2.07 - 4.2)^2$$

$$+ 15 \times 10^3 + 20\,000 \times (2.07 - 3.6)^2$$

$$= 682.1 \times 10^3 \text{ kg m}^2.$$

The first, third and fifth terms are the inertias of the block, engine and alternator about their own centres of gravity, while the second, fourth and sixth terms represent the inertias of their point masses about the centre of gravity of the complete system. Notice the use of the parallel-axes theorem. Similarly, the mass moment of inertia about the vertical axis is found to be

$$\theta_{\bar{z}} = 1139.5 \times 10^3 \text{ kg m}^2.$$

We mentioned earlier that it will be useful to calculate the rocking and pitching inertias about the centroid of the bearing surface (point C in Fig. 8.11). We shall only consider the rocking mode and therefore, transforming $\theta_{\bar{y}}$ about the Y axis, we obtain

$$\begin{aligned}\theta_Y &= 682.1\times 10^3 + 222\,488\times 2.07^2\\ &= 1635.4\times 10^3\ \mathrm{kg\,m^2}.\end{aligned}$$

Later we shall also require the ratio

$$\gamma = \theta_{\bar{y}}/\theta_Y = 0.417.$$

8.4 NATURAL FREQUENCIES

8.4.1 Vertical vibration

It will be assumed that the X and Y coordinates of the centre of gravity of the machinery and foundation block coincide with the centroid of the base area of the foundation, or very nearly so. The relevant British Standard Code of Practice (BSI, 1974) recommends that the discrepancy should not exceed 5% of either base dimension. If this is so it can be safely assumed that vertical vibration will occur as a simple mass on a spring and that coupling with rotation can be ignored. The natural frequency will be given by

$$f_z = \frac{1}{2\pi}\sqrt{\left(\frac{C_z A}{m}\right)} \tag{8.23}$$

where m is the total mass of block and machinery, and A is the area of the base.

8.4.2 Yawing vibration

It is possible for the dynamic loading of machinery to possess rotational components about a vertical axis. Therefore, frequencies of rotational, or yawing, vibration about the vertical axis are often calculated. The system has one degree of freedom but, in this case, the rotational stiffness and mass moment of inertia about the vertical axis are required. The natural frequency is given by

$$f_{\theta z} = \frac{1}{2\pi}\sqrt{\left(\frac{C_\psi I_z}{\theta_z}\right)} \tag{8.24}$$

where $C_\psi I_z$ represents the rotational stiffness given by Equation (8.16) and θ_z is the mass moment of inertia about the vertical axis.

8.4.3 Rocking and lateral mode

It was mentioned earlier that rocking motion about the Y axis forms a coupled system with horizontal motion in the X direction. This can be explained with reference to Fig. 8.14. By ignoring damping and considering the effects of forces relative to the centre of gravity at G, the equations of motion may be written

$$m\ddot{u} + C_x A u_0 = P_x(t) \tag{8.25a}$$

$$\theta_{\bar{y}}\ddot{\phi} + C_\phi I_y \phi = C_x A u_0 S + mgS\phi + M_y(t). \tag{8.25b}$$

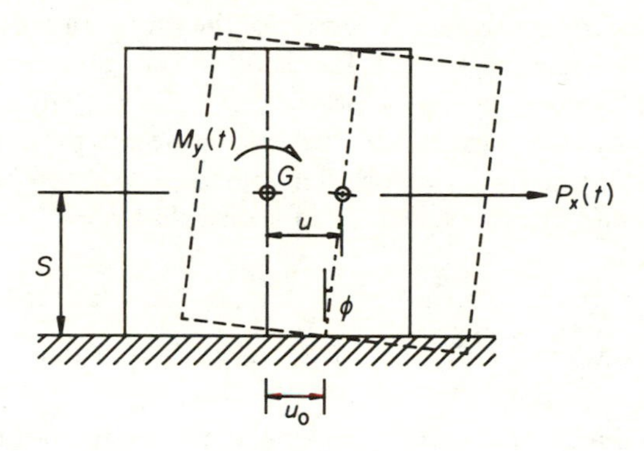

Fig. 8.14 Rocking and lateral mode.

Equation (8.25a) represents equilibrium of horizontal forces. The elastic restoring force is proportional to the horizontal displacement at the base, u_0. Equation (8.25b) represents angular equilibrium about G. The elastic rotational restoring force, $C_\phi I_y \phi$, acts in the same direction as the rotational inertia, $\theta_{\bar{y}}\ddot{\phi}$. However, the horizontal elastic force, $C_x A u_0$, will have a lever arm S and will supplement the externally applied moment, $M_y(t)$. In addition, the vertical weight, mg, will be offset from the centroid of the base by an amount $S\phi$, giving rise to a further couple $mgS\phi$. The displacement at the base, u_0, can be written in terms of the rotation and horizontal displacement of the centre of gravity, ϕ and u respectively. Thus, by substituting

$$u_0 = u - S\phi \tag{8.26}$$

into Equations (8.25) we obtain

$$m\ddot{u} + C_x A u - C_x A S\phi = P_x(t) \tag{8.27a}$$

$$\theta_{\bar{y}}\ddot{\phi} - C_x A S u + (C_\phi I_y - mgS + C_x A S^2)\phi = M_y(t). \tag{8.27b}$$

It can be seen that these are simultaneous second-order differential equations in the two degrees of freedom u and ϕ. The method of solution will follow the procedure for two-degree-of-freedom systems without damping outlined in Chapter 3.

The natural frequencies may be obtained by considering free vibration and setting the right-hand sides of Equations (8.27) to zero. The two limiting frequencies are defined as

$$\omega_x^2 = C_x A/m \tag{8.28a}$$

$$\omega_\phi^2 = \frac{C_\phi I_y - mgS}{\theta_Y} \tag{8.28b}$$

where I_y is the second moment of area of the base contact surface about the Y axis and θ_y is the mass moment of inertia of the block and machinery about the Y axis at the contact surface. Equation (8.28a) gives the frequency of lateral motion when rotation is prevented while (8.28b) corresponds to rocking about the base surface when lateral motion is prevented.

Substituting Equations (8.28), together with $\gamma = \theta_{\bar{y}}/\theta_Y$, into Equations (8.27) with the right-hand sides set to zero, and noting that $\theta_Y = \theta_{\bar{y}} + mS^2$, we obtain

$$\ddot{u} + \omega_x^2 u - \omega_x^2 S\phi = 0 \tag{8.29a}$$

$$S\ddot{\phi} - \omega_x^2 \frac{(1-\gamma)}{\gamma} u + \frac{1}{\gamma}[\omega_\phi^2 + \omega_x^2(1-\gamma)]S\phi = 0. \tag{8.29b}$$

These equations may be solved by assuming a periodic motion of the form

$$\begin{aligned} u &= U \sin \omega t \\ \phi &= \Phi \sin \omega t. \end{aligned} \tag{8.30}$$

Hence the frequency equation is obtained, which is given by

$$\omega^4 - \frac{(\omega_x^2 + \omega_\phi^2)}{\gamma}\omega^2 + \frac{\omega_x^2\,\omega_\phi^2}{\gamma} = 0. \tag{8.31}$$

This has two solutions, ω_1^2 and ω_2^2, whose frequencies will be on either side of the two limiting frequencies. It will also be found that the ratio of the amplitudes is given by

$$P = U/\Phi = \omega_x^2 S/(\omega_x^2 - \omega^2). \tag{8.32}$$

Thus the lower of the natural frequencies will give a positive ratio for the amplitudes and will result in the mode of vibration shown in Fig. 8.15(a), while the higher natural frequency will give a negative ratio resulting in the mode shown in Fig. 8.15(b).

EXAMPLE 8.2

We shall now calculate the natural frequencies of the engine–generator system with its foundation block shown in Fig. 8.13. First we shall need data

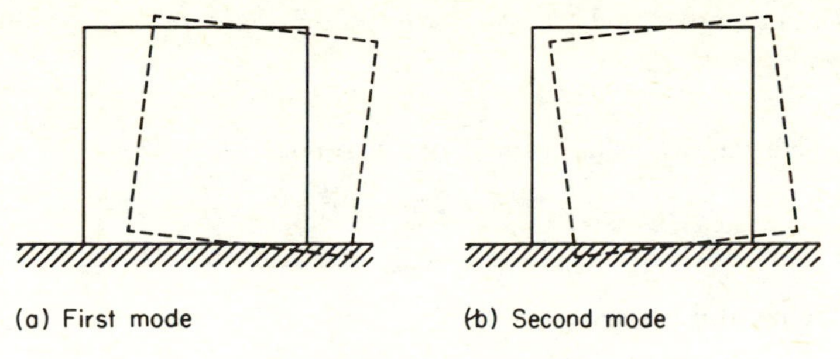

Fig. 8.15 Modes of vibration.

for the stiffness of the underlying material. Assume that it is a reasonably consistent volcanic rock with a Young's modulus of 4000 MN/m^2 and Poisson's ratio of 0.3. Then, using Equations (8.12) and (8.16), together with Table 8.1, the stiffness coefficients may be determined:

$$C_z = \frac{2000 \times 1.08}{(1-0.3^2)\sqrt{(4 \times 7)}} = 897 \text{ MN/m}^3$$

$$C_x = 0.5\ C_z = 448 \text{ MN/m}^3$$

$$C_\phi = 2\ C_z = 1794 \text{ MN/m}^3$$

$$C_\psi = 0.75\ C_z = 673 \text{ MN/m}^3.$$

The vertical natural frequency is

$$f_z = \frac{1}{2\pi}\sqrt{\left(\frac{897 \times 10^6 \times 28}{222\,488}\right)} = 53.5 \text{ Hz}.$$

The yawing frequency is

$$f_{\theta z} = \frac{1}{2\pi}\sqrt{\left(\frac{673 \times 10^6 \times 28(4^2+7^2)/12}{1.1395 \times 10^6}\right)} = 47.6 \text{ Hz}.$$

To obtain the rocking and lateral frequencies the limiting frequencies, Equations (8.28), must first be obtained. These are

$$\omega_x^2 = \frac{448 \times 10^6 \times 28}{222\,488} = 5.638 \times 10^4$$

$$\omega_\phi^2 = \frac{1794 \times 10^6 \times 7 \times 4^3/12 - 222\,488 \times 9.81 \times 2.07}{1635.4 \times 10^3}$$

$$= 4.095 \times 10^4.$$

Note that the *mgS* term is negligibly small and, in fact, is not even included in the relevant formula in the British Code of Practice (BSI, 1974).

We found earlier that $\gamma = 0.417$, substituting these values into Equation (8.31) we obtain

$$\omega^4 - 23.34 \times 10^4 \omega^2 + 55.36 \times 10^8 = 0,$$

from which the two solutions may be determined, these being

$$\omega_1^2 = 2.68 \times 10^4$$

$$\omega_2^2 = 20.66 \times 10^4.$$

Thus the natural frequencies are

$$f_1 = 26.0 \text{ Hz}, \quad f_2 = 72.3 \text{ Hz}.$$

8.5 AMPLITUDES OF FORCED VIBRATION

8.5.1 Vertical vibration

The equation of motion for the appropriate single-degree-of-freedom system, when loaded by an oscillatory vertical force, $P_z \sin \Omega t$, will be

$$m\ddot{w} + C_z A w = P_z \sin \Omega t. \tag{8.33}$$

By assuming that $w = w_p \sin \Omega t$ it may be verified that the amplitude will be given by

$$w_p = (P_z/m)/(\omega_z^2 - \Omega^2). \tag{8.34}$$

Yawing motions may be obtained in a similar way.

8.5.2 Rocking and lateral mode

We shall first consider a periodic force, $P_x \sin \Omega t$, acting transversely through the centre of gravity of the system. Thus, by referring to Equations (8.27), it may be seen that the equations of motion become

$$m\ddot{u} + C_x A u - C_x A S \phi = P_x \sin \Omega t \tag{8.35a}$$

$$\theta_{\bar{y}} \ddot{\phi} - C_x A S u + (C_\phi I_y - mgS + C_x A S^2)\phi = 0. \tag{8.35b}$$

Assuming that the ensuing motion is given by

$$u = u_p \sin \Omega t \tag{8.36a}$$

$$\phi = \phi_p \sin \Omega t \tag{8.36b}$$

then substitution in Equations (8.35) yields

$$(C_x A - \Omega^2 m)u_p - C_x A S \phi_p = P_x \tag{8.37a}$$

$$-C_x A S u_p + [C_\phi I_y - mgS + C_x A S^2 - \Omega^2 \theta_{\bar{y}}]\phi_p = 0. \tag{8.37b}$$

Eliminating ϕ_p we find that

$$u_p = (C_\phi I_y - mgS + C_x AS^2 - \Omega^2 \theta_{\bar{y}}) P_{(x)} / \Delta(\Omega^2) \tag{8.38}$$

where the denominator is given by

$$\Delta(\Omega^2) = (C_\phi I_y - mgS + C_x AS^2 - \Omega^2 \theta_{\bar{y}})(C_x A - \Omega^2 m) - C_x^2 A^2 S^2. \tag{8.39}$$

By introducing the limiting frequencies given by Equations (8.28) and noting that $\gamma = \theta_{\bar{y}}/\theta_Y$ and that $\theta_Y = \theta_{\bar{y}} + mS^2$, the denominator can be simplified to

$$\Delta(\Omega^2) = m\theta_{\bar{y}} \left[\Omega^4 - \frac{(\omega_x^2 + \omega_\phi^2)}{\gamma} \Omega^2 + \frac{\omega_x^2 \omega_\phi^2}{\gamma} \right]. \tag{8.40}$$

The expression within square brackets is the same as Equation (8.31) from which the natural frequencies were obtained. Therefore

$$\Delta(\Omega^2) = m\theta_{\bar{y}} (\omega_1^2 - \Omega^2)(\omega_2^2 - \Omega^2). \tag{8.41}$$

By eliminating u_p from Equations (8.37) the rotational amplitude may be found:

$$\phi_p = C_x ASP_x / \Delta(\Omega^2). \tag{8.42}$$

Similarly, when an oscillating moment, $M_y \sin \Omega t$, is applied to the system, the amplitudes of forced motion may be found. They are

$$u_m = C_x AS\, M_y / \Delta(\Omega^2) \tag{8.43a}$$

$$\phi_m = \frac{(C_x A - \Omega^2 m)\, M_y}{\Delta(\Omega^2)}. \tag{8.43b}$$

Finally, it is the usual practice to calculate the amplitudes at the top surface of the block or at its base. The maximum values of these amplitudes are given by

$$\begin{aligned} u_t &= u + (h - S)\phi \\ u_b &= u + S\phi, \end{aligned} \tag{8.44}$$

where h is the overall height of the block.

EXAMPLE 8.3

Using the same data as Examples 8.1 and 8.2 we shall now calculate the vibration amplitudes when the system is subjected to the following dynamic loads:

Operating speed of engine = 480 r.p.m.

Therefore

$$\Omega = 2\pi \times 480/60 = 50.3 \text{ rad/s}.$$

Primary vertical force at engine speed:

$$P_z = 25 \text{ kN at } 50.3 \text{ rad/s}.$$

Primary lateral force at engine speed:

$$P_x = 32 \text{ kN at } 50.3 \text{ rad/s}.$$

The engine is a ten-cylinder four-stroke diesel, and therefore there may be forces at $10/4 = 2.5$ times engine speed. This is estimated to be a torque only, of the following magnitude:

$$M_y = 185 \text{ kN m at } 125.7 \text{ rad/s}.$$

The remaining engine forces are adequately balanced.
Vertical amplitude:

$$w_p = \frac{25\,000/222\,488}{(2\pi \times 53.5)^2 - 50.3^2} = 1.02 \times 10^{-6} \text{ m}.$$

Rocking and lateral mode:

(a) Amplitudes resulting from lateral force P_x,

$$\Delta(\Omega^2) = 222\,488 \times 622.1 \times 10^3 \times (26\,800 - 50.3^2)(206\,600 - 50.3^2)$$
$$= 6.855 \times 10^{20}.$$

Therefore, using (8.38) and (8.42),

$$u_p = \frac{[66.98 \times 10^9 - 0.04 \times 10^9 + 53.75 \times 10^9 - 50.3^2 \times 622.1 \times 10^3] \times 32 \times 10^3}{6.855 \times 10^{20}}$$
$$= 5.56 \times 10^{-6} \text{ m};$$

$$\phi_p = \frac{448 \times 10^6 \times 28 \times 2.07 \times 32 \times 10^3}{685.5 \times 10^{18}}$$
$$= 0.15 \times 10^{-6} \text{ rad}.$$

(b) Amplitudes resulting from harmonic torque M_y,

$$\Delta(\Omega^2) = 222\,488 \times 622.1 \times 10^3 \times (26\,800 - 125.7^2)(206\,600 - 125.7^2)$$
$$= 290.5 \times 10^{18}.$$

Therefore, using (8.43),

$$u_m = \frac{448 \times 10^6 \times 28 \times 2.07 \times 185 \times 10^3}{290.5 \times 10^{18}}$$
$$= 16.54 \times 10^{-6} \text{ m};$$

$$\phi_m = \frac{(448 \times 10^6 \times 28 - 125.7^2 \times 222\,488) \times 185 \times 10^3}{290.5 \times 10^{18}}$$
$$= 5.75 \times 10^{-6} \text{ rad}.$$

The maximum displacements will then be

$$u = u_p + u_m = 22.1 \times 10^{-6} \text{ m}$$

$$\phi = \phi_p + \phi_m = 5.9 \times 10^{-6} \text{ rad},$$

and therefore the maximum horizontal displacements at top and bottom of the block will be

$$u_t = [22.1 + (3 - 2.07) \times 5.9] \times 10^{-6} = 27.6 \times 10^{-6} \text{ m}$$

$$u_b = [22.1 + 2.07 \times 5.9] \times 10^{-6} = 34.3 \times 10^{-6} \text{ m}.$$

8.6 CRITERIA FOR THE DESIGN OF MACHINE FOUNDATIONS

8.6.1 Resonant frequencies

In the previous section it was found that the amplitudes of forced vibration are controlled by the magnitude of the denominator $\Delta(\Omega^2)$ given in Equation (8.41). This term vanishes if the operating frequency, or an active harmonic of it, coincides with either of the natural frequencies of the system, resulting in theoretically infinite amplitudes. The same applies to modes with a single degree of freedom whose amplitudes are given by expressions such as Equation (8.34). In practice resonant amplitudes would be limited by the existence of damping in the system.

The simplified method recommended by Barkan (1962) and adopted in the British Code of Practice (BSI, 1974) depends for its reliability on the natural frequency of the system being significantly different from the operating frequency. In fact, it is recommended that the frequency of the disturbing force should be less than half or greater than twice the natural frequency. In order to avoid other harmonics being close to the natural frequency it is preferable to design a stiff foundation whose lowest natural frequency is always higher than the highest disturbing frequency. Unfortunately, it is not always possible to avoid the disturbing frequencies completely and it is often necessary to permit minor harmonics to be somewhat closer to resonance. Then the calculated amplitudes will give an indication of possible problems.

8.6.2 Human response to vibration amplitudes

Human beings are surprisingly sensitive to vibrations. This subject is discussed in detail in Chapter 11. At this moment it is sufficient to say that human tolerance to vibration is amplitude and frequency related. Curves defining recommended vibration limits are given in the British Code (BSI, 1974) and are shown in Fig. 8.16. The lines are defined as follows:

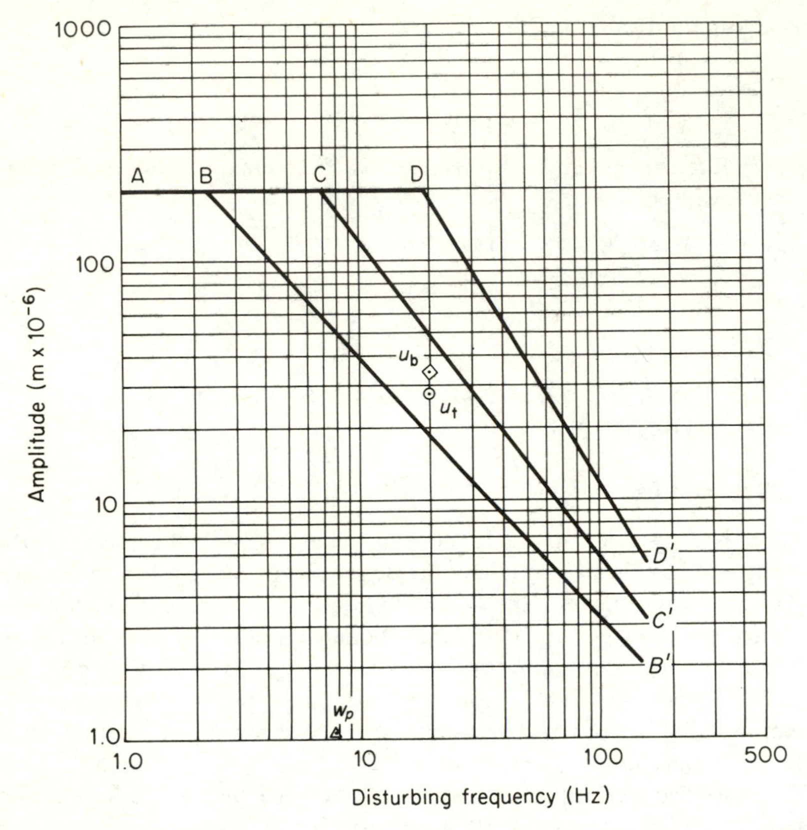

Fig. 8.16 Amplitude limits for foundation blocks (from British Standard CP2012, Part I; BSI, 1974).

ADD': limit to avoid damage to buildings
ACC': limit to avoid serious discomfort to persons
ABB': limit to ensure reasonable comfort to persons.

The amplitudes calculated in Example 8.3 are plotted on the graph. The vertical amplitude w_p is negligible. The horizontal vibration at the top of the block, u_t, is in the uncomfortable region for personnel who might be attending to the machinery. However, a generator building is an environment where high levels of vibration would be expected and therefore there should be little risk of complaints. The horizontal vibration at the base of the foundation block, u_b, is relevant to the risk of damage to the enclosing building. It is evidently well within the acceptable limit of 0.2 mm. It should be noted that in these two cases the higher harmonic frequency of 20 Hz was assumed to be relevant.

8.6.3 Transmission to adjacent buildings

Reference has already been made to the fact that vibration of a mass on an elastic half-space results in waves being radiated outwards over the free surface. If the dynamic disturbance is very large, as is often the case with many industrial processes, the waves may be great enough to cause damage to nearby buildings. This subject is dealt with in more detail in Chapter 10.

Another problem that has been considered by Warburton, Richardson and Webster (1971) is the interaction between oscillating masses close to each other on the surface of an elastic half-space. If a dynamic load is applied to one mass the waves propagated outwards are experienced as an oscillating ground motion by a second mass nearby. However, for practical values of the relevant parameters, the amount of interaction is small and could be safely neglected in most applications.

9

Moving loads

9.1 INTRODUCTION

The most obvious examples of structures subjected to moving loads are highway and railway bridges. Analysis is complicated by the fact that the mass of a moving vehicle or locomotive is usually large compared with that of the bridge itself. Furthermore, there is a problem of interaction between the motion of the bridge and the suspension of the vehicle. Stresses under moving loads are very much greater than under the same loads applied statically, especially if the riding surface is uneven. This is an important consideration in the design of suspension bridges which are intended to carry railway traffic.

Human induced forces come under the heading of 'moving' loads and are important sources of dynamic excitation. Pedestrians walking or running across footbridges have been known to induce large enough vibrations to be alarming. Vibrations of light floors, caused by heel impact in normal usage, can be disturbing to occupants of buildings. Surprisingly large dynamic forces can be exerted by running up or down stairs and, finally, there is the important problem of dynamic crowd loading which can affect dance hall floors and large sports stadia.

9.2 DYNAMIC ANALYSIS OF BEAMS UNDER MOVING LOADS

9.2.1 Simply supported beam under moving force

Many highway and railway bridges consist of simply supported girders. Therefore, useful insight into the dynamic behaviour of bridges may be obtained by studying the response of simply supported beams under moving loads. Timoshenko (1928) reviewed the analysis of some important illustrative examples. In the first instance we shall consider the passage of a constant force travelling across a beam at uniform speed as shown in Fig. 9.1(a). The equations of motion of a dynamically loaded beam in generalized coordinates

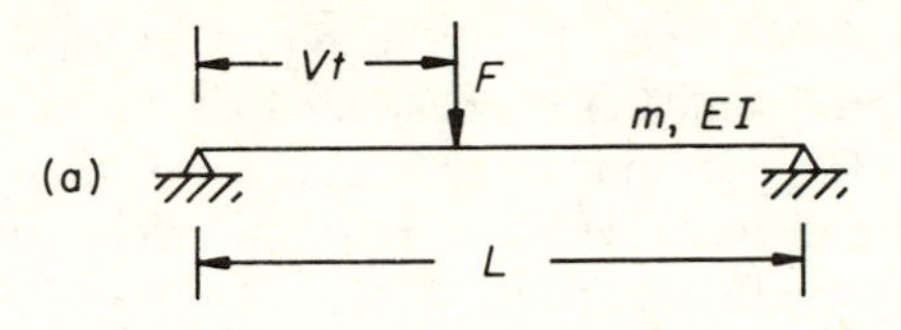

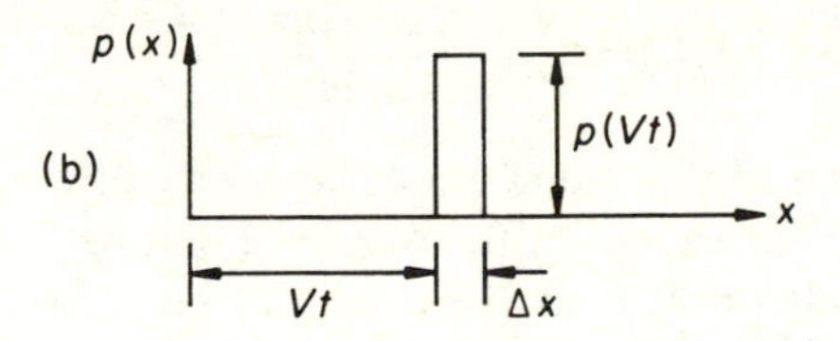

Fig. 9.1 Moving constant force.

were given in Chapter 3. Equation (3.82) stated that

$$\ddot{Y}_n(t)+\omega_n^2 Y_n(t)=Q_n(t)/M_n; \quad n=1, 2, \ldots, \infty \tag{3.82}$$

where

$$Q_n(t)=\int_0^L \phi_n(x)\,p(x,t)\,\mathrm{d}x \tag{3.83}$$

and

$$M_n=m\int_0^L \phi_n^2(x)\,\mathrm{d}x. \tag{3.84}$$

The mode shapes of a simply supported beam were given in Equation (3.49) as

$$\phi_n(x)=C_1 \sin(n\pi x/L). \tag{3.49}$$

The general form of the distributed load, $p(x, t)$, may be expressed as a force of intensity $p(Vt)$ applied over an infinitesimal length Δx as shown in Fig. 9.1(b) such that the product $p(Vt)\ \Delta x \to F$ as $\Delta x \to 0$.

Thus the integral in the expression for generalized force (3.83) will become

$$Q_n(t)=FC_1 \sin(n\pi Vt/L). \tag{9.1}$$

The generalized mass is given by

$$M_n=m\int_0^L C_1^2 \sin^2(n\pi x/L)\,\mathrm{d}x=(mL/2)C_1^2. \tag{9.2}$$

Making the value of the arbitrary constant $C_1=1$ and substituting (9.1) and (9.2) into (3.82) we find that

$$\ddot{Y}_n(t)+\omega_n^2 Y_n(t)=(2F/mL)\sin(n\pi Vt/L); \quad n=1, 2, \ldots, \infty. \tag{9.3}$$

The right-hand side of this equation is a periodic function with natural circular frequency $\Omega = n\pi V/L$. Thus the equation is similar to (2.20) in Chapter 2 which represented a single degree of freedom with a periodic forcing function $\sin \Omega t$. The solution to (9.3) consists of forced vibrations and free vibrations. The forced part may be obtained by analogy with (2.25), neglecting damping ($c=0$), and is given by

$$Y_n(t) = \frac{(2F/mL)\sin(n\pi Vt/L)}{\omega_n^2 - (n\pi V/L)^2}. \tag{9.4}$$

In Chapter 2 it was said that there would also be free vibrations that would die away after the first few cycles because of damping in the system. In the case of moving loads on bridges these cannot be ignored since the forced vibration consists effectively of only one half cycle. The free vibrations consist of two parts: those due to initial conditions which will be zero if the bridge is initially at rest; and those due to the sudden application of load. The latter consist of important transient vibrations which have to be included. The free vibrations are given by Equation (2.9). Thus, neglecting damping, the full solution to (9.3) is given by

$$Y_n(t) = A\cos\omega_n t + B\sin\omega_n t + \frac{(2F/mL)\sin(n\pi Vt/L)}{\omega_n^2 - (n\pi V/L)^2}. \tag{9.5}$$

The initial conditions with the beam at rest are

$$Y_n(0) = 0 \quad \text{and} \quad \dot{Y}_n(0) = 0. \tag{9.6}$$

Using these it is found that

$$A = 0$$

$$B = -\frac{2F}{\omega mL}\frac{n\pi V}{L}\frac{1}{\omega_n^2 - (n\pi V/L)^2}. \tag{9.7}$$

Remembering also that the full dynamic response is obtained by combining the individual modal responses, using Equation (3.77),

$$v(x,t) = \sum_{n=1}^{\infty} \phi_n(x) Y_n(t) = \sum_{n=1}^{\infty} Y_n(t)\sin(n\pi x/L), \tag{9.8}$$

we finally obtain for the dynamic response of the beam

$$v(x,t) = \frac{2F}{mL}\sum_{n=1}^{\infty}\frac{\sin(n\pi x/L)}{\omega_n^2 - (n\pi V/L)^2}\left\{\sin\left(\frac{n\pi Vt}{L}\right) - \frac{(n\pi V/L)}{\omega}\sin\omega_n t\right\}. \tag{9.9}$$

It has been suggested that modes higher than the fundamental may be neglected without serious loss of accuracy when calculating the deflections and bending moments in bridges under moving loads (Inglis, 1934; Biggs, Suer and Louw, 1959). It is therefore instructive to look at the relative importance of terms in Equation (9.9) with $n=1$. Using the formula for the fundamental natural frequency of a simply supported beam ($\omega^2 = \pi^4 EI/L^4 m$), given by Equation (3.48), the midspan deflection ($x = L/2$) may be obtained from (9.9) as follows:

$$v_c(t) = \frac{2FL^3}{\pi^4 EI}\left(\frac{\sin(\pi Vt/L) - (\pi V/L\omega)\sin\omega t}{1-(\pi V/L\omega)^2}\right). \tag{9.10}$$

The first term in the numerator within the large parentheses is due to the forced vibration and consists of a half sine wave as the force crosses the beam. The second term is the free vibration caused by the sudden application of the force. Its amplitude depends on the speed. The pseudo-static or crawl deflection occurs when $\pi V/L$ is negligibly small compared with ω and is given by

$$\delta_c(t) = \frac{2FL^3}{\pi^4 EI}\sin(\pi Vt/L). \tag{9.11}$$

Notice that this reaches a maximum when the load is at midspan ($Vt = L/2$) and is very nearly equal to $FL^3/48EI$, which is the deflection of a beam with static central load.

EXAMPLE 9.1

A beam of 30 m span, with a natural frequency of 2 Hz, is crossed by a force travelling at 15 m/s. Determine the dynamic deflection at the midspan of the beam while the force crosses the span. Compare with the crawl deflection.

Solution

The parameters are:

$$f = 2\text{ Hz};\quad V = 15\text{ m/s};\quad L = 30.$$

Therefore $V/L = 0.5$, $\omega = 4\pi$ and $\pi V/L\omega = 0.125$.

Substituting in (9.10) we obtain

$$v_c(t) = (\delta_c)_{\max}\left(\frac{\sin(0.5\pi t) - 0.125\sin 4\pi t}{1-(0.125)^2}\right).$$

This expression is evaluated at intervals of 0.2 s and is given in Table 9.1. The

Table 9.1. Dynamic response of beam under moving force

t (s)	0.2	0.4	0.6	0.8	1.0	1.2	1.4	1.6	1.8	2.0
$v_c/(\delta_c)_{\max}$	0.239	0.718	0.701	1.041	1.016	0.892	0.943	0.476	0.388	0
x/L	0.1	0.2	0.3	0.4	0.5	0.6	0.7	0.8	0.9	1.0
$\delta_c/(\delta_c)_{\max}$	0.309	0.588	0.809	0.951	1.0	0.951	0.809	0.588	0.309	0

crawl deflection curve in (9.11) is evaluated at intervals of $0.1x/L$ and included in Table 9.1 for comparison. Intermediate values may be calculated and plotted as in Fig. 9.2. The oscillation of the beam about the crawl deflection is evident.

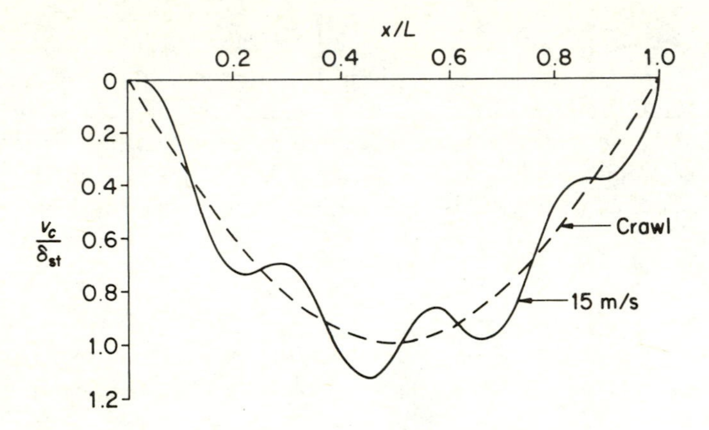

Fig. 9.2 Dynamic response of beam under moving force.

9.2.2 Maximum response

The dynamic load factor (DLF) was defined in Chapter 2 as the ratio of maximum dynamic response to the maximum static response. An estimate of this can be obtained from (9.10) by assuming that the maximum response occurs when the amplitudes of forced and free vibrations add together. Hence

$$\mathrm{DLF} = \frac{(v_c)_{\max}}{(\delta_c)_{\max}} \simeq \frac{1+(\pi V/L\omega)}{1-(\pi V/L\omega)^2} = \frac{1}{1-(\pi V/L\omega)}. \tag{9.12}$$

Taking the values of the previous example this would give DLF = 1.143.

It is clear that this dynamic magnification increases with speed. The loading frequency, $(\pi V/L)$, is generally quite low compared with the natural frequency, ω. However, dynamic magnification may be important in the design of bridges for very high-speed transit vehicles. It is natural to wonder what happens if the loading frequency and the natural frequency coincide. This is not true resonance since there is only one half cycle of loading as the force crosses the span. Substituting $\pi V/L = \omega$ into (9.10) yields an indeterminate result since both numerator and denominator vanish. However, a result may be obtained by considering the response when $\pi V/L = \omega + \delta\omega$ and eventually setting $\delta\omega$ to zero. By this method it may be shown that the maximum response is about 50% greater than the static response and occurs just as the force leaves the span. At greater speeds the response actually diminishes.

An interesting historical note is that a British government committee, enquiring into the safety of railway bridges in 1849, deliberated at length whether it was safer to go across a bridge fast or slow (HMSO, 1849). The matter could not be resolved until an adequate mathematical treatment was first achieved by Stokes (1883). The critical speed is generally very high and in the case of Example 9.1 would be 120 m/s $\simeq$ 270 miles/hour.

9.2.3 Moving pulsating force

An important load case is that of a force travelling at constant speed but with a pulsating amplitude. This is of historical interest in connection with the vibration produced by the unbalanced forces of the driving wheels of steam locomotives. Modern railway locomotives do not generate dynamic loads of this kind. However, footbridges may be analysed by assuming that pedestrian loading is equivalent to a moving pulsating force (BS 5400, 1978).

The problem is similar to that depicted in Fig. 9.1 except that the constant force F is replaced by the pulsating force $F_0 \sin \Omega t$. Equation (9.3) is therefore appropriately modified and becomes

$$\begin{aligned}\ddot{Y}_n(t)+\omega_n^2 Y_n(t)&=(2/mL)F_0 \sin \Omega t \sin(n\pi Vt/L); \quad n=1, 2, \ldots, \infty \qquad (9.13)\\ &=(F_0/mL)[\cos(\Omega-\Omega_v)t-\cos(\Omega+\Omega_v)t]; \quad n=1, 2, \ldots, \infty\end{aligned}$$

where $\Omega_v = n\pi V/L$.

The response of a single-degree-of-freedom system with a cosine forcing function is very similar to that with a sine forcing function. Therefore, adopting the method outlined in Chapter 3, and obtaining the coefficients of the free vibrations using (9.6), it is possible to show that the displacement of the beam is given by

$$v(x, t)=\frac{F_0}{mL}\sum_{n=1}^{\infty} \sin\frac{n\pi x}{L}\left(\frac{\cos(\Omega-\Omega_v)t-\cos\omega_n t}{\omega_n^2-(\Omega-\Omega_v)^2}-\frac{\cos(\Omega+\Omega_v)t-\cos\omega_n t}{\omega_n^2-(\Omega+\Omega_v)^2}\right). \qquad (9.14)$$

True resonance does not occur in this case but large vibrations can build up if the pulsating frequency coincides with the lowest natural frequency of the beam. If there is no damping, as in the example here, the forced vibrations get larger as the load crosses the span and are largest when it just leaves the span. Biggs (1964) showed that the maximum deflection at midspan is given approximately by

$$(v_c)_{\max}=\frac{2F_0}{m\pi V\omega_1}=v_{\mathrm{st}}\frac{\omega_1 L}{\pi V} \qquad (9.15)$$

where v_{st} is the static deflection under a load of F_0 at midspan. Some damping is usually present, which would reduce the vibrations.

Usually there will be a constant gravity force accompanying the pulsating load. Its additional effect may be obtained by combining (9.9) with (9.14).

9.2.4 Moving sprung mass

The analysis of a moving force on a single beam is a useful introduction to the vibration of bridges under moving vehicles. However, in practice the mass of a vehicle or locomotive may be significant compared with the mass of a bridge. Neglecting the combined inertia may lead to substantial error. Furthermore, vehicle suspension and roadway unevenness are important factors affecting the response of bridges and should be taken into account.

A simplified model is shown in Fig. 9.3. This was first suggested by Inglis (1934) and consists of a mass, m_s, representing the primary mass of the vehicle or locomotive, an unsprung mass, m_u, representing the wheels, and a spring and dashpot for the suspension. When the bridge and vehicle are in motion there will be an interacting force given by

$$F=(m_s+m_u)g-m_u\ddot{v}_u+c(\dot{v}_s-\dot{v}_u)+k(v_s-v_u). \tag{9.16}$$

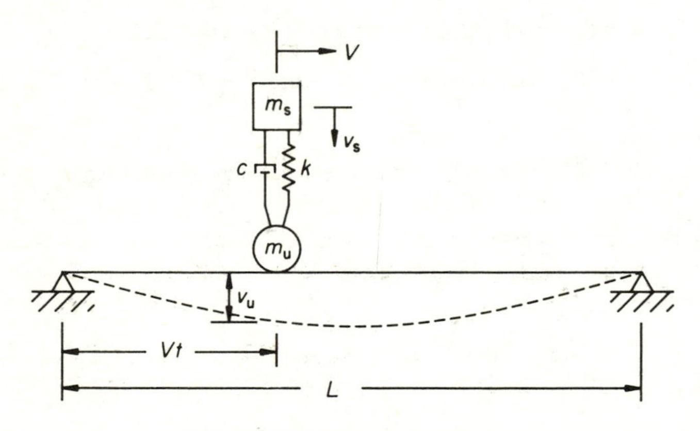

Fig. 9.3 Moving sprung mass.

The first term represents the weight of the vehicle. The second term arises from the negative inertia of the unsprung mass and assumes that the wheel remains in contact with the bridge. The remaining terms are caused by relative motion of the bridge surface and the sprung mass, noting that v_s is the displacement taken from the equilibrium position of the latter. The displacement under the unsprung mass is given by

$$v_u=\sum_{n=1}^{\infty} Y_n \sin(n\pi Vt/L). \tag{9.17}$$

Substituting (9.16) and (9.17) into (9.3) we obtain

$$\ddot{Y}_n(t)+\omega_n^2 Y_n(t) = \frac{2m_s}{mL}\sin\frac{2\pi Vt}{L}\left[\left(1+\frac{m_u}{m_s}\right)g-\frac{m_u}{m_s}\sum_{n=1}^{\infty}\ddot{Y}_n\sin\frac{n\pi Vt}{L} + 2\xi_s\omega_v\left(\dot{v}_s-\sum_{n=1}^{\infty}\dot{Y}_n\sin\frac{n\pi Vt}{L}\right) + \omega_s^2\left(v_s-\sum_{n=1}^{\infty}Y_n\sin\frac{n\pi Vt}{L}\right)\right]; \quad n=1, 2, \ldots, \infty \tag{9.18}$$

where ξ_s is the critical damping ratio in the vehicle suspension and ω_s is the natural frequency of the vehicle on its suspension.

There is an additional equation to take account of the movement of the sprung mass, given by

$$\ddot{v}_s+2\xi_s\omega_s\left(\dot{v}_s-\sum_{n=1}^{\infty}\dot{Y}_n\sin\frac{n\pi Vt}{L}\right) + \omega_s^2\left(v_s-\sum_{n=1}^{\infty}Y_n\sin\frac{n\pi Vt}{L}\right). \tag{9.19}$$

Equations (9.18) and (9.19) are coupled second-order equations with variable coefficients which can only be solved by direct numerical integration. Inglis (1934) included only the fundamental mode and then solved the equations manually. The computations were exceedingly protracted and tedious but he succeeded in obtaining solutions to several important load cases which were used in the studies by the British Bridge Stress Committee (HMSO, 1928).

Roadway unevenness can be taken into account in one of two ways. Biggs, Suer and Louw (1959) observed that the most significant effect of roadway unevenness is the bounce of a vehicle on its suspension as it entered the span. They included this in the initial conditions of (9.18) and (9.19) but ignored the effects of roughness on the span itself. Toledo-Leyva and Veletsos (1958) modified Equation (9.16) to take account of the effect of roadway unevenness on the interacting force.

An approximate solution to Equations (9.18) and (9.19) was obtained by Smith (1969) who noted that the oscillation of a vehicle as it entered the span of a bridge would involve a fixed amount of energy derived from initial bounce on its suspension. This energy would be shared between the vehicle and the bridge during the time it crossed the span. The upper bound of response was assumed to occur when the vehicle was close to midspan and all the periodic components of the oscillatory system added together for maximum effect. A formula for the dynamic load factor was derived and is given by

$$\mathrm{DLF}=\frac{1}{1-\alpha}+\frac{\beta\rho^2\exp(-\pi\xi_s/2\rho\alpha)}{\sqrt{[(\rho^2+2\mu+1)^2-4\rho^2]}} \tag{9.20}$$

where $\alpha=\pi V/L\omega$ is the speed parameter, β is the initial bounce of vehicle (defined below), $\rho=\omega/\omega_s$ is the frequency ratio, and $\mu=m_s/mL$ is the mass ratio. The initial bounce is defined by

$$\beta=\frac{\text{amplitude of oscillation of vehicle as it enters the span}}{\text{compression of suspension of vehicle under its static weight}}. \tag{9.21}$$

It should be noted that the first term in (9.20) is identical to (9.12) and is the DLF due to speed only. Smith (1969) showed that (9.20) provides a good estimate of the maximum response over a wide range of relevant parameters.

9.3 VIBRATION OF BRIDGES UNDER MOVING LOADS

9.3.1 Railway bridges

For many years the dynamic design of railway bridges was dominated by the problem of 'hammerblow' which consisted of pulsating forces generated by the balance weights on the driving wheels of steam locomotives. The investigations by Inglis (1934) included the effects of hammerblow and from his results an impact factor was derived. In principle this was the ratio of dynamic to static response and was therefore the dynamic load factor (DLF) referred to earlier. It was tacitly assumed that bending moments were magnified by the same ratio. Allowance for dynamic effects was therefore made by multiplying bridge loads by the impact factor and then designing a bridge to carry the static factored loads.

A combination of theoretical analysis, observations and experience have shown that impact factors decrease with span. Design codes often present impact factors as formulae or curves related to span (Lee, 1973). For example, the United Kingdom code (BS 5400, 1978) specifies a dynamic factor ranging from 2.0 for short spans (<3.6 m) to 1.0 for long spans (>67 m) by which the static bending moments must be multiplied.

Diesel and electric traction have replaced steam locomotives in most countries and therefore 'hammerblow' is no longer a significant problem. However, speeds have gone up and dynamic effects are largely due to irregularities in the track.

More careful attention has to be paid to suspension bridges that are intended to carry railway traffic because of their considerable flexibility. Hirai and Ito (1967) developed a dynamic analysis for suspension bridges which took account of the distributed forces under a locomotive, the mass effect, hammerblow, springing, rail joints and wave propagation. Their calculations, together with experimental results, enabled them to derive values of parameters to be used in the design of suspension bridges to take rail traffic.

9.3.2 Highway bridges

The factors affecting the vibration of highway bridges under the action of rapidly moving heavy vehicles include speed, bounce on suspension and roughness of road surfaces. There has been considerable research into methods of analysis for various kinds of bridges under moving vehicles. Much work has been done at the University of Illinois (Veletsos and Huang, 1970) to extend the early theoretical work of Inglis and Biggs. Studies of the performance of various kinds of bridges under highway traffic have included: multi-span bridges by Nieto-Ramirez and Veletsos (1966) and by Kishan and Traill-Nash (1973); multi-girder and slab bridges by Walker (1968) and Smith (1973b); cantilever bridges by Wen and Toridis (1962); and horizontally curved bridges by Vashi, Schelling and Heins (1970) and Christiano and Culver (1969).

The great amount of research into the vibration of highway bridges does not appear to be prompted by any serious design problems. If we consider the effect of vibration on maximum stresses, we should note that the dynamic load factor increases with speed. However, it is known that the mean spacing of vehicles also increases with speed (Salter, 1974) according to the formula

$$S = S_0 V_f / (V_f - V) \tag{9.22}$$

where S_0 is the minimum spacing of stationary vehicles, V is the mean speed of traffic, and V_f is the free speed. Therefore, since bridge girders are designed to carry the weight of closely spaced stationary vehicles, it is evident that the increase in dynamic load factor with speed is easily compensated by the increased spacing between vehicles when in motion. At high speed there would usually be a maximum of one vehicle per lane on a short span bridge. On the other hand, it has been suggested by Schilling (1982) that dynamic load factors should be applied to stresses under isolated vehicles for fatigue design.

It is possible that vibrations of a highway bridge could be disturbing to people walking or standing on the footpath. Human response to vibration is considered in Chapter 11. However, calculations by Blanchard, Davies and Smith (1977) indicated that, under normal conditions, traffic-induced vibrations would not exceed recommended limits. However, the presence of sharp discontinuities in the roadway surface could induce unpleasant vibrations; but this would be an unusual situation.

9.4 HUMAN-INDUCED VIBRATION

9.4.1 Pedestrian loading on footbridges

The custom of troops breaking step when marching over a bridge may be traced to the collapse of a cast iron bridge at Broughton in 1831 under the resonance effect of 60 soldiers (Tilly, Cullington and Eyre, 1984). Until recently there has been little guidance in design codes relating to vibrations caused by pedestrians. Consequently, there have been isolated examples of footbridges that were found to be too lively when built and required remedial action in the form of additional damping (Brown, 1977).

Pedestrian loading has been determined with the aid of an orthopaedic 'gait' machine (Skorecki, 1966). Force–time curves giving the vertical component of typical foot impacts are shown in Fig. 9.4. Two peaks occur characteristically under 'heel strike' and 'toe off'. The mechanics of walking is very complicated but a simple simulation was proposed by Blanchard, Davies and Smith (1977), who assumed that the worst condition occurs when a pedestrian walks in resonance with the natural frequency of a bridge with a stride length of 0.9 m. The pacing frequency of normal walking lies between about 1.5 and 3.0 Hz, whereas frequencies above 3.0 Hz are representative of running or jogging. It is difficult to excite a footbridge with a frequency above 4 Hz.

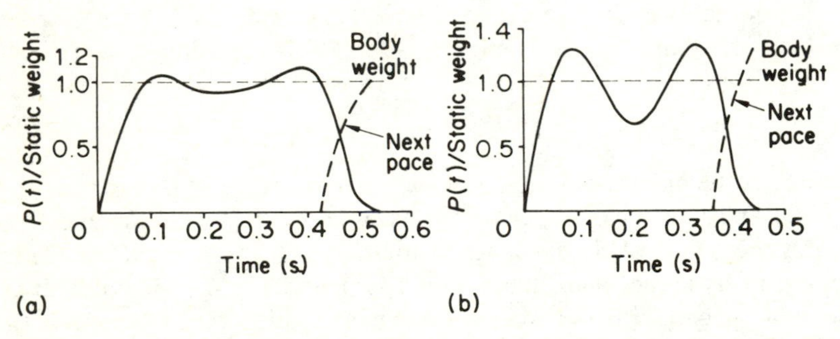

Fig. 9.4 Force–time curves for walking (vertical component). (a) Normal walk; (b) fast walk.

Simulated pedestrian loading of a footbridge is shown in Fig. 9.5. A simple solution may be obtained by ignoring all modes above the fundamental. Therefore, the displacement at any point is given by

$$v(x, t) = Y_1(t)\phi_1(x). \tag{9.23}$$

The pedestrian forcing function is represented by a series of point loads, with a force–time curve of the form shown in Fig. 9.4, applied at successive time intervals equal to the period of vibration, T, as shown in Fig. 9.5(b). Thus,

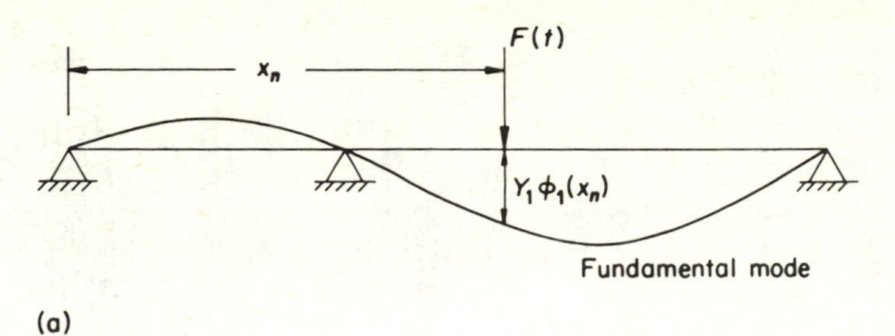

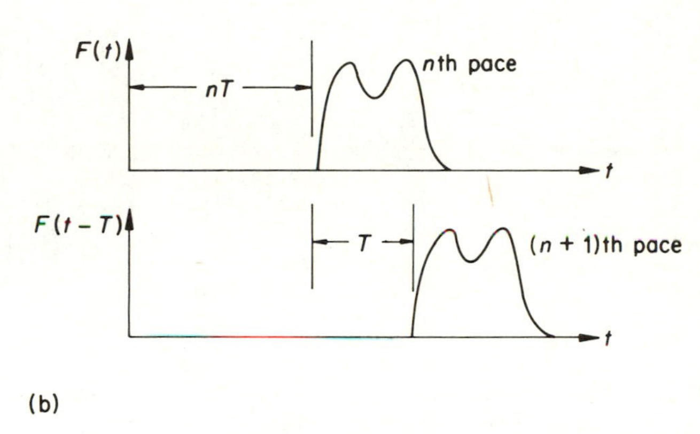

Fig. 9.5 Simulated pedestrian loading of a footbridge. (a) Deflection curve of footbridge; (b) pedestrian loading function.

from Equation (3.82), the equation of motion is given by

$$\ddot{Y}_1 + \omega_1^2 Y_1 = \frac{1}{M_1} \sum_{n=1}^{N} \phi_1(x_n) F(t - nT) \tag{9.24}$$

where N is the number of paces required to cross the bridge. The natural frequency ω_1, mode shape $\phi_1(x)$, and generalized mass M_1 of the fundamental mode may be obtained by any of the methods described in Chapters 3 or 4. Numerical integration is required to obtain a solution to (9.24). An example of the calculated and observed vibration of a slender steel box girder footbridge under pedestrian loading is shown in Fig. 9.6.

Blanchard, Davies and Smith (1977) used the above method to analyse footbridges with different configurations. It was found that the dynamic deflection could be expressed conveniently in the form

$$v_{max} = v_{st} K \psi \tag{9.25}$$

where v_{st} is the static deflection under the weight of a pedestrian at the point

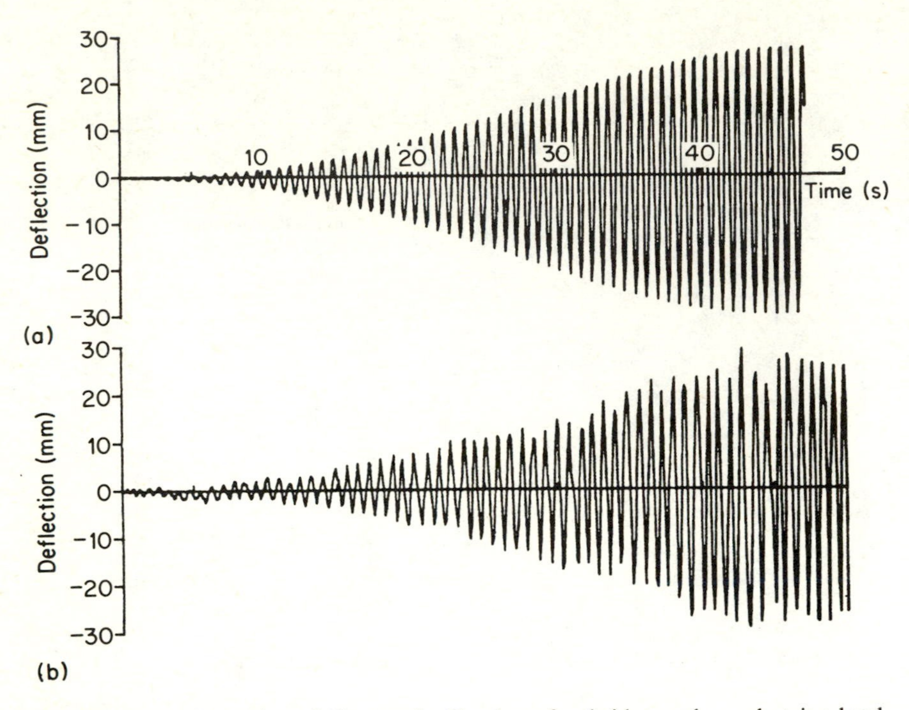

Fig. 9.6 Dynamic motion of Clapton-in-Gordano footbridge under pedestrian loading. (a) Calculated deflection; (b) Observed deflection.

of greatest deflection, K is the configuration factor for the type of structure as given in Table 9.2; and ψ is the dynamic response factor.

Blanchard, Davies and Smith (1977) included bridge damping in the equations of motion (see §3.3) and showed that the dynamic response factor varied with span and damping as shown in Fig. 9.7.

A simplified loading function was proposed by Blanchard, Davies and Smith (1977) to permit analysis of bridges of more general configuration. It consists of a pulsating force of amplitude F_0 moving along the span with a velocity of $0.9\,f$ and in resonance with the bridge. Therefore, the equation of motion will be

$$\ddot{Y}_1 + \omega_1^2 Y_1 = (F_0/M_1)\phi(0.9\,ft)\sin\omega_1 t. \tag{9.26}$$

It was found that the amplitude F_0 should be approximately 25% of the static weight of a pedestrian to produce the same response as the rigorous method. A further simplification is available if the mode shape, $\phi(x)$, of the main span is assumed to be a half sine curve, in which case the solution given in §9.2.3 may be used since $\phi(0.9\,ft) = \sin(0.9\,\pi ft/L)$ and $0.9\,f$ may be substituted for the speed V.

Table 9.2 Configuration factor

Configuration	*a/L*	*K*
L	–	1.0
a *L*	1.0	0.7
	0.8	0.9
	<0.6	1.0
a *L* *a*	1.0	0.6
	0.8	0.8
	<0.6	0.9

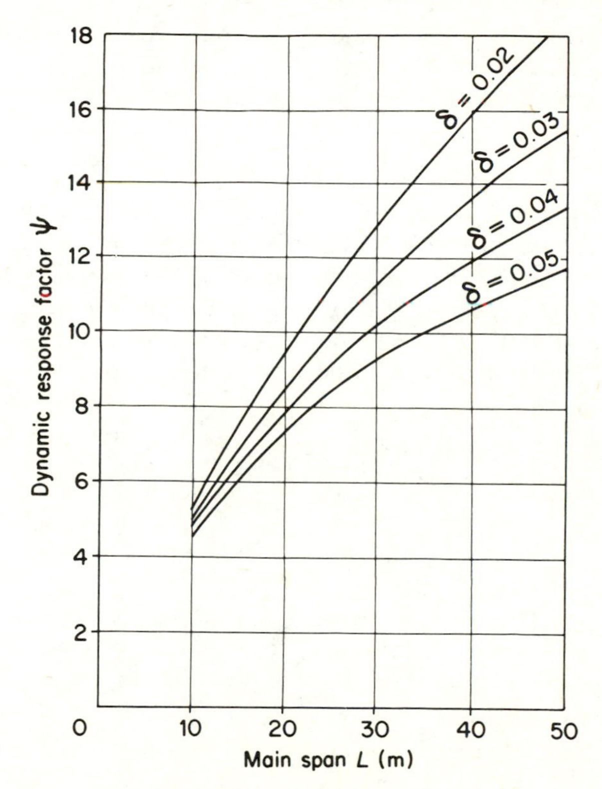

Fig. 9.7 Dynamic response factor.

The most important criterion for dynamic design of footbridges is that they should not vibrate to an extent that users would be disturbed or alarmed. This is a human response criterion, a topic which will be discussed in detail in Chapter 11. Meanwhile it is sufficient to note that vertical acceleration is a useful measure of human response to bridge vibration. The UK bridge code (BS 5400, 1978) recommends a maximum acceleration of footbridge decks of $\pm\frac{1}{2}\sqrt{f}\,\mathrm{m/s^2}$ when one pedestrian walks over the main span in step with the natural frequency, f. This 'one-pedestrian' test was calibrated against some real bridges which were known to be only just acceptable. Analysis by the Blanchard, Davies and Smith (1977) method is recommended.

It has been confirmed by Matsumoto, Nishioka, Shiojiri and Matsuzake (1978) and Wheeler (1982) that the 'one–pedestrian' test is realistic. It is possible for two pedestrians to walk in step with the natural frequency of a footbridge with the assistance of a metronome and amplitudes are approximately twice the single pedestrian case (Tilly, Cullington and Eyre, 1984). However, such a condition is difficult to maintain.

The well-known problem of groups of vandals deliberately exciting footbridges in resonance is perhaps less serious than it may seem. Reports of damage due to this form of loading are rare. Wheeler (1982) found experimentally that it was difficult for three people together to produce a greater motion than the 'one-pedestrian' case.

EXAMPLE 9.2

A simply supported footbridge of 18 m span has a total mass of 12 600 kg and a flexural stiffness (EI) of $3.0 \times 10^8\ \mathrm{N\,m^2}$. Determine the maximum amplitude of vibration and vertical acceleration caused by a pedestrian walking across in step with the frequency. It may be assumed that the pedestrian weighs 700 N, has a stride of 0.9 m and produces an effective pulsating force of 180 N amplitude. The damping in the bridge is given by $\delta = 0.05$.

Solution

The natural frequency of the bridge is

$$f = \frac{\pi}{2 \times 18^2}\sqrt{\left(\frac{3 \times 10^8}{12\,600/18}\right)} = 3.17\ \text{Hz}.$$

The static deflection with the pedestrian at midspan is

$$v_{st} = \frac{700 \times 18^3}{48 \times 3 \times 10^8} = 0.2835 \times 10^{-3}\ \text{m}.$$

Therefore, from (9.25)

$$v_{max} = 0.2835 \times 10^{-3} \times 1.0 \times 6.8 = 1.93 \times 10^{-3}\ \text{m}.$$

Maximum acceleration is given by

$$a_{max} = \omega^2 v_{max} = 4\pi^2 \times 3.2^2 \times 1.93 \times 10^{-3} = 0.78 \text{ m/s}^2.$$

Notice that this is less than the recommended limit of $\frac{1}{2}\sqrt{f} = 0.89$ m/s^2.

Using the simplified moving pulsating force method we may employ Equation (9.15). However, the static deflection should now be calculated under the pulsating amplitude of 180 N. Therefore

$$v_{st} = 0.0729 \times 10^{-3} \text{ m}.$$

Hence, using (9.15) we obtain

$$(v_c)_{max} = 0.0729 \times 10^{-3} \times \frac{2\pi \times 3.2 \times 18}{\pi \times 0.9 \times 3.2} = 2.92 \times 10^{-3} \text{ m}.$$

This calculation neglects damping and therefore it should be expected that the maximum response would be overestimated.

9.4.2 Foot impact on light floors

It is common experience that perceptible vibrations can be induced in lightweight flooring systems by a variety of normal activities. Polonsek (1970) considered the effects of normal walking, dancing, children playing, domestic appliances, door slam and other sources of vibration. Of these, a single person walking was not only the most frequently occurring but also one of the activities that gave rise to the greatest nuisance. It is not a question of overstress of the structural system but a problem of vibrations being perceptible or annoying to other seated persons in the vicinity. Especially susceptible are timber joist floors (Polonsek, 1970) and long-span lightweight concrete floors supported on steel joists (Lenzen 1966; Pernica and Allen, 1982).

Human response to vibration is considered in detail in Chapter 11. In the case of floors, vibration occurs in short bursts at each footstep consisting of peak amplitude under heel strike followed by rapidly decaying motion due to damping. It has been established that, with this kind of transient vibration, human response depends not only on frequency and amplitude of motion, but also on the rate of decay and hence damping (Wiss and Parmelee, 1974). A guide to limits of floor acceleration are given in an appendix to the Canadian code for steel structures (CSA, 1984) and are shown in Fig. 9.8. These are based on the work of Allen and Rainer (1976) and apply to normal building occupancies such as schools, offices and residential. A more stringent curve is appropriate for continuous vibration such as might be caused by domestic appliances.

For the purpose of determining the dynamic response, Lenzen and Murray (1969) recommended the 'heel drop' test. This consists of a person of average weight rising up on his toes and then dropping suddenly on his heels near

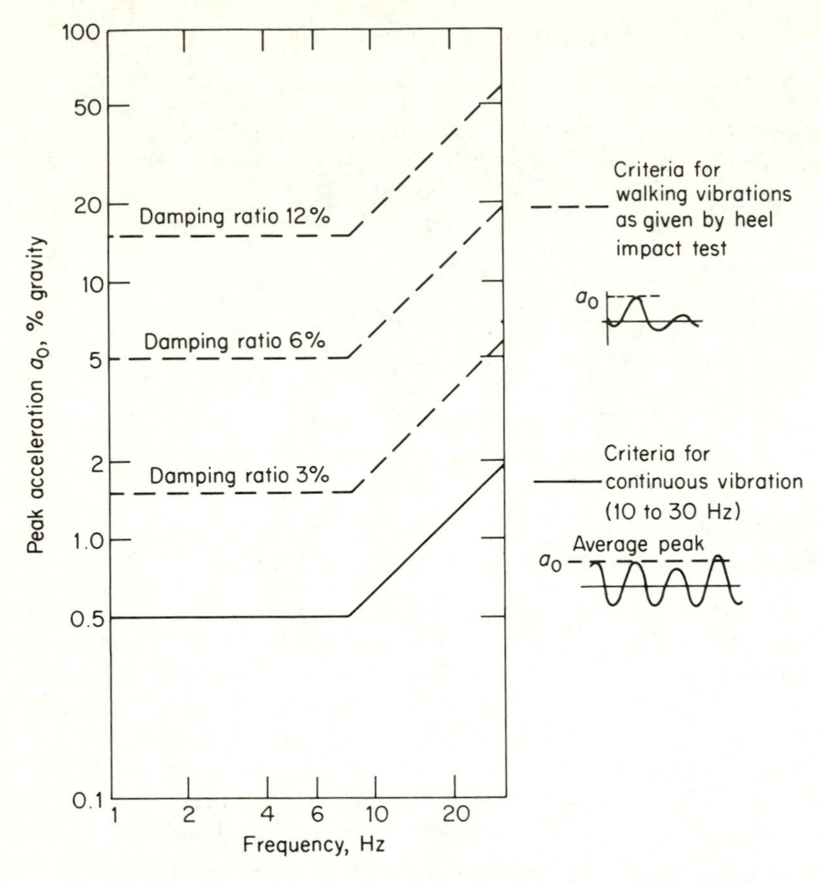

Fig. 9.8 Recommended vibration limits for light floors (CSA, 1984).

midspan. A typical force–time curve is shown in Fig. 9.9(a) and the ensuing dynamic response is shown in Fig. 9.9(b). This is considered to be representative of foot impact when walking. Simple calculations may be performed by treating the floor as an equivalent one-degree-of-freedom system and applying a triangular impulse varying from 2.7 kN to zero in 1/20 second as shown. The Duhamel integral, given by Equation (2.32), could be used to obtain an explicit response history. However, an even simpler formula was obtained by Allen and Rainer who assumed that for floors with a frequency of less than 10 Hz the velocity of the floor under the impact could be calculated by momentum considerations in the form given by Equation (2.30). The impulse would then be the integral under the curve in Fig. 9.9(a), i.e. 70 N s, and the velocity after t seconds, using (2.10), is given by

$$\dot{u} = \frac{1}{M} \exp(-2\pi f \xi t) \cos 2\pi f t \tag{9.27}$$

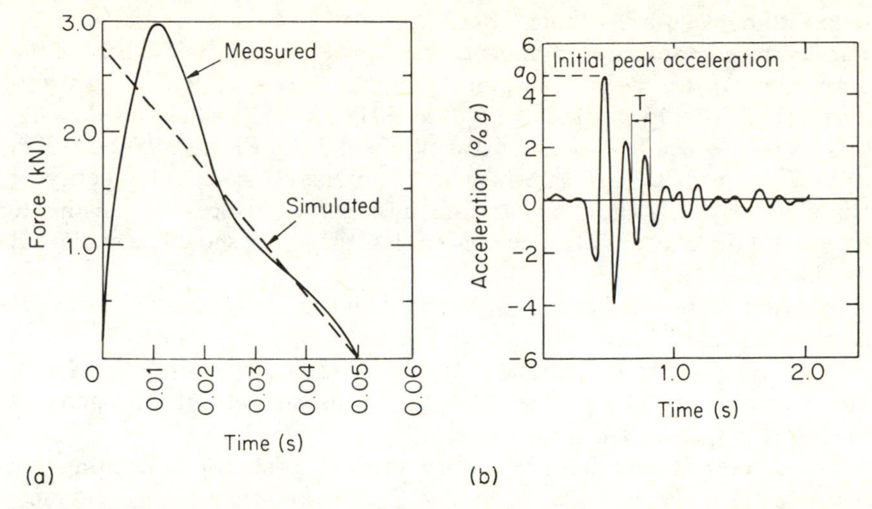

Fig. 9.9 Heel drop test. (a) Average force v time curve for heel impact (from Lenzen and Murray, 1969); (b) Typical floor response to heel impact (from Allen and Rainer, 1976).

where I is the impulse and M is the equivalent mass of the floor. Thus the maximum acceleration is approximately given by

$$a_0 = (0.9)\, 2\pi f I/M \tag{9.28}$$

where the factor of 0.9 allows for the loss of amplitude due to damping in the first half cycle. On the basis of heel drop tests on 42 floors the equivalent mass was recommended to be taken as 0.4 × the distributed mass of the floor. The Canadian code gives more detailed guidance on the calculation of equivalent mass for floors with different supporting systems.

The above calculations were verified with observations on long-span floors (greater than 7 m). Whale (1983) has reviewed experiments on timber floors of shorter span. Existing static deflection limitations were found to give poor control over dynamic performance of floors. The work of Onysko and Bellosillo (1978) indicated that a deflection/span limit of 1/615 under a distributed load of 2 kN/m^2 would be required to achieve satisfactory stiffness.

9.4.3 Dynamic crowd loading

In most design codes crowd loading is covered by specifying a uniformly distributed live load equivalent to people very tightly packed together, e.g. 5 kN/m^2 in the UK code. However, substantial dynamic effects occur in dance halls, on grandstands and at pop concerts. Irwin (1981a) reported extreme conditions at a pop concert in Edinburgh when crowd compaction close to the stage was resulting in static live load 1.74 times the design value. Furthermore, coordinated jumping of the crowd in time with the beat of the music generated a dynamic response factor of 1.97 at a predominant frequency of 2.5 Hz. Frequencies as high as 5 Hz can be generated by dancing though the frequency is usually between 2 and 3 Hz. Heins and Yoo (1975) investigated a dance hall in which the floor had a natural frequency of approximately 3 Hz and the vibrations during 'rock' dances were distinctly unpleasant. Frequencies of a floor system less than 5 Hz should be avoided if possible.

A similar problem may occur in sports stadia or grandstands when sports fans sway to and fro rhythmically. The Canadian Building Code (NBC, 1985) recommends that the structure of grandstands should be designed for forces due to swaying of 0.3 kN per metre length of seats parallel to the row and half this value perpendicular to each row.

At the present time no satisfactory method exists for calculating the response due to dynamic crowd loading. It is not known, with any accuracy, what loads one person would generate when dancing or jumping and it is not known how many could act in unison for a sustained period of time. Some guidance is offered in the Supplement to the National Building Code of Canada (NBC, 1985) in which it is suggested that the dynamic load due to a rhythmic activity be represented by $\alpha w_p \sin 2\pi ft$, where w_p is the weight per unit area of participants, f is the forcing frequency and α is a dynamic factor which depends on the activity. Values of the dynamic factor ranging from 0.25 for sports events to 1.5 for jumping gymnastic exercises are given. The dynamic response may then be determined using the single-degree-of-freedom formula for forced vibration.

9.4.4 Loads on staircases

Staircases are normally designed to carry the same static live loads as the floors to which they give access. This is nearly always satisfactory for staircases of conventional construction. However, there have been instances of staircases with light or unusual supporting structures, e.g. designed for dramatic architectural effect, that have proved to be susceptible to the dynamic effects of human loading.

It is known that the human body can generate very substantial dynamic overloads chiefly due to heel strike. Energetic walking can give rise to a peak load of up to twice body weight (BW). In the heel drop test of Fig. 9.9(a) the peak load is roughly 4 × BW. Even greater impacts may occur when a person runs up or down stairs. Some experiments were carried out by the author using an orthopaedic force plate. Subjects ran up or down a short flight of steps. Examples of the vertical component of foot impact are shown in Fig. 9.10. When running up the peak load is generated by 'toe off' and is about

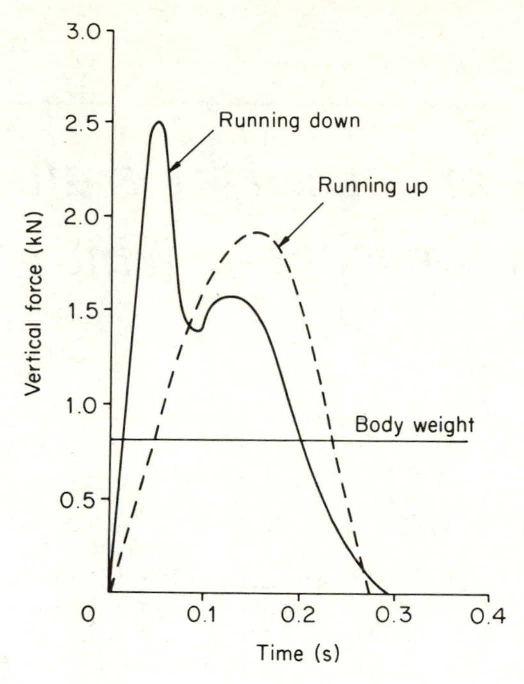

Fig. 9.10 Dynamic loading on staircase.

$3 \times$ BW. When running down the peak load occurs under 'heel strike' and is generally about $4-5 \times$ BW. Evidently dynamic loads of this magnitude should not be ignored. A simple method of dynamic analysis, such as Equation (9.28), could be employed when checking sensitive structures.

10

Vibration caused by traffic, blasting and pile-driving

10.1 INTRODUCTION

Complaints about vibration caused by traffic, pile-driving, blasting and sonic boom are very widespread. For example, a survey by Sando and Batty (1974) indicated that 8% of residents of the United Kingdom were seriously bothered by traffic-induced vibrations. Steffens (1974) reviewed a large number of instances of vibration nuisance from a variety of sources including forge hammers, machinery, pile-driving, blasting and aircraft.

Complaints about vibration nuisance are often accompanied by reports of damage, e.g. cracks, to properties. Steffens (1974) and others have observed that some of the reported damage was pre-existing and was probably caused by other factors such as shrinkage and settlement. Damage to buildings can occur in serious cases, but frequently the direct causes of complaint are disturbance from windows and crockery rattling, startle or alarm at sudden impacts, or householders' fear of damage to their own property. The effects of ground-transmitted vibration are therefore not simple, but they are so extensive that some means of assessment is required.

10.2 GENERATION AND TRANSMISSION OF VIBRATION

10.2.1 Impulsive sources

Forge hammers, pile-drivers and blasting generate impacts on the ground surface. A simplified description of the process is that of an elastic half-space subjected to a point load having an impulsive time variation. This problem was analysed in Chapter 3, in which it was explained that surface waves are propagated outwards from the source of the disturbance. Nearby buildings will experience the waves as a local ground motion (see Fig. 10.1). Expres-

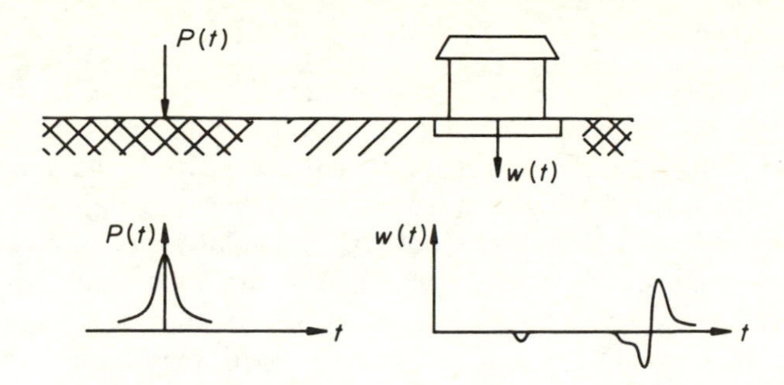

Fig. 10.1 Ground vibration caused by impulsive loading.

sions for estimating the ground motion were given by Equations (3.160) and (3.161), though the input parameters would normally be so uncertain that the predicted response should be regarded as very tentative.

The propagation of vibration caused by blasting has been studied by the US Bureau of Mines (Siskind, Stachura, Stagg and Kopp, 1980) and others (Dowding and Corser, 1981; Medearis, 1978). Blasting operations in opencast mining are performed by detonating a row of charges with very short time delays (25 to 60 ms) between each detonation. Ground motions at a nearby location, e.g. a residence, are related to the distance and the mass of each charge. Medearis (1978) obtained data from 74 commercial blasts including quarrying, strip mining and construction work. By analysis of the results he derived a relationship between the observed peak ground velocity and the distance of a single charge and its mass, as follows:

$$V = 466(L/\sqrt{M})^{-1.34} \tag{10.1}$$

where V is the peak ground velocity in mm/s, being the resultant of motion in three orthogonal directions, L is the distance from the charge in metres, and M is the mass of charge in kg.

The frequencies of ground motion resulting from blasting cover a wide band. Siskind, Stachura, Stagg and Kopp (1980) reported observed frequency ranges from different forms of blasting activity as given in Table 10.1. Dowding and Corser (1981) reported similar experiences.

Table 10.1 Ground vibration frequencies due to blasting

Activity	*Range (Hz)*	*Mean (Hz)*
Surface coal mining	5–30	20
Quarrying	10–40	25
Construction	15–100	40

10.2.2 Traffic vibration

As a heavy vehicle progresses over an irregular road surface, an interacting dynamic force is induced in the tyres and suspension (Trott and Whiffin, 1965). The ensuing dynamic load generates surface waves in the same manner as impulsive loads which were described in the previous section. The differences are that the load is in motion and is an irregular function of time. This problem was analysed by Smith (1973a), who assumed that the irregular forcing function could be treated as a succession of rectangular impulses. The response to each impulse, at a point near the road, may be obtained using Equation (3.161). The total response is the sum of the separate responses taking due account of the fact that the impulses are applied in succession along the path of the vehicle. The numerical results showed that an irregular surface gives rise to significant vibration only at points close to the road. An isolated irregularity such as a ridge or a ramp can give rise to a more serious level of vibration.

Vibrations generated by traffic have been measured experimentally by the Road Research Laboratory (Burt, 1972). The observations were in agreement with the theoretical analysis of Smith in that there should be no significant vibration resulting from heavy lorries travelling over roads in reasonable condition. However, when there are large irregularities, such as pot holes or those caused by poorly backfilled transverse cable or pipe trenches, it is possible to generate annoying vibrations in nearby buildings. The solution is to repair the road surface.

10.2.3 Airborne vibration

It has now been established that the principal mechanism of traffic-induced vibration is low-frequency noise resulting in structural vibration of windows and floors (Martin, 1978). Sound pressure waves from the engines of heavy lorries, with frequencies predominantly below 100 Hz, can excite window panes and the air within a room into vibration, with consequent structural response of walls and floor. However, measured floor vibrations in houses close to heavily trafficked roads were relatively small compared with known human response (see Chapter 11). Therefore, other factors were contributing to the annoyance, particularly psychological ones such as alarm or fear of damage.

Blasting in open-cast mining results in an air pressure pulse generated by the sudden displacement of the rock. The shape of the air pressure pulse is reasonably consistent, an example being shown in Fig. 10.2. It consists of a sharp initial peak followed by a more gradual decay with some suction before returning to normal (Dowding, Fulthorpe and Langan, 1982). Sharper spikes occur if the ground cover is poor, allowing expanding gas to be released. This

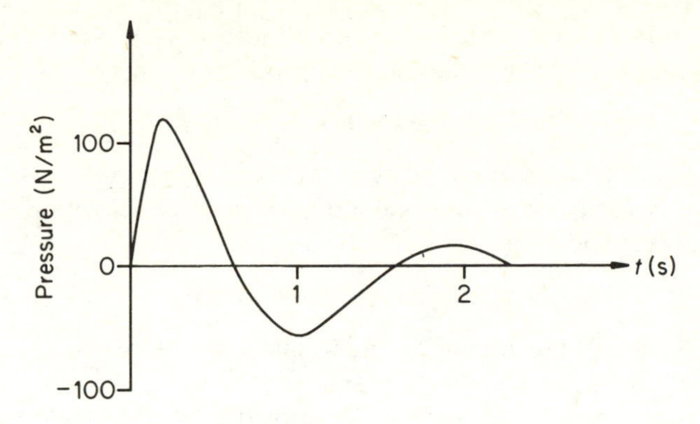

Fig. 10.2 Typical air pressure pulse from blasting.

tends to happen in construction blasts which are often close to the surface. Air blast due to sonic boom tends to be very similar, though the shape varies with the type of aircraft (Clarkson and Mayes, 1972). Peak air pressure varies with distance but 100 N/m^2 is not uncommon for both blasting and sonic boom.

10.2.4 Assessment of vibration intensity

Intensities of ground vibration from the sources described here are notoriously difficult to predict by calculation. If it is necessary to estimate ground vibrations resulting from a planned project, it is generally preferable to make use of measured vibrations in a comparable situation. This procedure was adopted by Bean and Page (1976) during the planning of a large cut and cover road tunnel in a heavily populated location.

Ground-transmitted vibration comes more frequently from sources that did not exist when the affected buildings were designed. Therefore the problem is more often one of legal dispute after the event than one of foresight and prevention. Fortunately, transmitted vibration intensities can be measured accurately with modern instrumentation and sophisticated computer-based equipment is readily available for recording and analysis.

10.3 EFFECTS OF GROUND AND AIRBORNE VIBRATION

10.3.1 Response of buildings

The response of buildings to ground vibration depends on whether they are close to the source or far from it. If they are close (<25 m) Dowding and Corser (1981) suggest that foundations and buried structures will be subjected to ground strain. This is caused by the amplitude of the ground wave varying over the dimension of the building, thus producing a net strain given by

$$\varepsilon = V/C \tag{10.2}$$

where V is the peak particle velocity of the ground and C is the wave propagation velocity. V is the resultant of measured velocities along three orthogonal axes and is given by

$$V = |\sqrt{[V_x^2(t) + V_y^2(t) + V_z^2(t)]}|_{\max}. \tag{10.3}$$

Alternatively, if only the amplitude and frequency of vibration are known, the velocity may be estimated from $V = 2\pi fA$.

Buildings that are far from the source of disturbance (>65 m) respond freely to the support motion, which may be assumed to be uniform over the base of the structure. Medearis (1978) used the response-spectrum method of analysis to determine the dynamic response of single- or two-storey residences to input support motion from blasting. He found that the amplification factor varied with frequency, as shown in Fig. 10.3. These results were in agreement with observations of real buildings by Siskind, Stachura, Stagg and Kopp (1980), who distinguished between midwall bending and racking of an entire building. Racking tended to occur at frequencies below 12 Hz and midwall bending at higher frequencies. The motions are illustrated in Fig. 10.4.

Structural damage may occur under racking deformation with diagonal cracks at the corners of window frames sometimes being observed.

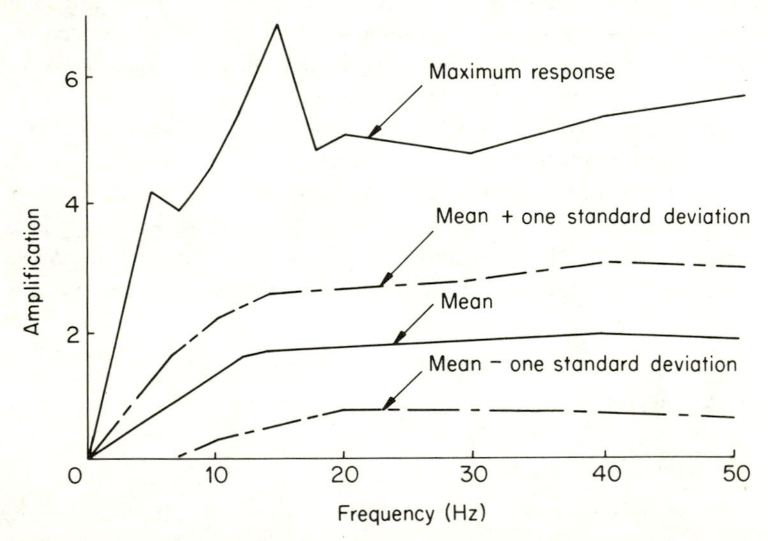

Fig. 10.3 Dynamic amplification of single-degree-of-freedom system subjected to blasting ground motions (after Medearis, 1978).

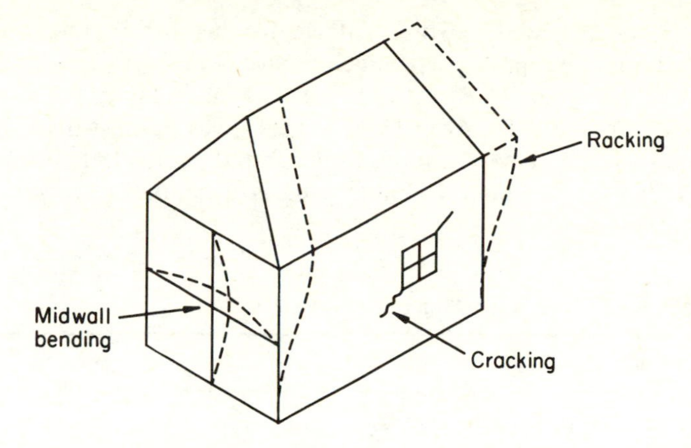

Fig. 10.4 Deformation due to blasting.

Dynamic response due to airblast may be calculated by the methods of Chapters 2 and 3. Siskind, Stachura, Stagg and Kopp (1980) obtained empirical relations between airblast pressures and the ensuing response. They suggested that damage was unlikely with airblast pressures of less than 200 N/m^2.

Peak particle velocity, as given by Equation (10.3), has been established as being the best measure of the damage potential of ground vibrations (Nicholls, Johnson and Duvall, 1971). The US Bureau of Mines originally suggested a limit of 50 mm/s as being safe for residences but in fact problems occur below this level and Siskind, Stagg, Kopp and Dowding (1980) now suggest 12.5 mm/s as a safe limit for older properties. Splittgerber (1978) made the following observations of shock and blast vibration damage in the 3 to 100 Hz range:

threshold damage	3–5 mm/s
minor damage	5–30 mm/s
major damage	>100 mm/s.

'Minor damage' represents cracking of plaster and other superficial architectural damage requiring little more than routine maintenance, whereas 'major damage' could entail a certain amount of building work for repair.

Historic structures require a greater degree of control over potential ground vibration because of their sensitivity and value. Crockett (1965) reported a survey of 40 cathedrals and abbeys indicating that where there was a heavily trafficked road adjacent to such a building it tended to lean towards the road. Where there was no nearby road there was no lean. Konon and

Schuring (1985) reviewed existing criteria for historic buildings and recommended a limiting peak particle velocity of 6–12.5 mm/s.

A good general guide to effects of vibrations on buildings is the German provisional standard DIN 4150 (1975). This gives guide values for peak particle velocity at foundation level, under impulsive loading, below which significant damage is unlikely. Table 10.2 is a simplified version of the recommendations.

Table 10.2 Guide values for peak particle velocity

Type of building	V_{max} *(mm/s)*
Residential buildings and similar structures in good repair	8
Well-braced structures made of heavy structural elements and frames: well-maintained	30
Other structures and historic structures	4

10.3.2 Human response

Human response will be covered in detail in the next chapter. Meanwhile, it is sufficient to say that humans are generally more sensitive than structures. Therefore the limits for structural damage should apply to infrequent events. It has been found that in residential areas there are complaints if frequently occurring vibrations exceed the threshold of perception (Okada, Yamashita and Fukuda, 1971).

It has already been noted that traffic-induced vibration results principally from low-frequency noise generated by vehicles. Furthermore, complaints about the vibration do not correlate very well with measured vibration because of other factors such as fear of damage and annoyance caused by rattling objects. A study by Watts (1984) showed that there was better correlation between subjective rating of vibration nuisance and 18-hour noise exposure measures. The measurement and prediction of noise levels due to traffic is well established (HMSO, 1975) and therefore this is a potentially useful means of assessment.

10.3.3 Sensitive equipment

It is well known that sensitive experimental apparatus such as balances and galvanometers can be affected by quite small vibrations and therefore have to be sited away from possible disturbances. There are many examples of even more sensitive equipment, including electron microscopes, mass spectrometers, astronomical telescopes, microelectronic manufacturing plant, and testing precision accelerometers. Some examples of the sensitivity of laboratory instruments were obtained by Whiffin and Leonard (1971). The smallest features of some microelectronic circuits can be as small as 2×10^{-7} m, therefore requiring that the maximum amplitude of vibration of the manufacturing equipment should not exceed one tenth of this. Similarly, precision accelerometers may require an environment of less than $g \times 10^{-5}$ for testing purposes. It is evident that vibration isolation of the equipment is required. This normally consists of a heavy concrete block mounted on compressed air springs. Vibration isolation was discussed in Chapter 2.

11

Human response to vibration

11.1 INTRODUCTION

Human beings are surprisingly sensitive to vibration and are often disturbed by intensities that are well below those required to overstress the structures they inhabit. Therefore, in the design of structures, human response to vibrations caused by frequent sources of dynamic loading, e.g. wind, should be regarded as a serviceability limit state. Extreme loading, e.g. due to earthquakes, would be considered as an ultimate limit state, and dynamic stresses should be kept below the levels likely to cause collapse.

Human response to vibration varies greatly with the particular circumstances. One extreme is the pleasure obtained from violent fairground rides, while the other extreme is the near panic that can be caused by the much gentler motion of long-span bridges (Irwin, 1983). It is obvious that the response depends on expectations. In the first example violent motion is not only expected but paid for! In the second case rigidity and firmness is often expected and the sensation of motion may undermine confidence in the structure. There are many examples in between these extremes where vibration may be annoying, disturbing or detrimental to working efficiency. Examples include the swaying motion of tall buildings in wind, floor vibrations caused by machinery, and vibration of bridges under wind or traffic. Vibrations are most commonly experienced in various forms of transport, such as passenger cars, trains and aircraft.

The earliest systematic study of human response to vibration was carried out by Reiher and Meister (1931). They used a vibrating platform that was able to oscillate vertically or horizontally with a wide range of frequencies and amplitudes. Subjects either stood or reclined on the platform and were exposed to a constant-amplitude motion for about 10 minutes. They were then asked to give their subjective opinions on the intensity of the vibration by classifying it in one of the following categories: (1) imperceptible; (2) just perceptible; (3) clearly perceptible; (4) annoying; (5) unpleasant; (6) intolerable. By plotting the amplitude against frequency, using different notation for each category, they were able to draw boundaries between the six levels of comfort. Their results for standing subjects exposed to vertical vibration are shown in Fig. 11.1. It is evident that sensitivity, as measured by amplitude, is a rapidly decreasing function of frequency.

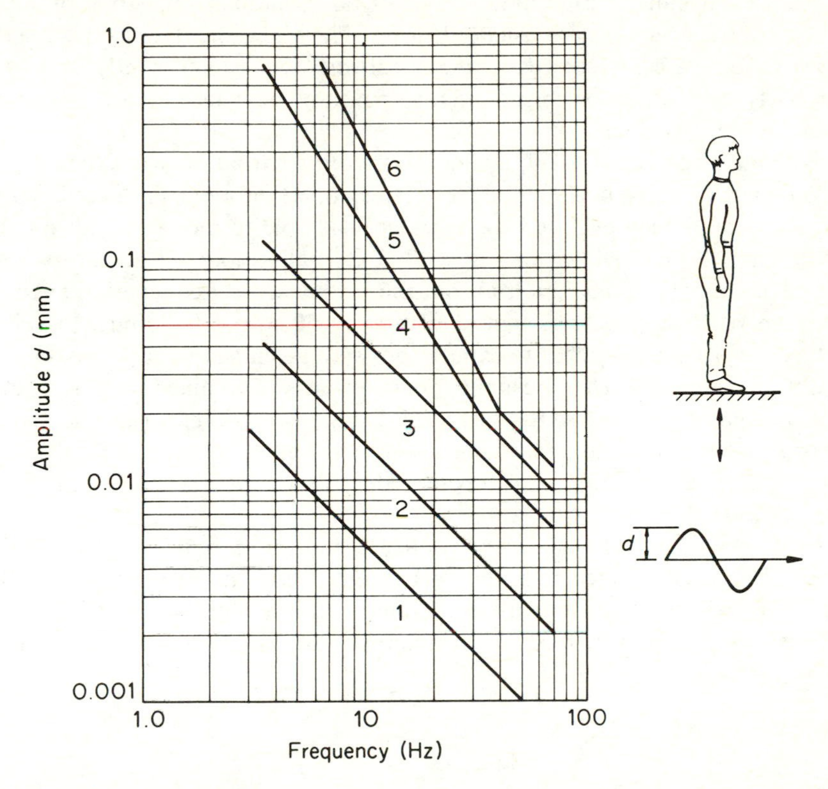

Fig. 11.1 Equal sensation contours for standing position with vertical vibration (after Reiher and Meister, 1934).

Reiher and Meister exposed subjects to vertical, transverse and longitudinal vibration separately when standing or lying down. Thus they obtained six sets of results similar to those shown in Fig. 11.1. They found that transverse or longitudinal horizontal motion was less disturbing than vertical vibration for standing subjects. The difference in sensation had the effect of moving the boundaries up by approximately one level. They found that, when reclining, subjects tended to be more tolerant of the motion. The curves were further refined by Meister (1935).

The growth of high-speed transport, and aircraft in particular, prompted a great amount of research into human response to vibration (reviews by Goldman and von Gierke, 1961; van Eldik Thieme, 1961; Guignard and Guignard, 1970). However, most of the data apply to seated subjects, often with armrests or other restraints, being exposed to prolonged vibration in the 1 Hz to 100 Hz frequency range. Occupants of buildings and other civil engineering structures are generally exposed to a significantly different vibration environment. First, frequencies of tall buildings, long-span bridges and offshore structures are usually below 1 Hz where the data from vehicle research is scarce. Secondly, peak vibrations occur infrequently and for relatively brief durations. And finally, psychological factors are often of overriding importance.

Human response to low-frequency horizontal vibration was studied by Chen and Robertson (1972) using a horizontal motion simulator. This consisted of a small enclosed room which was able to move on rails along orthogonal axes as shown in Fig. 11.2. Externally mounted actuators imposed simple harmonic motion on the room with frequencies ranging between 0.067 and 0.2 Hz, typical of 40- or 50-storey buildings. Initially subjects were not told what to expect and were given simple tasks involving walking or standing. The threshold of perception was obtained when subjects commented on peculiar sensations. Other subjects were informed about the purpose of the test and some others were subjected to repeat tests. It was found that the threshold of perception decreased when subjects had their attention drawn to the vibration.

Further studies of human response to the motion of buildings, and other large structures, were carried out by Hansen, Reed and Vanmarcke (1973) and Irwin (1978). The latter used a similar motion simulator to Chen and Robertson but later refined it to include rotational or yaw components (Irwin, 1981b).

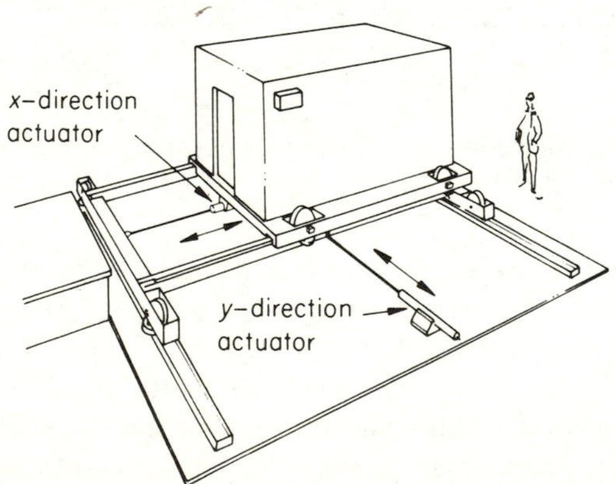

Fig. 11.2 Horizontal motion simulator (after Chen and Robertson, 1972).

11.2 FACTORS AFFECTING HUMAN RESPONSE TO VIBRATION

11.2.1 The mechanics of vibration sense

The human body possesses a variety of apparatus for detecting vibration and movement. The four most important mechanisms of motion sense are as follows: (1) tactile sense, (2) vestibular system or inner ear, (3) deep sensibility, i.e. the stomach and other internal organs, and, (4) visual sense.

The tactile sense of human beings is perhaps the most important mechanism for the detection of vibration. Perception occurs as a result of the distortion of sensory receptors in the tissue underlying the skin. The soles of the feet and the buttocks provide substantially reliable motion sense for whole-body vibration in the frequency range from 3 to 100 Hz.

The motion-sensing apparatus of the inner ear is usually referred to as the vestibular system. This apparatus is shown diagrammatically in Fig. 11.3. The semicircular canals are sensitive to angular accelerations. Fluid rotates in the canals and deflects hair-like cells in proportion to the acceleration. The otolith organ consists of lumps of calcium carbonate surrounded by gelatinous material and suspended between sensory receptors. It is thought to behave like a seismic accelerometer and detects translational motion. The vestibular system induces reliable impressions of dynamic motion in the frequency range between 1 and 3 Hz (Walsh, 1960).

Deep sensibility is the terminology used to describe the sensory mechanism of the stomach, the intestines and other internal viscera. Internal receptors detect the relative movements between various visceral organs but the information conveyed to the brain is not always reliable. However, deep

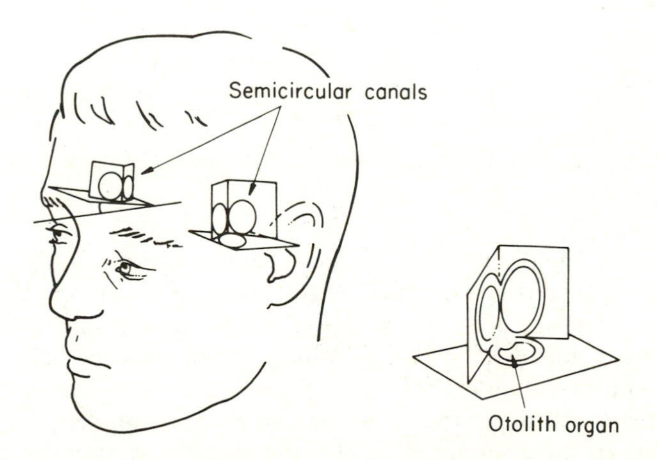

Fig. 11.3 The vestibular system.

sensitivity plays a significant part in determining the discomfort of a human being undergoing vibration, especially at low frequencies. Motion sickness arises partly as a result of prolonged oscillation of the internal viscera.

Vision plays a very important part in the general sense of equilibrium of a human being and without it this sense is seriously impaired. As far as vibration is concerned visual impressions are of importance at low frequencies when the amplitude of oscillation is generally large. In this case relative movements may be observed without necessarily being felt and it is known that disturbing visual impressions help to induce motion sickness.

11.2.2 The human body as a mechanical system

It is useful to think of the body as a mechanical system consisting of lumped masses, springs and damping elements. We can then study the dynamic response of the human body when subjected to vibration and consider the effects of frequency and posture. An example of an appropriate mechanical model is shown in Fig. 11.4 (Goldman and von Gierke, 1961).

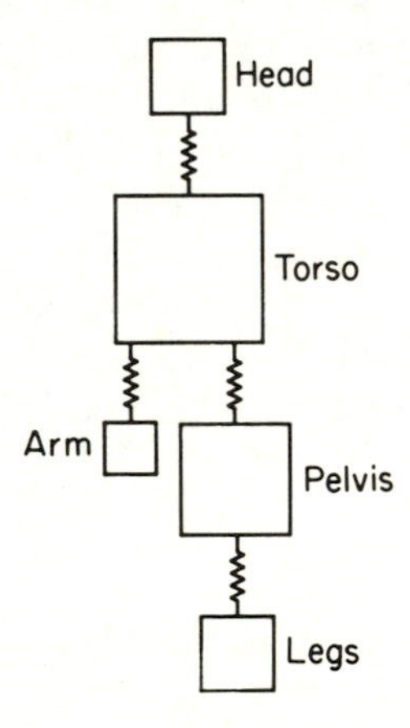

Fig. 11.4 The human body as a mechanical system.

It has been found that the frequency range between 2 and 30 Hz is particularly uncomfortable because of the existence of important resonant frequencies of major body parts. Resonant frequencies are indicated by transmissibility tests in which a subject is seated or standing on a vibrating platform. Transmissibility is the ratio of motion of a specific body part to motion of the platform. The most significant resonant frequencies of a standing subject are given in Table 11.1, together with a brief description of the nature of the motion.

Table 11.1 Human body resonant frequencies

Frequency (Hz)	*Nature of motion*
2	No resonance. Body acts as a unit mass
3–3.5	Oscillation of abdominal viscera coupled with air oscillation of the mouth and chest
5–7	In-phase motion of torso and pelvis
10–14	Upper torso with a forward nodding motion of the upper spine
20–30	Relative movement between head and shoulders
60–90	Eyeball resonance

The resonances are significant, especially those below 10 Hz, because the relative displacements of parts of the body contribute to human discomfort. When subjected to horizontal vibration, the human body is much more flexible, with bending of the neck and spine in a swaying mode. Various forms of restraint tend to suppress the lower-frequency resonances and are therefore beneficial in terms of human comfort.

11.2.3 Effect of amplitude and frequency

We have already noted that Reiher and Meister found that human response varied with amplitude and frequency. The results were plotted on logarithmic scales for the simple reason that the human sensory system tends to respond logarithmically to various forms of input. Mack, Stevens and Stevens (1960) showed that sensation grows as a power function of stimulus. The best example of this is human response to sound. The loudest noise that the human ear can tolerate is roughly 10^{14} times as intense (sound pressure level) as the feeblest that can be heard. Yet it is difficult to distinguish between sounds differing by a factor of as much as two. Think about the noise made by a motorcycle accelerating from the traffic lights. Surprisingly, two motorbikes do not seem to be twice as painful. This is why noise is interpreted on the decibel scale, which is logarithmic.

Response to frequency is also logarithmic. The well-known system of octave bands in music is based upon ranges of frequency going up in factors of two. For example, middle C is 256 Hz and the C above is 512 Hz, spanning eight notes which sound natural to the human ear. Human response to vibration follows the same law, though the eight octave bands between 0.063 Hz and 16 Hz are the most important for large civil engineering structures.

11.2.4 Measures of vibration exposure

Efforts have been made to define suitable measures of vibration exposure. Postlethwaite (1944) suggested a unit called the 'trem' which was supposed to be analogous to units used in human response to sound. Displacement, velocity, acceleration and jerk (time derivative of acceleration), as functions of frequency, have been proposed as suitable parameters. This has sometimes been motivated by the desire to find a parameter that is reasonably independent of frequency. Over small frequency ranges this has sometimes been possible and indeed, at high frequencies, peak velocity has been usefully employed as a measure of vibration nuisance arising from ground-borne vibration.

Most researchers favour acceleration against frequency as the most useful measure of vibration exposure. This has two advantages. One is that the curves are tolerably horizontal with a marked reduction at the major resonant frequency of the human body. The other is that results can be presented in terms of g, the acceleration due to gravity, thus harmonizing units. Some examples of human response curves are shown in Fig. 11.5.

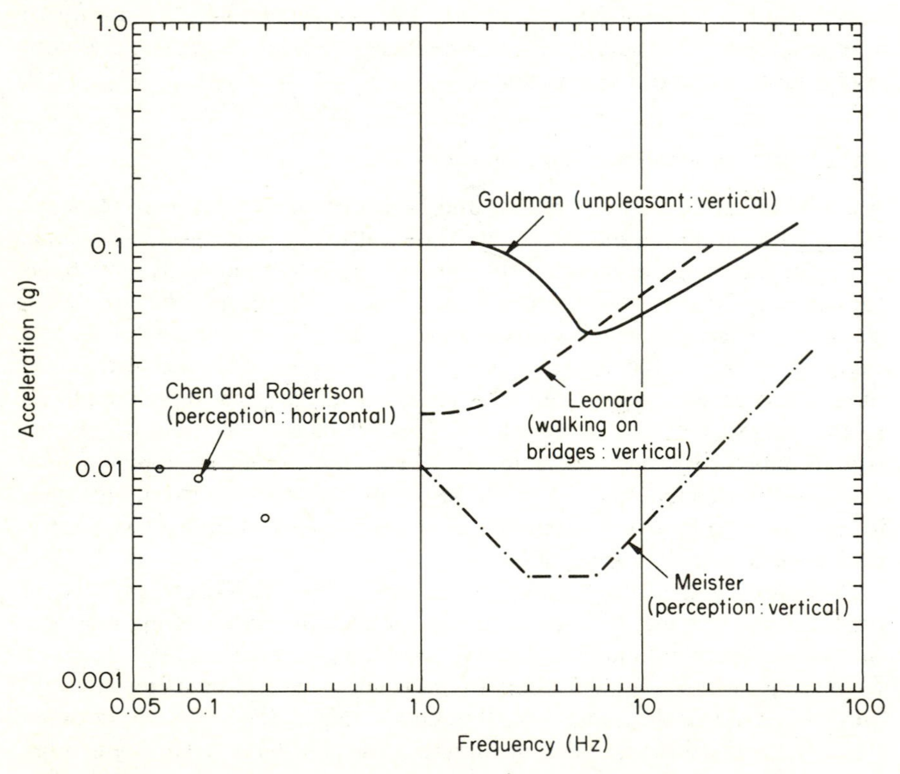

Fig. 11.5 Human tolerance to vibration: acceleration versus frequency diagram.

A further point is that, in reality, vibration is usually random or of complex waveform. Therefore, while it is natural to present constant-amplitude laboratory data as zero to peak displacement or acceleration, it is more generally useful in the form of root mean square amplitude. There is a growing body of data on variable-amplitude human response for which this form of presentation is available.

11.2.5 Other factors affecting human response to vibration

So far we have observed that human response to vibration is affected by amplitude, frequency and direction of motion, and the posture of the subject. Other factors are also very important, including: (1) duration of exposure, (2) waveform, (3) type of activity, and (4) psychological factors.

Much of the early research into human response was concerned with the comfort of aircraft pilots and passengers in transport vehicles. Therefore periods of exposure to vibration tended to be long. Prolonged exposure to vibration can result in inattentiveness or loss of performance, or even motion sickness if the frequency is low. Allen (1974) suggested a time dependence curve based on a review of published low-frequency results together with experience obtained at the Royal Aircraft Establishment, Farnborough. His curve is shown in Fig. 11.6. A similar exposure–time relationship exists in the frequency range above 1.0 Hz based upon data obtained by Holland (1967).

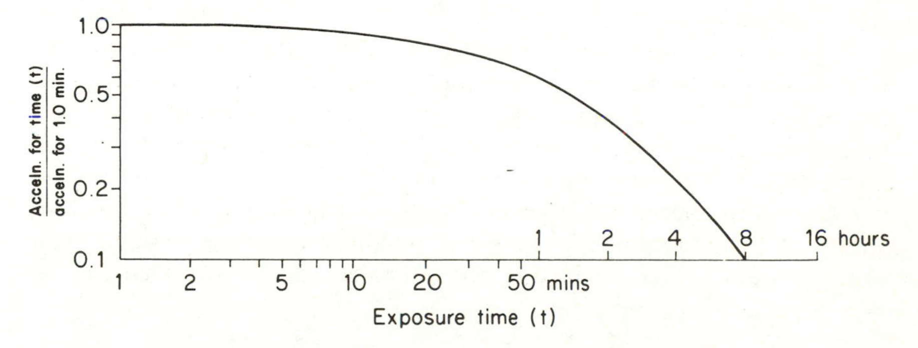

Fig 11.6 Exposure time curve (after Allen, 1974).

Wind-induced oscillation of tall buildings is irregular in waveform and therefore it is necessary to obtain root mean square data averaged over a specified period. Results of human response studies with random vibration correlate reasonably well with the corresponding root mean square amplitude of simple harmonic motion provided that there are not sharp spikes or discontinuities in the waveform. Griffin and Whitham (1980) showed that human response thresholds were reduced if there were sudden spikes in the waveform. Pile-driving produces repeated impulsive shocks characterized by a rapid rise to a peak followed by a slower decline at intervals of about 2 to 10 s depending on the type of hammer used.

It is to be expected that people doing precision work, such as in hospital operating theatres, will be affected by vibration at the threshold of perception. Other activities, such as sleeping, general office work and manufacturing could tolerate higher levels of vibration without seriously impeding the activity. Ashley (1978) compiled a table of weighting factors to take account of diverse activities in different types of vibration exposure. The application of these will be discussed in the next section. Irwin (1983) reported the effects of vibration on activities ranging from computer assembly to fairground rides and demonstrated a very wide range of human response thresholds.

It is almost impossible for people to be entirely objective about their response to vibration. Human comfort is, after all, subjective and is bound to be conditioned by the psychology of each individual. Indeed, Guignard and Guignard (1970) suggested that differences in personality, experience and attitude to the testing environment are more important factors affecting human response than physical variations between individuals.

Human tolerance experiments tend to focus attention on the vibration, which has the effect of reducing the threshold curves. Extraneous noises, such as rattling of windows and rattling of china in crockery cabinets, also focus attention on vibration and lead to a more severe attitude (Onysko and Bellosillo, 1978). Tall buildings sometimes creak audibly in high winds. This can be disconcerting.

People expect large buildings and structures to be rigid. Perceived motion undermines people's feelings of safety and security. This is particularly true of tall buildings and long-span bridges where unexpected motion, however small, can cause alarm. However, after a time people sometimes become accustomed to a certain amount of motion.

Age and experience are important factors. It is therefore essential that studies of human response to motion of buildings should be based on data obtained from a wide population. Considerable scatter of results is to be expected.

11.3 CRITERIA AND ASSESSMENT

A criterion of human comfort is a specification of an unwanted effect caused by vibration. This can then be used as a standard for judging the effects of vibration under diverse conditions. There are three basic criteria which are relevant to vibration exposure in addition to other forms of environmental stress.

(a) *Comfort or amenity*. This would be associated with the threshold of perception for people engaged in delicate tasks. A slightly higher level may be disturbing in a residential area. Significant adverse comment, disturbance or alarm come into this category.

(b) *Preservation of working efficiency*. Vibration should not interfere with normal working activities. Office workers may be affected by quite low intensities especially if the vibration is frequent or continuous. Operators in offshore structures are able to work effectively in surprisingly adverse conditions, but there must be a limit. Prolonged vibration is known to induce fatigue, inattentiveness and other undesirable effects even when the intensity does not seem uncomfortable initially.

(c) *Health and safety*. Upper boundaries of vibration exposure are characterized by the need to protect personnel from injury or disability. This may require the provision of handholds or restraint harness. Duration of exposure is again very important since levels of vibration that are not necessarily uncomfortable for short periods can induce headaches, sickness and other disabilities if experienced for longer periods.

It has already been suggested that the nature of the vibration event is significant in relation to assessment of its effects on human performance. Three types of event may be envisaged:

(a) *Continuous vibration*. This could arise from the effects of machinery in plant rooms of tall buildings, or in factories. Intermittent periods of continuous vibration can be almost as disturbing, such as might be caused by the motion of fork-lift trucks. Repeated impulsive shocks caused by pile-driving are in this category.

(b) *Occasional impulsive motion or shock*. Vibrations caused by blasting operations would come into this category and are characterized by long periods between impacts. The effects of sonic boom or earthquakes are even rarer events of this type.

(c) *Motion induced by storms*. Vibration caused by storm winds occurs infrequently but peaks of motion may persist for periods ranging from five minutes to about one hour. By interviewing the occupants of two very tall (40-storey) buildings, Hansen, Reed and Vanmarcke (1973) found that the percentage of people objecting to a certain amplitude depended on the return period of the storm causing the motion. The more frequent the storm, the greater was the percentage of people objecting.

11.3.1 Human response to vibration in buildings and other large structures

Horizontal motion

The motion of tall buildings and offshore structures in storm conditions is predominantly horizontal. Recommendations for the evaluation of the comfort of occupants of such structures are given in the British Standards BS 6472 (1984) and BS 6611 (1984). The latter standard applies to the low-frequency range from 0.063 Hz to 1.0 Hz and is technically equivalent to the International Standard ISO 6897 (1984). The standards are based substantially on the work of Irwin (1978) who suggested the form of a base curve for the threshold of perception of sensitive humans. This is shown in Fig. 11.7.

Hansen, Reed and Vanmarcke (1973) found that owners and occupants of tall buildings believed that not more than 2% of people in the upper floors should object to the motion induced by storms with a return period of about six years. On the basis of further data, the British Standard (BS 6611, 1984) applies to a storm with a return period of five years. Furthermore, the dynamic motion is averaged over the worst ten minutes of the storm peak. Irwin (1978) suggested that briefer peaks did not have a significant impact on human recollection unless they were associated with more violent events like earthquakes. Irwin's 2% complaint level for the worst ten minutes of storms with a 1 in 5 year return period is shown in Fig. 11.7. This refers to horizontal motion applied to standing, seated or reclining occupants of buildings used for general purposes, e.g. offices, hotels. If there is any torsional motion visual cues have a tendency to reduce this threshold (Irwin, 1981a).

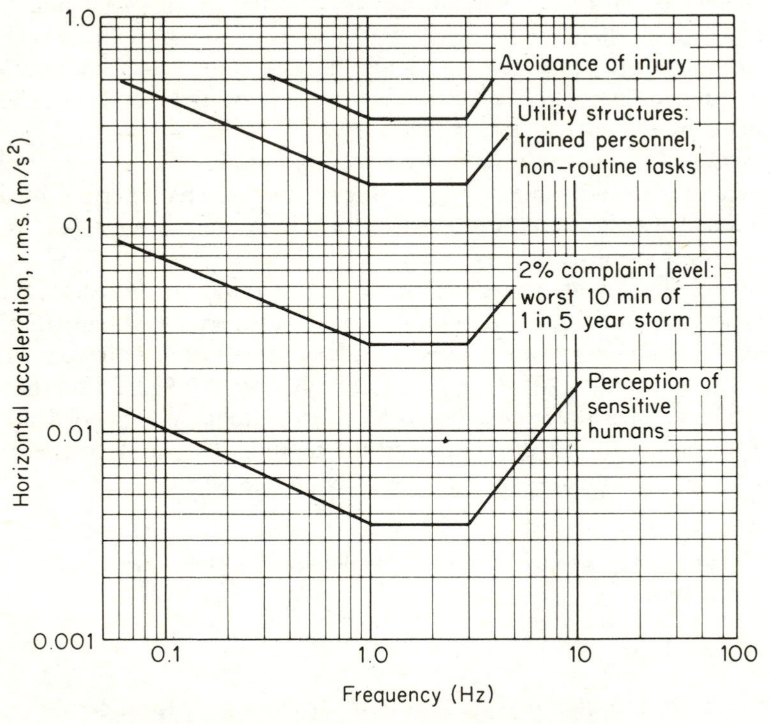

Fig. 11.7 Human response to horizontal motion.

Higher intensities of motion are acceptable in utility structures such as offshore platforms and bridges. The curve shown in Fig. 11.7 was suggested by Irwin (1978) as being appropriate for trained personnel executing non-routine tasks requiring skill and decision making. This would apply to maintenance workers on bridges and certain operations on offshore structures. Drilling is considered to be a routine task and is often carried out in very rough weather, being limited only by the ability of the equipment to function correctly. Irwin (1984) suggested a higher level still for avoidance of injury.

Vertical vibration

A base curve representing the threshold of perception for vertical vibration is given in BS 6472 (1984) and is shown in Fig. 11.8. This is appropriate for

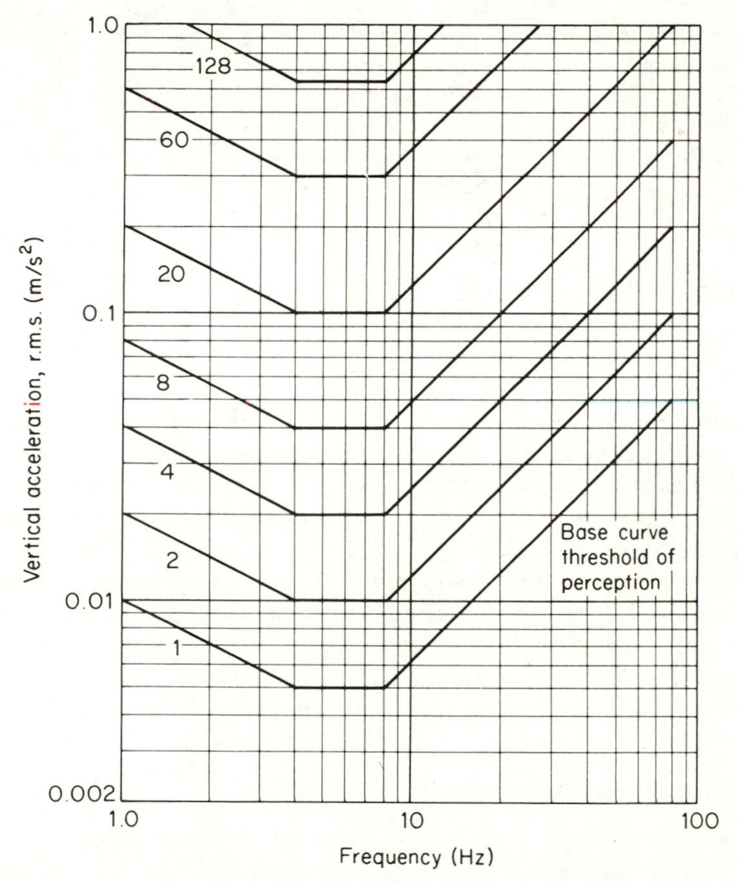

Fig. 11.8 Human response to vertical motion.

delicate work, e.g. hospital operating theatres, precision manufacturing or laboratories. Higher thresholds are acceptable for different activities. Ashley (1978) suggested that the base curve for perception of motion could be multiplied by weighting factors to take account of different types of building occupancy and different kinds of dynamic event, ranging from continuous vibration to impulsive vibration and shock. A complete table of factors, together with detailed notes and explanation is given in BS 6472 (1984). A brief version of the table is given as Table 11.2.

Table 11.2 Weighting factors above threshold of perception for acceptable building vibration

Place	*Time*	*Continuous or intermittent vibration and repeated shock*	*Impulsive shock with several occurrences per day*
Critical working areas (e.g. hospital operating theatre)	Day Night	1 1	1 1
Residential	Day Night	2 to 4 1.4	60 to 90 20
Office	Day Night	4 4	128 128
Workshops	Day Night.	8 8	128 128

Appropriately factored curves are included in Fig. 11.8. The same system of factors may be applied to the horizontal vibration base curve shown in Fig. 11.7. Impulsive shocks caused by pile-driving or blasting may be more acceptable if prior warning is given and construction work is to take place over a limited period of time.

A very similar procedure is adopted in the provisional German code DIN 4150: Part 2 (1975). Formulae are used to calculate perception intensity (KB values), according to the imposed frequency and acceleration (or velocity or displacement). Then the KB values are compared with recommendations which depend on the building use, time of day and duration of vibration.

11.3.2 Human response to vibration of bridges

Because of their structural form, being wide but shallow in depth, bridges are predominantly susceptible to vertical vibration. The only exception is the wind-induced horizontal oscillation of very long suspension bridges.

Higher values of vertical motion are acceptable in bridges, in contrast to buildings, because users are in the open and aware of the presence of wind or traffic. Furthermore, people crossing are exposed to vibration for short periods of time only. Irwin (1978) suggested a base curve for acceptable human response to vibration of bridges under frequent forms of loading. This is shown in Fig. 11.9.

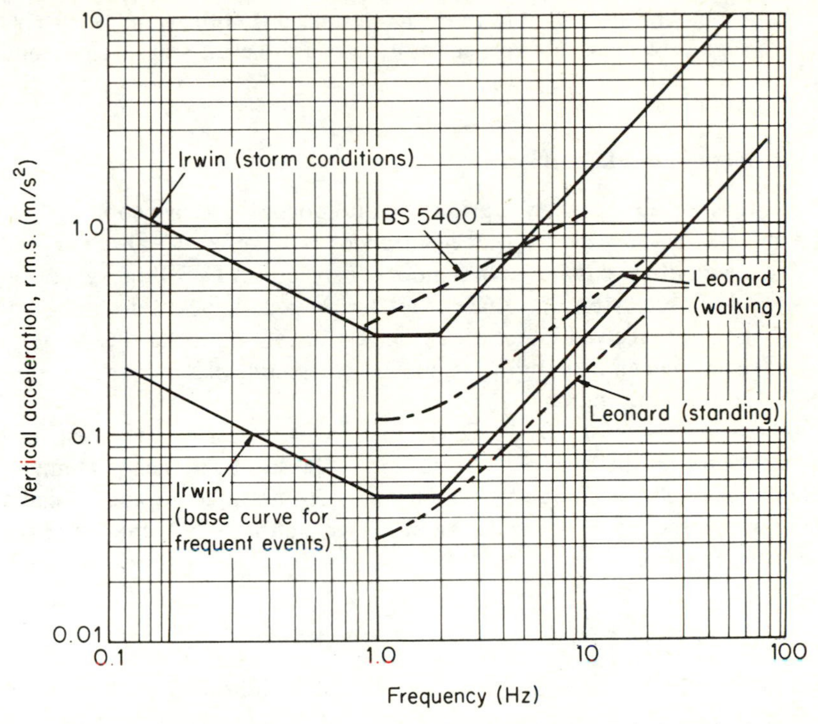

Fig. 11.9 Human response to vibration of bridges (vertical motion).

A survey of users of highway bridges in the USA (Siess and Vincent, 1958) indicated that reports of disturbing vibration came from pedestrians in the majority of cases. It appeared that the reason for this was that drivers and passengers in vehicles seldom noticed the oscillations of bridges because these were obscured by the normal vibration of vehicles. It has been shown by Leonard (1966) that pedestrians are less susceptible to vertical vibration when walking than when standing. This is because the body motion and required muscular coordination tends to obscure the surface vibration, particularly at frequencies above 3 Hz. At lower frequencies displacements are usually much greater and can interfere with normal gait, as on the deck of a ship. Leonard's curves for walking and standing on bridges (converted to root mean square motion by multiplying by $1/\sqrt{2}$) are also shown in Fig. 11.9. It has been found that the majority of pedestrian users of any bridge will come to accept a higher threshold after a period of acclimatization. There is obviously an upper limit above which there is difficulty in walking. A less stringent recommendation is suggested in the British Standard for bridge design (BS 5400, 1978). This recognizes the need for economy and relates specifically to footbridges but is relevant to footways on other bridges. The curve is also shown in Fig. 11.9 converted to r.m.s. acceleration. It approximates closely to Irwin's suggested curve for vertical motion of bridges in storm wind conditions.

11.3.3 Vibration of light floors

The disturbing effect of the vibration of light floors caused by people walking, and other everyday usage, was mentioned in Chapter 9. This can be a problem with long-span floors of slender construction. The vibration consists of short bursts of vibration under footstep impact, quickly decaying with damping. The perceived vibration can give rise to a disconcerting feeling of flimsiness to the person in motion and can be a source of annoyance to other people in the same room.

Allen and Rainer (1976) devised a method of assessing the intensity of vibration and, on the basis of a large number of tests, recommended limits for acceptable levels. Their recommendations were adopted by the Canadian Standards Association for the design of steel structures (CSA, 1984) and were shown in Fig. 9.8.

EXAMPLE 11.1

Determine the acceptable horizontal motion of a tall office building, with a natural period of vibration of 4 s, in the worst 10 minutes of the 1 in 5 year storm wind. Assume that the peak motion is approximately four times the r.m.s. motion.

Solution

The natural frequency of the building is 0.25 Hz. Therefore, by referring to the curve of the 2% complaint level in Fig. 11.7, it will be seen that the acceptable r.m.s. motion is an acceleration of 0.046 m/s^2. Therefore, the peak acceleration will be $4 \times 0.046 = 0.184$ m/s^2. Assuming that the motion is simple harmonic at the peak vibration, the peak dynamic component of deflection should not exceed

$$U_{\max} = \frac{0.184 \times 1000}{4\pi^2 \times 0.25^2} = 74.6 \text{ mm.}$$

EXAMPLE 11.2

Refer to Example 8.3 and compare the calculated amplitude of the top of the engine–generator block with acceptable limits recommended in this chapter.

Solution

The engine/generator machinery will require the attendance of operators for lengthy periods and the situation could therefore be described as a workshop. Referring to Table 11.2 the weighting factor for continuous vibration is 8.

The natural frequency of the engine–generator block in the rocking and translational mode was 20 Hz. The motion is predominantly horizontal and therefore the horizontal base curve in Fig. 11.7 should be used. The perceptible r.m.s. acceleration at 20 Hz is 0.035 m/s^2. Using the weighting factor, the acceptable acceleration is $8 \times 0.035 = 0.28$ m/s^2. Converting to peak displacement we obtain

$$U_{\max} = \frac{0.28 \times \sqrt{2}}{4\pi^2 \times 20^2} \times 1000 = 0.025 \text{ mm.}$$

The calculated displacement was 0.0275 mm. Therefore, the conditions on the top of the generator block may be slightly uncomfortable for a person working in the vicinity all day.

12

Fatigue

12.1 INTRODUCTION

The phenomenon of fatigue is usually associated with a progressive form of failure of metal structures under cyclic or repeated loading. The long-term and somewhat insidious nature of the phenomenon was used for dramatic effect by Nevil Shute in his novel *No Highway*, which was about an airliner which crashed due to a fatigue failure in its tailplane. Fatigue life assessment is important in the design of civil engineering structures such as offshore platforms, tall masts and bridges.

Fatigue affects components and connections subjected to millions of cycles of dynamic stress. Failure is in the form of progressively lengthening cracks beginning at sharp discontinuities or defects. Fatigue is a very common problem in welded steel structures because of the presence of microscopic flaws in the weld metal. Offshore structures are typically subjected to 10^8 cycles of wave loading in a life of about 30 years. Premature fatigue cracking of welds at the connections of tubular bracing members is a well-known hazard and was the primary cause of collapse, and subsequent overturning, of the *Alexander Kielland* semi-submersible platform in the North Sea in 1980. Tall masts and other slender structures may undergo almost continual oscillation in the wind. Therefore, structural details are likely to be subjected to many millions of cycles of stress, depending on the natural frequency. Bridges on major trunk roads may be loaded by as many as a million heavy lorries per year which would amount to about 5×10^8 axle loads during a life of 120 years. In the case of bridges, however, fatigue is mainly a static or pseudo-static effect caused by the repeated applications of load. The dynamic components under traffic or wind are usually quite small.

12.1.1 The fatigue process

There are two distinct stages in the fatigue process, these being (1) crack initiation and (2) crack propagation. In metallic materials the atoms are arranged in geometrical patterns to form individual crystals. Defects in the patterns, called dislocations, tend to move and extend under repetitions of stress. Eventually several dislocations may join up into a micro-crack. The presence of a micro-crack gives rise to very high stresses locally, and as the micro-cracks extend they join up into macro-cracks as shown in Fig. 12.1(a). At this stage they could be about 10^{-4} m in length and on the threshold of visibility.

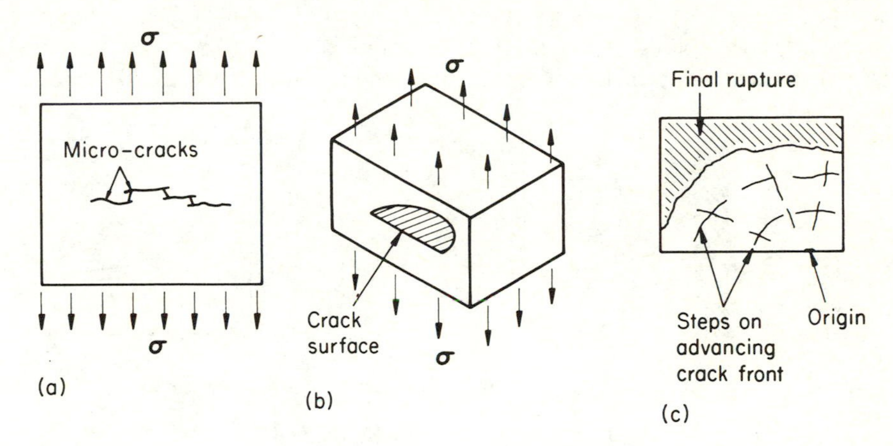

Fig. 12.1 Characteristic fatigue failure. (a) Macro-crack; (b) crack growth; (c) rupture surface.

The second stage is the growth or propagation of macro-cracks under repeated loading until the stressed section is so reduced in area that sudden rupture occurs. Fatigue cracks usually advance through a body of solid material with a circular or elliptical crack front perpendicular to the direction of stress (Fig. 12.1(b)), and, after failure, display a characteristic stepped rupture surface as shown in Fig. 12.1(c).

The above stages describe the progress of fatigue failure in an ideal homogeneous material. However, while it is true that fatigue cracking may eventually occur in smooth virgin material, it more often emanates from a flaw or defect. The most common example of this is the fatigue of welded steel structures where inclusions, gas bubbles and other minute defects act as incipient cracks. Thus the first phase of the process is effectively missed out and fatigue life occurs almost entirely during the crack propagation phase (Gurney, 1979; Berge, 1978). Stress concentrations due to sudden changes in geometry usually determine the location of fatigue cracks. Other factors contributing to the incidence of fatigue are rubbing or fretting actions and the existence of corrosion.

12.1.2 *S–N* curves

The classic experiment on fatigue was carried out by Wohler (1871) who was studying the strength of railway carriage axles. This consisted of a rotating cantilever shaft with an end load, as shown in Fig. 12.2(a), such that any fibre on the surface of the shaft at X would be subjected to an alternating stress as in Fig. 12.2(b). The stress amplitude, S, was changed by using different loads for each sample. The number of rotations, N, until rupture of the shaft was recorded, and the results plotted on a graph of S against N (Fig. 12.2(c)). It was found that by decreasing the stress amplitude, the cycles to failure increased, there being evidence of a minimum stress S_0, below which no fatigue failure occurred. This was therefore a constant-amplitude fatigue test.

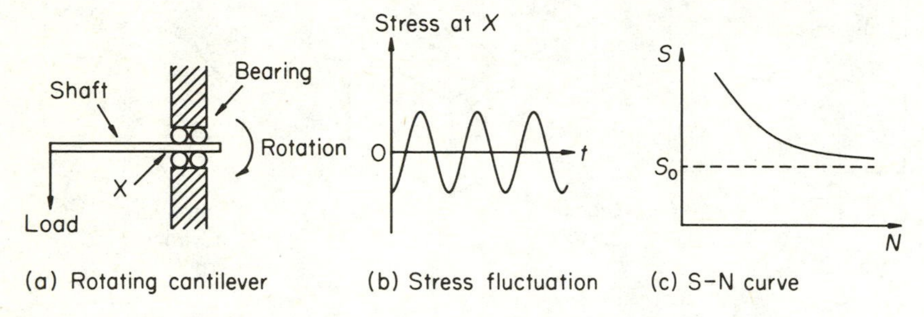

Fig. 12.2 Rotating bending fatigue test.

Constant-amplitude fatigue tests provide the most basic data on the fatigue endurance of materials or structural components. It is not surprising that this is the most common test, though the shapes of samples and the way they are loaded may vary widely according to the structural application. Some examples of fatigue specimens are shown in Fig. 12.3. Axially loaded samples can be tested in general-purpose machines at frequencies up to 150 Hz, hence

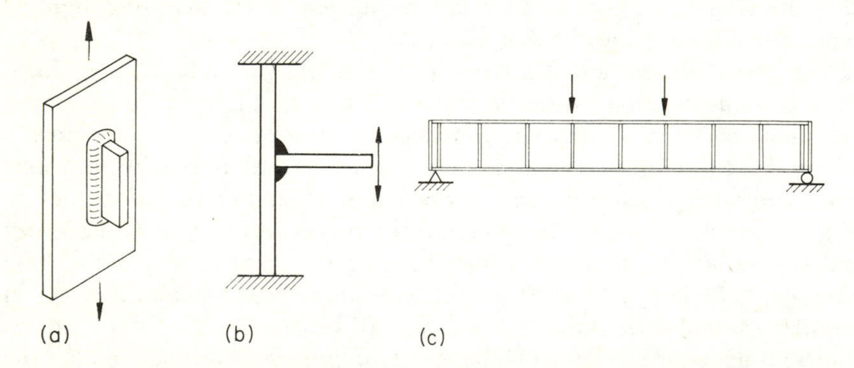

Fig. 12.3 Examples of fatigue specimens. (a) Axial; (b) bending; (c) bending of a beam.

obtaining nearly 10^8 cycles of load in a week. Specimens subjected to bending, such as fillet welded connections (Fig. 12.3(b)) or full-scale beams (Fig. 12.3(c)), would have to be loaded at slower rates because of their greater flexibility. It is often difficult to apply more than 10^5 cycles in a day with large specimens.

An example of a large-scale fatigue test is shown in Fig. 12.4. The specimen, which is a device used for repair of offshore structures, consists of a fabricated steel sleeve clamped over the original tubular connection. Grout is injected to fill any remaining annular cavities and ensure a tight fit between the parts. Dynamic loading is being applied through one of the diagonal members and

Fig. 12.4 Large scale fatigue test on a repair sleeve for offshore structures (reproduced with permission of Wimpey Laboratories, UK).

stresses are being monitored by means of strain gauges attached to the surface of the fabrication. It is easy to see in Fig. 12.3 that it is not always possible to test samples so that the stress oscillates between tension and compression. In order to interpret the results of tests in which there is a mean stress S_m, it is useful to adopt the notation illustrated in Fig. 12.5. In many applications, especially welded structures, the stress range S_r, is the most significant parameter, and would be plotted against N.

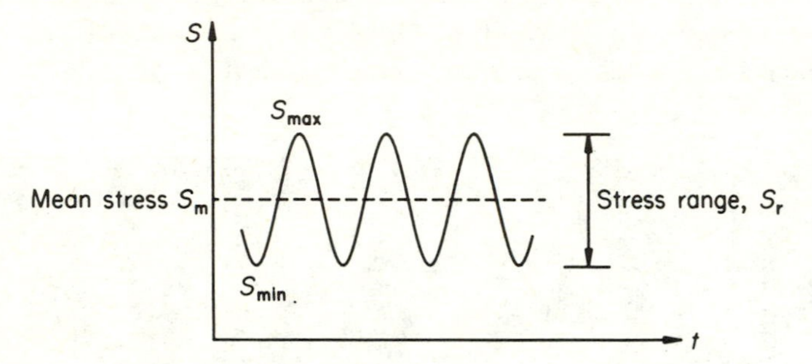

Fig. 12.5 Constant-amplitude fatigue notation.

It has been found empirically that the results of fatigue tests follow a power-law relationship between stress and cycles. Therefore the curve in Fig. 12.2(c) can be expressed as

$$N = aS^m, \tag{12.1}$$

and by taking logarithms the linear form

$$\log N = \log a + m \log S \tag{12.2}$$

may be obtained, where m is the slope of the curve when plotted on logarithmic scales. This is indeed the usual practice and in Fig. 12.6 some results of tests on welded samples are shown (Maddox, 1974). The straight line, given by the mean, has a slope of $m = -3.42$.

Because of the scatter of experimental results it is necessary to define a lower limit for design purposes. This is usually obtained statistically by determining a lower confidence limit below which only a small percentage of samples would be likely to fail. If it is assumed that the experimental scatter in the fatigue lives of samples, all tested at the same stress level, obeys a normal distribution as shown in Fig. 12.6, then the standard deviation may be obtained by calculating the root mean square error between the experimental results and the mean curve.

The probability that the fatigue life of a sample will be less than N_c is equal to the shaded area under the normal distribution. This depends on the difference between the mean N_m and the confidence limit N_c. If the confidence limit is taken to be two standard deviations below the mean, then the failure

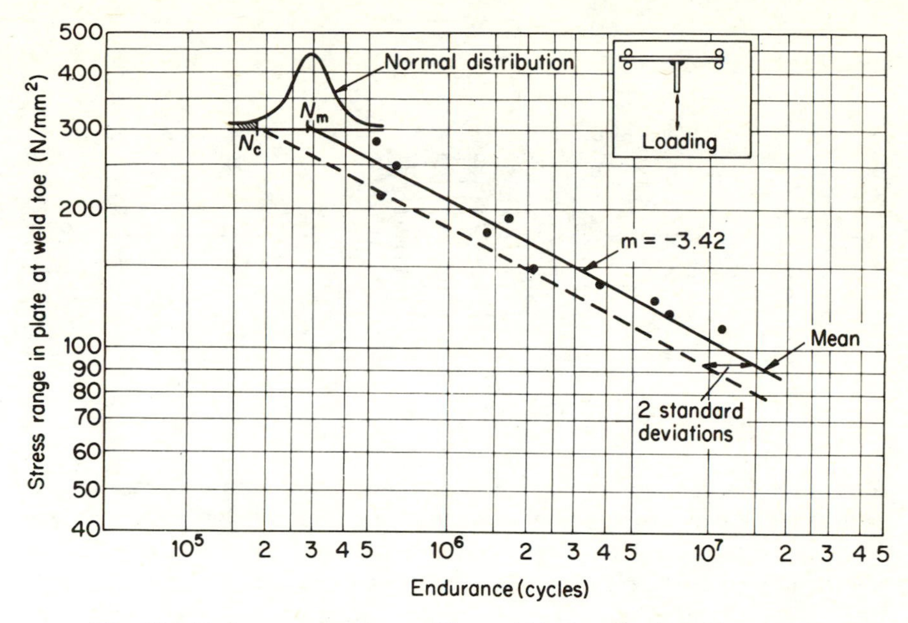

Fig. 12.6 Fatigue test results on fillet weld in bending (Maddox, 1974).

probability can be shown to be only 2.3%. This is the usual procedure adopted for obtaining design curves (Gurney, 1979).

12.1.3 Stress concentration

A factor that is known to have a profound influence on fatigue life is stress concentration (Gurney, 1979). Sudden discontinuities such as cracks, holes or changes of section act as stress raisers which increase the stress locally. An example is illustrated in Fig. 12.7 where the stress contours in a fillet-welded connection are shown. It is evident that there is a concentration of stress at the weld toe and experience has shown that this is where fatigue cracks usually begin.

It is possible to calculate stress concentration factors by elasticity theory or by finite element analysis. However, the detailed geometry in the vicinity of welded connections is generally uncertain and it has been found much better to obtain the fatigue lives of various classes of structural detail experimentally rather than theoretically. This will be discussed further in §12.4.

12.2 FRACTURE MECHANICS

So far we have considered fatigue from an empirical viewpoint. That is, we have discussed the observed relationship between fatigue endurance and

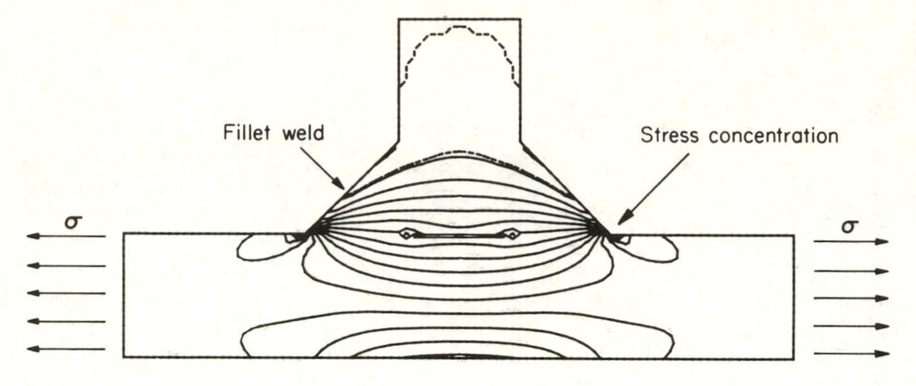

Fig. 12.7 Stress concentration at weld toe.

applied stress, and we have mentioned that other factors influence fatigue endurance. We have noted that, in welded structures, the bulk of the fatigue life occurs during crack propagation. The knowledge that cracks may exist in a structure raises the questions of stability in crack growth and acceptable sizes of cracks in service. To answer these questions we now turn to the subject of fracture mechanics in which attention is focused on the behaviour of materials in the vicinity of cracks.

12.2.1 Stress intensity factors

It is evident that there is a stress concentration at a crack tip. Calculations using the theory of elasticity have been carried out to determine the stress field around an elliptical crack (Sneddon, 1946). Unfortunately, the stress at a sharp crack tip becomes theoretically infinite when the radius of curvature diminishes to zero. This problem is avoided in linear elastic fracture mechanics (LEFM) by introducing a quantity known as the *stress intensity factor* which is a measure of the stress field in the close proximity of the crack tip.

Consider a crack of length $2a$ transverse to the direction of tensile stress in a remotely loaded two-dimensional sheet (Fig. 12.8). The components of stress at some point (r, θ) near the crack tip were shown by Sneddon to be given by

$$\begin{aligned}
\sigma_x &= \frac{\sigma\sqrt{(\pi a)}}{\sqrt{(2\pi r)}}\cos\frac{\theta}{2}\left(1-\sin\frac{\theta}{2}\sin\frac{3\theta}{2}\right)\\
\sigma_y &= \frac{\sigma\sqrt{(\pi a)}}{\sqrt{(2\pi r)}}\cos\frac{\theta}{2}\left(1+\sin\frac{\theta}{2}\sin\frac{3\theta}{2}\right)\\
\tau_{xy} &= \frac{\sigma\sqrt{(\pi a)}}{\sqrt{(2\pi r)}}\sin\frac{\theta}{2}\cos\frac{\theta}{2}\cos\frac{3\theta}{2}.
\end{aligned} \tag{12.3}$$

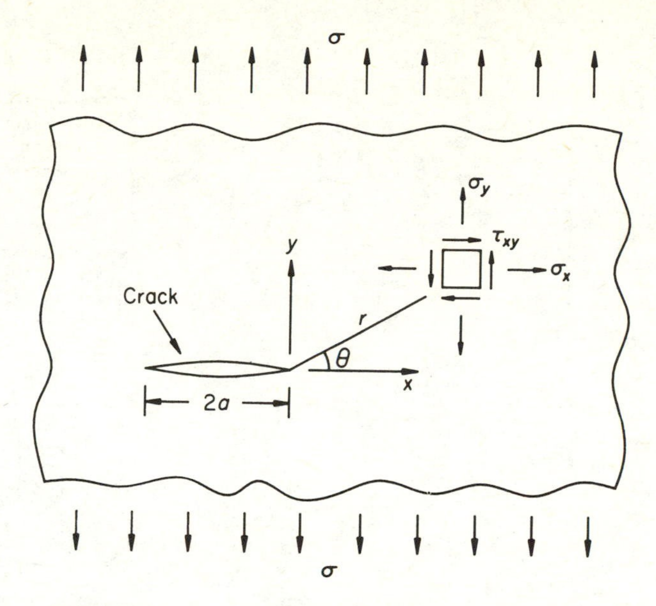

Fig. 12.8 Stresses in the vicinity of an elliptical crack.

Irwin (1958) noticed the existence of a common factor $\sigma\sqrt{(\pi a)}$ which scaled the equations according to stress and crack length. He defined this term as the stress intensity factor K, given by

$$K = \sigma\sqrt{(\pi a)}, \tag{12.4}$$

and observed that with a knowledge of crack length and applied stress, the stress field in the proximity of any crack could be accurately determined.

Equations (12.3) are only applicable to the simple case of a transverse crack in an infinite stress field. The stress distribution will be affected by the local geometry of an object of finite size. This is dealt with by introducing a geometry factor α so that the stress intensity factor becomes

$$K_1 = \alpha\sigma\sqrt{(\pi a)}. \tag{12.5}$$

The factor α can be determined by elasticity theory for some simple cases such as those shown in Fig. 12.9. A large number of standard solutions are available, covering a wide range of geometries and loading conditions (Rooke and Cartwright, 1976). Compounding of two or more standard solutions has also been suggested for more complex geometries (Rooke and Cartwright, 1975). The finite element method can be used for geometries where no suitable standard solution can be found. A fine mesh is required around the crack tip for calculation of the stress in an element close to the crack tip. The stress intensity can then be calculated directly from the stress field equations

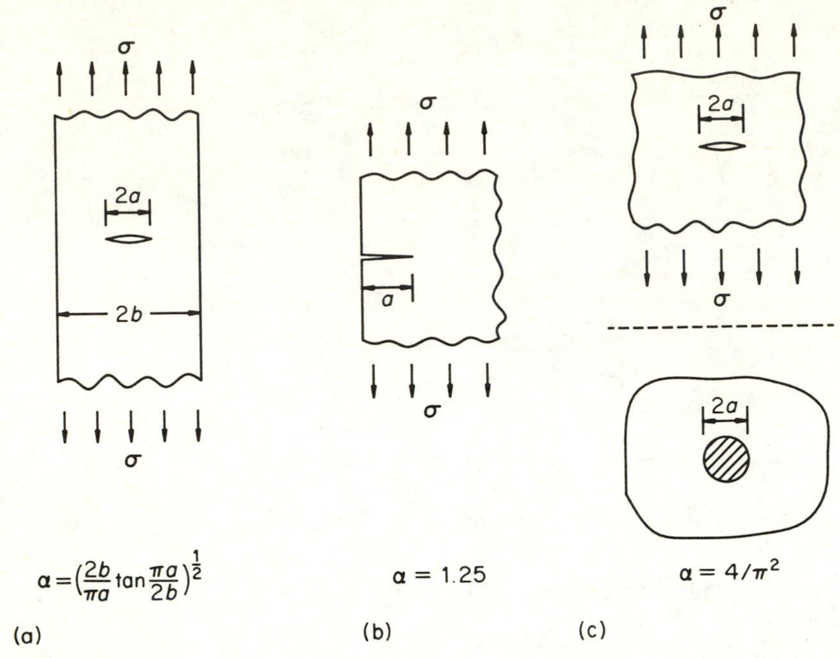

Fig. 12.9 Geometric factors. (a) Sheet of finite width; (b) edge crack; (c) disc crack in infinite solid.

by writing them in the form

$$K_1 = \sigma_{ij}\sqrt{(2\pi r)}/f_{ij}(\theta) \tag{12.6}$$

where σ_{ij} is one of the stress components σ_x, σ_y or τ_{xy} obtained by the finite element analysis and $f_{ij}(\theta)$ is the appropriate function in Equation (12.3). The calculation should be repeated using different elements close to the crack tip and the same value of K_1 should result. The scatter in values is an indication of the error involved. Extremely fine meshes are required in order to obtain reasonably accurate values of K_1. Crack tip finite elements have been developed and they have been shown to give more accurate estimates of K_1 (Tracey, 1974).

12.2.2 Fracture toughness

Fracture mechanics originated in the work of Griffith (1920) who considered the stability of a crack in a brittle material such as glass. He took the view that when a crack extended, elastic strain energy would necessarily be released. However, energy would be required to create the new surfaces (the ultimate condition is when all the atoms of the body have been vapourized).

Normally more work is required to create new surfaces than is released elastically as a crack extends. However, a critical crack length exists when the balance of energies is reversed and unstable propagation of the crack occurs. This combination of applied stress and crack length has been shown by Irwin (1956) to correlate well with the stress intensity factor. Evidently this critical combination is determined by the toughness of the material in its resistance to the formation of new surfaces. This critical stress intensity factor, denoted by K_{1c}, is called the fracture toughness and is a material property. It is important to note that this value is found to depend on the thickness of the samples used for testing. This is because failure tends to occur in shearing at the edges whereas the true brittle rupture assumed in the theory only occurs if the sample is thick enough to ensure plane strain conditions over the middle portion. This is shown in Fig. 12.10.

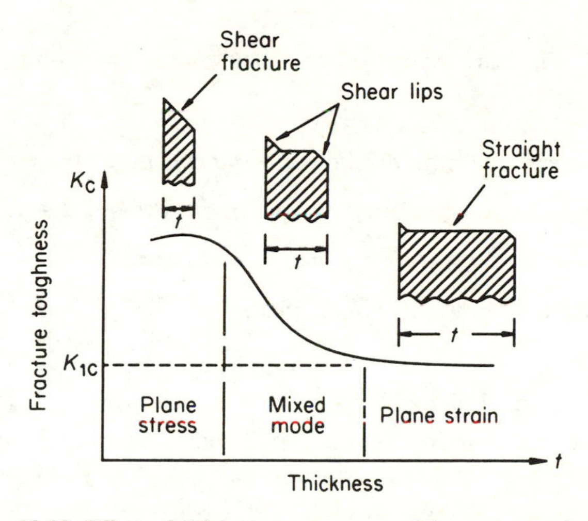

Fig. 12.10 Effect of thickness on measured fracture toughness.

The foregoing discussion has applied only to cracks that open in a direction normal to the crack surfaces. There are other modes of crack opening as indicated in Fig. 12.11. Mode I crack opening is the one most commonly encountered.

No material is perfectly elastic, even glass, and there will always be a small region of plasticity at the crack front as shown in Fig. 12.12. An allowance is usually made for this plastic region by assuming an effective crack length given by

$$a_1 = a + r_p \tag{12.7}$$

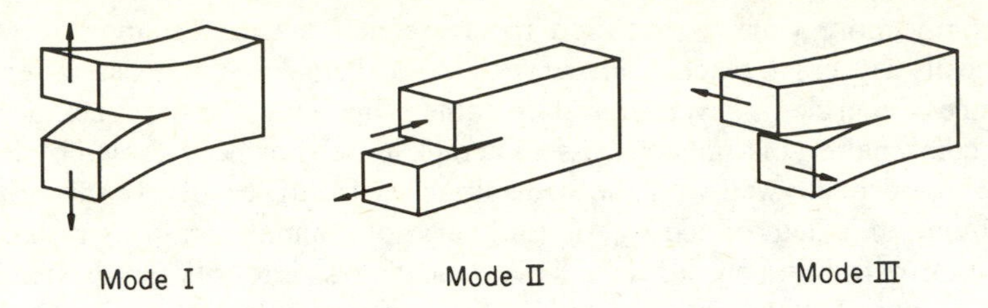

Fig. 12.11 Modes of crack opening.

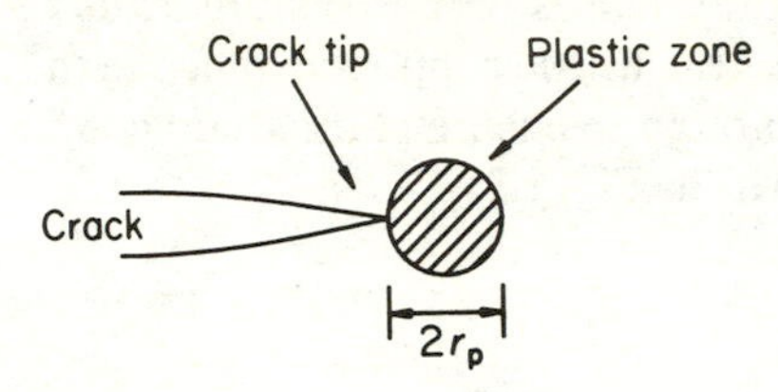

Fig. 12.12 Allowance for plastic zone at crack tip.

where r_p is the radius of the plastic region and is given by

$$r_p = (1/2\pi)(K/\sigma_y)^2; \quad \text{plane stress}$$
$$r_p = (1/6\pi)(K/\sigma_y)^2; \quad \text{plane strain} \qquad (12.8)$$

where σ_y is the yield stress (Irwin, 1958).

12.2.3 Fatigue crack propagation

A knowledge of the fracture toughness of a material makes it possible to use Equation (12.5) to estimate the critical crack length beyond which unstable propagation, and therefore rupture, will occur. The equation should be written in the form

$$a_c = (1/\pi)(K_{1c}/\alpha\sigma)^2 \qquad (12.9)$$

where σ is the applied stress. This value is most significant in the design of large and important welded steel structures such as offshore platforms. It gives an upper limit on the acceptable crack or flaw size.

Normally flaws incurred during fabrication will be much smaller and it is necessary to know how long it will take for a fatigue crack to propagate to the critical size. This will determine the life of the structure. The crack propagation rate, da/dN, is the rate at which the crack front advances per cycle. Attempts have been made to relate the rate of crack propagation to the range of applied stress, but the most satisfactory parameter has been found to be the range of stress intensity ΔK_1. This is embodied in the Paris equation (Paris,

1964) which states that

$$\mathrm{d}a/\mathrm{d}N = C(\Delta K_1)^m \tag{12.10}$$

where

$$\Delta K_1 = (\sigma_{max} - \sigma_{min})\alpha\sqrt{(\pi a)}. \tag{12.11}$$

In cases where σ_{min} is compressive the crack will close and the accepted practice is to ignore this part of the cycle (ESDU, 1976b).

Experimental observations of crack propagation give a curve similar to the one shown in Fig. 12.13. Thus the Paris equation is applicable only over small ranges of the curve when C and m are approximately constant. However, it is usually assumed that the curve is straight over the intermediate range of stress intensity, ignoring the threshold effects at very low stress intensity and stopping short below the critical value.

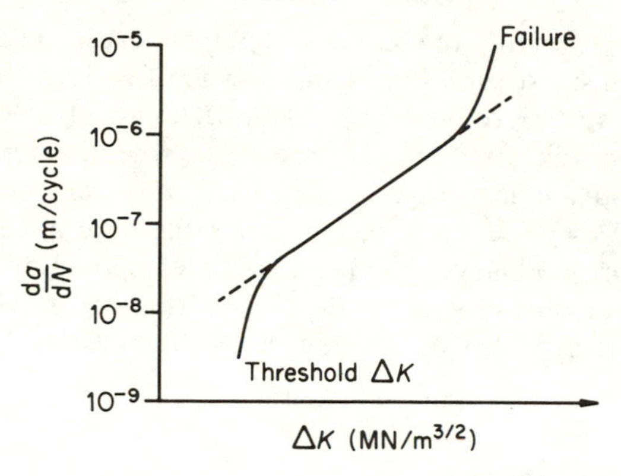

Fig. 12.13 Crack growth under cyclic loading.

Since the fatigue life of a welded component is taken up almost entirely by propagation of a crack originating from a small crack-like defect, an estimate of the number of cycles required to extend the crack to an unacceptable size may be obtained by integration of the Paris equation. Thus by substituting Equation (12.5) into (12.10) and integrating we obtain

$$N = \int_{a_i}^{a_f} 1/C(\Delta K_1)^m \,\mathrm{d}a$$

$$= \frac{1}{C\alpha^m \Delta\sigma^m \pi^{m/2}(m/2-1)} \left\{ \frac{1}{a_i^{m/2-1}} - \frac{1}{a_f^{m/2-1}} \right\} \tag{12.12}$$

where a_i is the initial crack length and a_f the final crack length. Cracks propagate very rapidly just prior to failure, as can be seen in Fig. 12.13, and therefore there is little error incurred in neglecting the term in a_f. It is just as if the integration were taken to infinity giving

$$N = [C\alpha^m \Delta\sigma^m \pi^{m/2}(m/2-1)\, a_i^{m/2-1}]^{-1}. \tag{12.13}$$

This can be further simplified to

$$N = a_i/[C\Delta K_i^m(m/2-1)] \tag{12.14}$$

where ΔK_i is the stress intensity factor with the initial crack length.

Experimental values of the exponent m have been found to vary widely from as low as 2 to as high as 10 (Ritchie and Knott, 1973). The average value is about 3. The range of values for structural steel is much narrower, being between 2.4 and 3.6 (Gurney, 1979). Notice that (12.14) predicts the same slope for the S–N curve as is found empirically for the majority of welded joints (see Fig. 12.6).

12.3 VARIABLE-AMPLITUDE FAILURE

Fatigue loading is seldom of constant amplitude, except in the laboratory. Examples of realistic stress histories are shown in Fig. 12.14. The first would be typical of the stresses recorded in a bridge deck under traffic loading. It is a mixture of sharp spikes and broader peaks. The second is typical of stresses induced by aerodynamic oscillation of a mast in turbulent wind. The history is characterized by a steady frequency, being the natural frequency of the mast, with a more gradual variation in amplitude imposed by the turbulence. Evidently some method of assessing the effect of variable amplitude loading is required. This is provided by a study of the fracture mechanics of crack propagation.

12.3.1 Cumulative damage

Crack propagation under variable-amplitude loading may be determined by considering a mean rate of crack propagation. From Equation (12.10), the

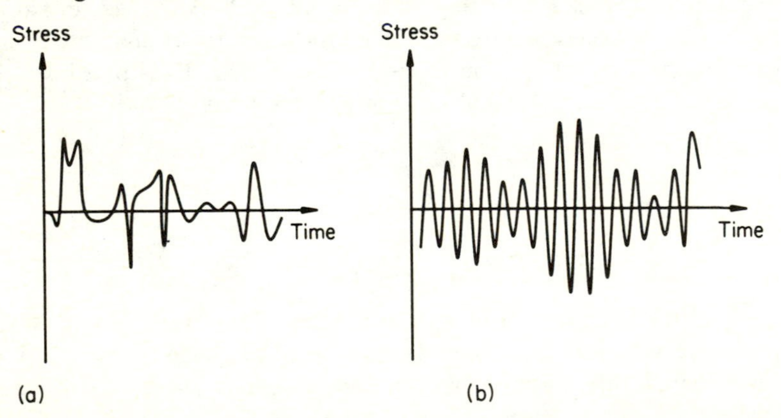

Fig. 12.14 Examples of practical stress histories. (a) Bridge deck due to traffic; (b) mast due to wind.

crack length advance due to n_j cycles of a stress intensity range ΔK_j may be written as

$$(\mathrm{d}a/\mathrm{d}N)_j n_j = Cn_j(\Delta K_j)^m. \tag{12.15}$$

Then the total crack front advance under a combination of stress ranges of different numbers of cycles is given by

$$(\mathrm{d}a/\mathrm{d}N)_{\mathrm{av}} N = \sum_{j=1}^{\infty} C n_j (\Delta K_j)^m \tag{12.16}$$

where $N = \Sigma n_j$ is the total number of cycles.

If we use Equation (12.14) to determine the total number of cycles to failure for each stress range we have

$$N_j = a_i/[C(m/2-1)\Delta K_j^m]. \tag{12.17}$$

Substituting in Equation (12.16) we obtain

$$(\mathrm{d}a/\mathrm{d}N)_{\mathrm{av}} \cdot N = [a_i/(m/2-1)] \sum_{j=1}^{\infty} n_j/N_j. \tag{12.18}$$

Turning now to the left-hand side of (12.18), if N is the total number of cycles to failure under the equivalent stress intensity range ΔK_0 required to produce the average rate of crack advance, then the left-hand side becomes

$$(\mathrm{d}a/\mathrm{d}N)_{\mathrm{av}} N = C(\Delta K_0)^m a_i/[C(m/2-1)(\Delta K_0)^m] = a_i/(m/2-1). \tag{12.19}$$

Hence Equation (12.18) reduces to

$$\Sigma n_j/N_j = 1.0. \tag{12.20}$$

Since the stress intensity ranges ΔK_j are also proportional to the corresponding stress ranges ΔS_j, it follows that (12.20) is identical to the empirically derived Palmgren–Miner rule (Palmgren, 1924; Miner, 1945). The Palmgren–Miner rule is effectively a law of cumulative damage, because the ratios on the left-hand side of (12.20) represent the damage accumulated by the various stress ranges. It is generally written as an inequality so that the sum of the cumulative damage ratios should not exceed 1.0. Wide variations in the cumulative damage sum have been reported but the bulk of the evidence is that the Palmgren–Miner rule is satisfactory for fatigue appraisal (Tilly and Nunn, 1980). Consequently it has been adapted widely for fatigue design rules (BS 5400, Part 10, 1980).

Under random loading there is evidence to suggest that the fatigue limit observed in constant-amplitude tests does not apply. One possible reason for this is that as the crack is advanced by large stress cycles the stress intensity at the crack tip increases, with the result that small cycles become progressively more damaging. Tilly and Nunn (1980) demonstrated that the Palmgren–Miner rule gives a good prediction of failure if the slope of the S–N curve is taken to be $(2m-1)$ below the fatigue limit.

12.3.2 Cycle counting

One of the difficulties in assessing fatigue life under variable amplitude or random loading is to determine the number of cycles occurring within each stress range. In the case of narrow-band variable loading, such as in Fig. 12.14(b), it is relatively simple because each cycle at the natural frequency of the structure is clearly identifiable. All that is required is to count the number of cycles that the stress range is within a certain band of magnitude and obtain a histogram of the numbers of cycles of different stress ranges. An example is shown in Fig. 12.15. If the loading has known statistical properties, as in the case of wind, it is often possible to calculate the frequency of cycles at various stress levels.

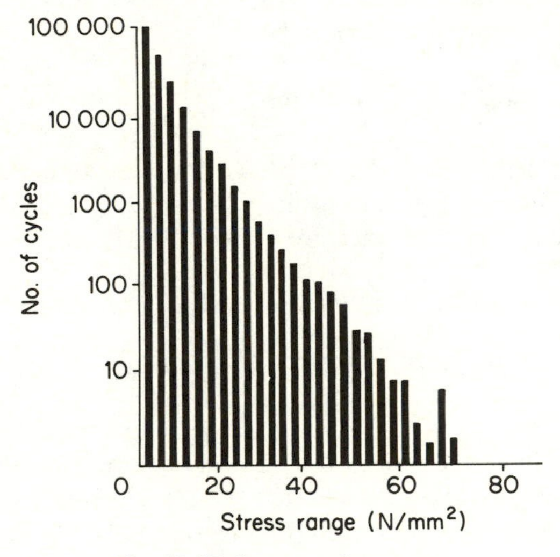

Fig. 12.15 Stress cycles count.

An irregular stress history (Fig. 12.14(a)) is more difficult because individual cycles are not immediately obvious. It would appear that the record is composed of small cycles superimposed upon larger ones. A number of different methods of counting cycles have been devised (Gurney, 1979) and it is uncertain which is the best. However, one that is widely adopted in practice is the 'rainflow' method (Matsuishi and Endot, 1968). It will be described here because of its convenience of use and its well established reputation (BS 5400, 1980).

The 'rainflow' method can best be described by rotating the time axis of a fluctuating stress signal through 90°, as shown in Fig. 12.16(a). The name is derived from the idea of rain dripping over the edges of the roofs of a Japanese pagoda. Rain is imagined to flow from the start of the record and

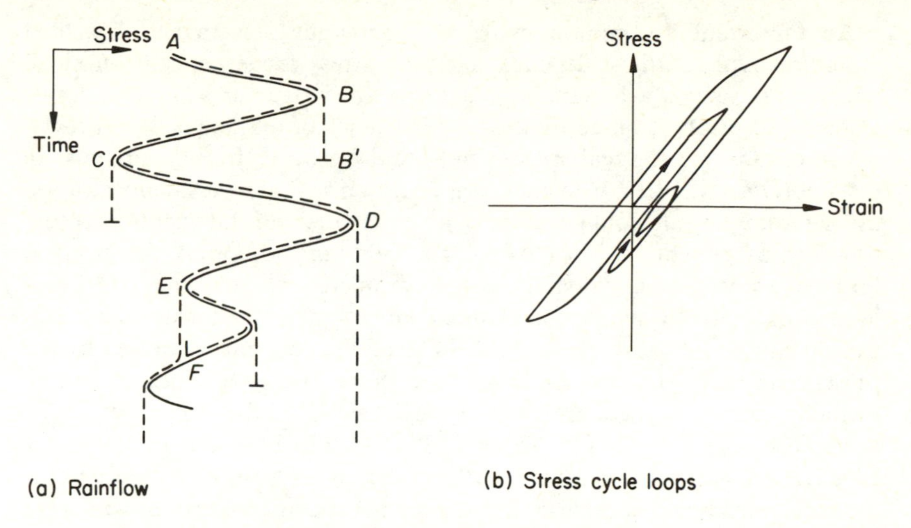

Fig. 12.16 The rainflow cycle counting method.

from the inside of every peak (*B*). When a change of direction is encountered flow is imagined to drip downwards (*B*–*B*′) as off the edge of a roof. Flow stops when it meets flow from a higher level (*F*) or when it is opposite a point which is arithmetically greater than (or equal to) the point from which it started (*B*′). Each separate flow is counted as a half cycle. Thus the irregular stress waveform is reduced to a number of half cycles. It can be seen that half cycles always come in pairs except for a limited number that are not completed at the end of a record. The same record is shown as a number of hysteresis loops in Fig. 12.16(b), which indicates the existence of individual cycles.

Extraction of stress cycles manually is a laborious process, though it is quite feasible and about 100 cycles could reasonably be counted in an hour. However, it is much more practical to do the cycle counting by computer. Data recorders exist which are capable of counting stress cycles in real time, with the means of producing a frequency spectrum for different stress ranges (Wastling and Smith, 1987).

12.4 FATIGUE OF STRUCTURES

12.4.1 Welded steel structures

The reader is referred to Gurney (1979) for a comprehensive reference on the fatigue assessment of welded steel structures. One of the most important applications is in the design of steel highway bridges (Tilly, 1978), for which a detailed code of practice is available (BS 5400, 1980).

An important assumption made in the fatigue assessment of welded structures is that fatigue depends solely on stress range and not on mean stress. This sounds a bit drastic but for welded structures there is ample evidence to justify it on account of the existence of tensile residual stresses which ensure that the real stresses pulsate downwards from yield stress in tension (Gurney, 1979). A further simplification is that stress concentration arising from the joint configuration is not calculated. It is taken into account, however, by presenting *S–N* curves for a wide range of joint types grouped into classes using nominal stress as the parameter. The nominal stress at a joint is obtained by simple calculations which ignore the localized concentration caused by the shape of the weld. Stress concentration caused by the presence of holes, or other features of the shape of the structure, should be included, however. There are eight classes of detail in the British Standard. Some examples are given in Table 12.1. It should be noted that Class A effectively applies to virgin metal with no defects and can be discounted for practical purposes. The table does not include tubular connections which are more relevant to offshore structures and require special techniques which will be discussed later. Reference should be made to BS 5400: Part 10 (1980) for the full classification. *S–N* curves for the mean minus two standard deviations are shown in Fig. 12.17. The slopes of curves D to W are taken as $m=-3$, curve C as $m=-3.5$, and curve B as $m=-4$.

The sketches in Table 12.1 give an approximate indication of where fatigue cracking may be expected to occur. It is usually found that fatigue cracking begins at the toe of a weld (McDonald and Thomson, 1980; Yamada and Hirt, 1982) though in certain configurations cracking tends to begin in the weld itself (Maddox, 1974).

12.4.2 Concrete structures

The fatigue endurance of concrete structures is of interest because of the use of concrete in highway bridges and offshore structures, both types of structure being subjected to many millions of cycles of loading during their lives. Fortunately, concrete, both reinforced and prestressed, has a good record of fatigue performance.

Nordby (1958) found that reinforced concrete beams generally had an endurance limit of 60 to 70% of the static ultimate load, based on tests up to 10^6 cycles. Failure occurred in bars near cracks or due to stress concentration. There was generally no problem with prestressed concrete beams, which had a factor of safety of approximately two against fatigue failure. These findings have been confirmed by more recent studies (Snowdon, 1970).

Plain concrete itself will eventually fail in fatigue (Raithby, 1979), but in reinforced concrete the failure nearly always takes place in the reinforcing bars. Tilly (1979) found that fatigue strength of the bars was not affected

Table 12.1 Classification of welded joints

A	Plain steel with all surfaces machined and polished and of uniform or smoothly varying cross section. Generally inappropriate for civil engineering structures.	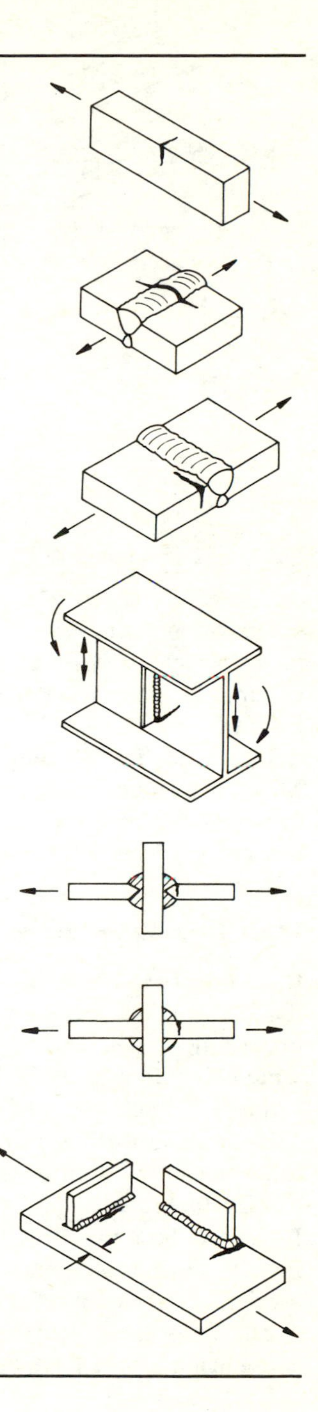
B	Plain steel in as-rolled condition, or with cleaned surfaces and any flame-cut edges subsequently ground or machined	
B	Full penetration butt weld, dressed flush, machined and ultrasonically tested	
C	Butt or fillet welds, loaded parallel with their length and made by automatic process with no stop–start positions	
D	Butt welds loaded perpendicular to their length and made in the shop in the flat position	
E	Butt welds made other than as above	
E	Parent metal at the end of a weld connecting a stiffener diaphragm or other attachment to a girder web in a region of combined bending and shear	
F	Fillet weld or T-butt weld with full penetration and any undercutting at the corners of the member dressed out by local grinding	
F_2	Joint made with partial penetration welds with any undercutting dressed out by local grinding as above	
G	Parent metal at the toes or ends of butt or fillet welded attachments on or within 10 mm of the edges or corners of a stressed member	
W	Weld metal in load-carrying joints made with fillet or partial penetration welds, with the welds either transverse or parallel to the direction of applied stress (based on nominal shear stress on the minimum weld throat area)	

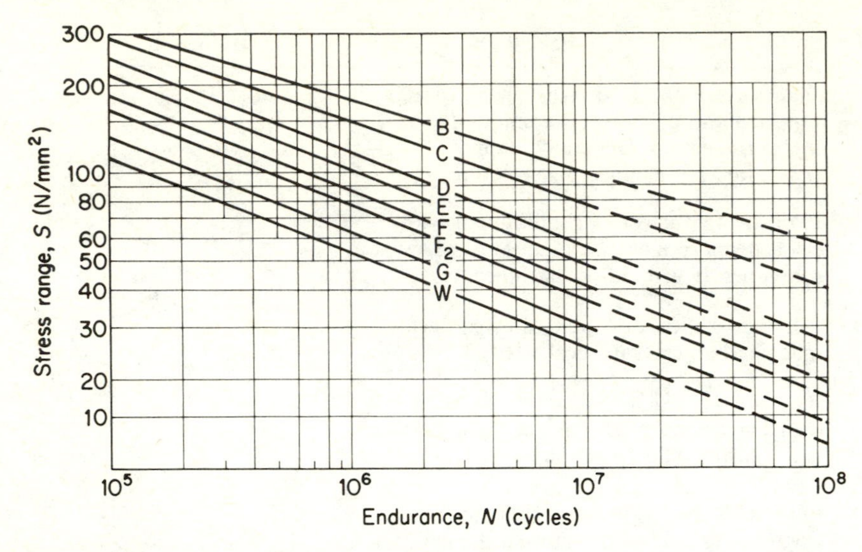

Fig. 12.17 Mean minus two standard deviations *S–N* curves (BS5400: Part 10: 1980).

much by bar strength but was reduced by 40 to 50% by the presence of ribs on high yield bars or manufacturers' markings. Larger bars had a lower endurance. Increase in the mean stress from 160 to 275 N/mm^2 reduced fatigue life by about 40%. This is significantly different from the behaviour of welded structures. Welding of reinforcing bars reduced their life by 20 to 55% Moss (1982) found that, when embedded in concrete beams, the performance of reinforcing bars was generally better than in air. Embedment tended to reduce the effects noted above.

12.4.3 Offshore structures

It has long been known that the presence of a corrosive environment reduces the fatigue endurance of materials significantly. It is also known that combined fatigue and corrosion is more damaging than either of the effects separately (Gough, 1932). This is because corrosion has the effect of sharpening the crack tip.

In civil engineering the corrosive medium of most interest is sea water or salt water spray. Johnson, Bretherton, Tomkins *et al.* (1978) found that crack propagation rates in steel structures were up to three times greater than in air. It has also been found that spraying with salt water is more damaging than immersion in brine since the former permits air to reach the crack and accelerate the oxidation (Duquette and Uhlig, 1968).

The importance of corrosion fatigue in offshore structures is obvious and consequently there have been considerable efforts in recent years to obtain further data. Research supported by the European Offshore Steel Research programme has produced further valuable data, confirming previous knowledge and extending it to offshore applications (EOSR, 1978).

A specific problem in the fatigue appraisal of steel offshore structures is the stress analysis of large tubular connections. First, large connections tend to be relatively more flexible than small tubular joints and considerable errors are introduced if rigidity is assumed (Fessler, Mockford and Webster, 1986). Secondly, stress concentration factors need to be evaluated carefully in order to ensure reliable prediction of fatigue life (Fisher, 1981). The procedure is to calculate maximum joint stresses, or *hot-spot* stresses (excluding the effect of weld profile itself), from the nominal member stresses by using parametric equations which take account of joint geometry. Appropriate parametric equations have been obtained by finite element analysis of tubular connections (Wordsworth and Smedley, 1978). *S–N* curves for hot-spot stresses are given in a design code published by the American Petroleum Institute (API, 1980). Using this procedure, reasonable correlation has been obtained between fatigue predictions and discovered cracks on actual offshore structures (Fisher, 1981).

EXAMPLE 12.1

A steel plate with a non-load-carrying fillet-welded attachment is loaded axially and has dimensions of 10 mm × 100 mm perpendicular to the direction of the load. The attachment is within 10 mm of the edge of the plate. The plate is loaded dynamically with 60% of the load cycles ranging from 0 to 20 kN, 30% from 0 to 30 kN and 10% from 0 to 70 kN. Determine the safe number of cycles that may be applied.

Solution

The stress ranges may be calculated from the applied loads and the cross-sectional area, and are

$$\Delta S_1 = 20\ \text{N/mm}^2, \quad \Delta S_2 = 30\ \text{N/mm}^2, \quad \Delta S_3 = 70\ \text{N/mm}^2.$$

According to Table 12.1 the weld class is G and therefore the safe fatigue lives, from Fig. 12.17, are

$$n_1 = 3 \times 10^7, \quad n_2 = 9.5 \times 10^6, \quad n_3 = 7.2 \times 10^5.$$

Therefore, using the Palmgren-Miner Rule (12.20) we obtain

$$\frac{n_1}{30} + \frac{n_2}{9.5} + \frac{n_3}{0.72} < 10^6.$$

But, by substituting $n_2 = 3n_3$, $n_1 = 6n_3$, we find that

$$n_3(0.2 + 0.316 + 1.389) < 10^6.$$

Therefore $n_3 < 0.525 \times 10^6$ and the total number of cycles must be

$$N = n_1 + n_2 + n_3 < 5.25 \times 10^6 \text{ cycles.}$$

References

Abdel-Ghaffar, A. M. and Rubin, L. I. (1984) Torsional earthquake response of suspension bridges, *J. Eng. Mechs.*, ASCE, **110**, 1467–84.

ACI Committee 318 (1983) *Building Code Requirements for Reinforced Concrete*, ACI 318–83, American Concrete Institute, Detroit.

Algermissen, S. T. and Perkins, D. M. (1976) A probabilistic estimate of maximum acceleration in rock in the contiguous United States, *US Geological Survey Open File Report*, 76–416.

Allen, C. R., St Amand, P., Richter, C. F. and Nordquist, J. M. (1965) Relationship between seismicity and geological structure in the Southern California region, *Bull. Seism. Soc. Amer.*, **55**, 753–97.

Allen, D. E. and Rainer, J. H. (1976) Vibration criteria for long span floors, *Can. J. Civil. Engng.* **3**, 165–73.

Allen, G. R. (1974) Proposed limits for exposure to whole body vertical vibration, 0.1 to 1.0 Hz, AGARD meeting, Oslo, April.

Altinisik, D., Karadeniz, H. and Severn, R. T. (1981) Theoretical and experimental studies on dynamic structure–fluid coupling, *Proc. ICE*, **71**, 675–704.

Altinisik, D. and Severn, R. T. (1981) Multiple-support and asynchronous base excitation of structures, *Proc. ICE*, **71**, 407–26.

Anagnostopoulos, S. A. (1982) Wave and earthquake response of offshore structures: evaluation of modal solutions, *J. Struct. Div. ASCE*, **108**, 2175–91.

Ang, A. H-S. (1974) Probability concepts in earthquake engineering, in *Applied Mechanics in Earthquake Engineering* (ed. W. D. Iwan), AMD-v8, ASME, NY.

API (1980) *Planning, Designing and Constructing Fixed Offshore Platforms.* American Petroleum Institute, API RP2A, 11th edition.

Arnold, R. N., Bycroft, G. N. and Warburton, G. B. (1955) Forced vibrations of a body on an infinite elastic solid, *J. Appl. Mech.*, **22**, *Trans. ASME, series E*, **77**, 391–400.

ASCE (1984) Chinese display ancient technology, *News*, *Civil Engineering* (*ASCE*), Nov., p. 14.

Ashley, C. (1978) Proposed international standards concerning vibration in buildings, Conf. on Instrumentation for Ground Vibration and Earthquakes, ICE, pp. 153–162.

ASME (1980) *Boiler and Pressure Vessel Code, Section III, Nuclear Components.*

ATC (1974) *An Evaluation of a Response Spectrum Approach to Seismic Design of Buildings*, Applied Technology Council, report for National Bureau of Standards, ATC-2, San Francisco.

ATC (1976) *A Methodology for Seismic Design and Construction of Single-Family Dwellings*, Applied Technology Council, ATC-4, San Francisco.

ATC (1978) *Tentative Provisions for the Development of Seismic Regulations for Buildings*, Applied Technology Council, ATC 3-06, National Bureau of Standards, SP510.

Bagchi, G. and Noonan, V. S. (1985) Seismic capability of equipment in operating plants in the light of experience data, *Trans. 8th Int. Conf. Struct. Mech. in Reactor Tech.*, SMIRT–8, vol. D9, no. 8, pp. 547–52.

Balendra, T. and Heidebrecht, A. C. (1987) A foundation factor for earthquake design using Canadian Code of Practice, *Canadian J. Civil Engng.*, vol. 14, no. 4, Aug.

Barkan, D. D. (1962). *Dynamics of Bases and Foundations*, McGraw-Hill, New York.

Bathe, K.-J. (1971) *Solution Methods of Large Generalized Eigenvalue Problems in Structural Engineering*, Report UC SESM 71-20, Civil Eng. Dept, Univ. of California, Berkeley.

Bathe, K.-J. (1982) *Finite Element Procedures in Engineering Analysis*, Prentice-Hall, Englewood Cliffs, New Jersey.

Bathe, K.-J. and Ramaswamy, S. (1980) An accelerated subspace iteration method. *J. Comp. Meth. Appl. Mech. Engng.*, **23**, 313–31.

Bathe, K.-J. and Wilson, E. L. (1973) Stability and accuracy analysis of direct integration methods, *Int. J. Earthqu. Engng. Struct. Dyn.*, **1**, 283–91.

Bathe, K.-J., Wilson, E. L. and Iding, R. H. (1974) *NONSAP: A Structural Analysis Program for Static and Dynamic Response of Nonlinear Systems.* Report No. UC SESM 74-3, Structural Engineering Laboratory, University of California, Berkeley.

Bathe, K.-J., Wilson, E. L. and Peterson, F. E. (1974) *SAPIV: A Structural Analysis Program for Static and Dynamic Response of Linear Systems*, Report No. EERC 73-11, Univ. of California, Berkeley.

Bean, R. and Page, J. (1976) *Traffic-Induced Ground Vibration in the Vicinity of Road Tunnels*, TRRL Supplementary Report 218 UC, Transport and Road Research Laboratory, Crowthorne, UK.

Bearman, P. W. (1980) Aerodynamic loads on buildings and structures. Paper 5, *Conference on Wind Engineering in the Eighties*, CIRIA, London.

Berge, S. (1978) Fatigue crack initiation in weldments of a carbon–manganese steel, Paper 6, *European Offshore Steels Research Seminar*, Welding Institute, UK.

Biggs, J. M. (1964) *Introduction to Structural Dynamics*, McGraw-Hill, New York.

Biggs, J. M., Suer, H. S. and Louw, J. M. (1959) Vibration of simple span highway bridges, *Trans. ASCE*, **124**, 291–318.

Blakeley, R. W. G., Cooney, R. C. and Megget, L. M. (1975) Seismic shear loading at flexural capacity in cantilever wall structures, *Bull. NZ Nat. Soc. Earthqu. Engng.*, **8**, 278–90.

Blanchard, J., Davies, B. L. and Smith, J. W. (1977) Design criteria and analysis for dynamic loading of footbridges, *Symposium on Dynamic Behaviour of Bridges*, Paper 7, *TRRL Supplementary Report* 275, Transport and Road Research Laboratory, Crowthorne, UK.

Blevins, R. D. (1979) *Formulas for Natural Frequencies and Mode Shapes*, Van Nostrand Reinhold, New York.

Boussinesq, J. (1885) *Application des Potentiels a l'Etude de l'Equilibre et du Mouvement des Solides Elastiques*, Paris.

Brock, J. E. (1946) A note on the damped vibration absorber, *Trans. ASME*, **68**, A-284.

Brown, C. W. (1977). An engineer's approach to dynamic aspects of bridge design,

Symposium on Dynamic Behaviour of Bridges, paper 8, *TRRL Supplementary Report* 275, Transport and Road Research Laboratory, Crowthorne, UK.

BSI (1972) *Code of Basic Data for the Design of Buildings, CP3*, Chapter V, Part 2, *Wind Loads*, British Standards Institution, London.

BSI (1974) *Code of Practice for Foundations for Machinery*: CP 2012: Part 1. *Foundations for Reciprocating Machines*, British Standards Institution, London.

BS 5400 (1978) *Steel, Concrete and Composite Bridges*. Part 2: *Specification for Loads*, British Standards Institution, London.

BS 5400 (1980) *Steel, Concrete and Composite Bridges*. Part 10: *Code of Practice for Fatigue*, British Standards Institution, London.

BS 6472 (1984) *Guide to Evaluation of Human Exposure to Vibration in Buildings* (*1 Hz to 80 Hz*), British Standards Institution, London.

BS 6611 (1984) *Guide to Evaluation of the Response of Occupants of Fixed Structures, Especially Buildings and Offshore Structures, to Low Frequency Horizontal Motion* (*0.063 Hz to 1.0 Hz*), British Standards Institution, London (technically equivalent to ISO 6897: 1984).

Burt, M. E. (1972) *Roads and the Environment*, *TRRL Report* LR 441, Transport and Road Research Laboratory, Crowthorne, UK.

Burton, R. W., McGonigle, R. W. and Neilson, G. (1981) *Seismicity and Seismic Risk Evaluations: Nigg Bay and St. Fergus*, Report No. 145, Global Seismology Unit, Edinburgh.

Bycroft, G. N. (1956) Forced vibrations of a rigid circular plate on a semi-infinite elastic space and on an elastic stratum, *Philos. Trans. Roy. Soc. Series A*, **248**, 327–68.

Canby, T. Y. (1973) California's San Andreas Fault, *Nat. Geog. Mag.*, **143**, Jan., 38–53.

CEB (1984) *Model Code for Seismic Design of Concrete Structures*, Bulletin d'Information No. 160, Comité Euro-International du Beton, Paris.

Chang, F. K. (1973) Human responses to motion in tall buildings, *Proc. ASCE*, **99**, ST6, 1259–72.

Chen, P. W. and Robertson, L. E. (1972) Human perception thresholds of horizontal motion, *J. Str. Div., ASCE*, **98**, ST8, 1681–95.

Cheung, Y. K. and Yeo, M. F. (1979) *A Practical Introduction to Finite Element Analysis*, Pitman, London.

Christiano, P. P. and Culver, C. G. (1969) Horizontally curved bridges subject to moving loads, *Proc. ASCE*, **95**, ST8, 1615–43.

Clarkson, B. L. and Mayes, W. H. (1972) Sonic-boom-induced building structure responses including damage, *J. Acoustical Soc. America*, **51**, 742–57.

Clough, R. W. and Penzien, J. (1975) *Dynamics of Structures*, McGraw-Hill, New York.

Cole, D. G. and Spooner, D. C. (1965) The damping capacity of concrete, *Int. Conf. on the Structure of Concrete and its Behaviour Under Load*, C & CA, London, Sept., pp. 217–225.

Contractor, G. P. and Thompson, F. C. (1940) The damping capacity of steel and its measurement, *J. Iron and Steel Inst.*, **141**, 157–218.

Cook, N. J. (1982) Towards better estimation of extreme winds, *J. Wind Engng. and Indust. Aerodyn.*, **9**, 295–324.

Cornell, C. A. (1968) Engineering seismic risk analysis, *Bull. Seism. Soc. Amer.*, **58**, 1583–606.

Crockett, J. H. A. (1965) Some practical aspects of vibration in civil engineering, *Symposium on Vibration in Civil Engineering*, Imperial College, April.

CSA (1984) *Steel Structures for Buildings*, Canadian Standards Association, CAN3-S16.1-M84.

Davenport, A. G. (1961) The application of statistical concepts to the wind loading of structures, *Proc. ICE*, **19**, 449–72.

Davenport, A. G. (1962) The response of slender, line-like structures to a gusty wind, *Proc. ICE*, **23**, 389–408.

Davenport, A. G. (1964) Note on the distribution of the largest value of a random function with application to gust loading, *Proc. ICE*, **28**, 187–96.

Davenport, A. G. (1983) Wind engineering, Chap. 9 in *Engineering Structures* (eds. P. S. Bulson, J. B. Caldwell and R. T. Severn), Thomas Telford, London.

Deinum, P. J., Dungar, R., Ellis, B. R., Jeary, A. P., Reed, G. A. L. and Severn, R. T. (1982) Vibration tests on Emosson arch dam, Switzerland, *Int. J. Earthqu. Engng. Struct. Dyn.*, **10**, 447–70.

Den Hartog, J. P. (1947) *Mechanical Vibrations*, 3rd ed., McGraw-Hill, New York.

De Salvo, G. J. and Swanson, J. A. (1985) *ANSYS: Engineering Analysis System: User Manual*, Swanson Analysis Systems, Houston, PA.

DIN 4150 (1975) Part 1, 2 and 3: 1975 *Vibrations in Buildings: Effects on Structures: Provisional Standards Revised Draft* Part 3: 1983, Deutsches Institut für Normung e.V., Berlin.

Dowding, C. H. and Corser, P. G. (1981) Cracking and construction blasting, *Proc. ASCE*, **107**, Co1, 89–106.

Dowding, C. H., Fulthorpe, C. S. and Langan, R. T. (1982) Simultaneous airblast and ground motion response, *Proc. ASCE*, **108**, ST11, 2363–78.

Dumanoglu, A. A., Severn, R. T., Brownjohn, J. M. W. and Taylor, C. A. (1987) Modal combination methods in the seismic analysis of the Humber Bridge, *Conf. on Steel and Aluminium Structures*, Cardiff. Elsevier, London, pp. 623–632.

Dungar, R. (1978) An efficient method of fluid–structure coupling in the dynamic analysis of structures. *Int. J. Num. Meth. Engng.*, **13**, 93–107.

Duquette, D. J. and Uhlig, H. H. (1968) Corrosion fatigue in salt water and spray, *Trans. Am. Soc. Metals*, **62**, 839.

EEFIT (1986) *The Mexican Earthquake of* 19 *September* 1985: *A Field Report by EEFIT*, SECED, Institution of Civil Engineers, London.

Ellis, B. R. (1980) An assessment of the accuracy of predicting the fundamental natural frequencies of buildings and the implications concerning the dynamic analysis of structures, *Proc. ICE*, **69**, 763–76.

Ellis, B. R. (1986) The significance of dynamic soil–structure interaction in tall buildings, *Proc. ICE*, **81**, 221–242.

EOSR (1978) *European Offshore Steels Research Seminar*, vols 1 and 2, Welding Institute, Cambridge.

Epstein, B. and Lomnitz, C. (1966) A model for the occurrence of large earthquakes, *Nature*, **211**, 954–6.

EQE (1984) *Seismic Experience Data Base*, EQE Inc., San Francisco, prepared for Seismic Qualification Utilities Group.

Ergatoudis, J. G., Irons, B. M. and Zienkiewicz, O. C. (1968) Curved, iso-parametric, quadrilateral elements for finite element analysis, *Int. J. Solids and Structures*, **4**, 31–42.

Eringen, A. C. (1953) Response of beams and plates to random loads, *J. Appl. Mech.*, **20**, 461.

ESDU (1976a) *The Response of Flexible Structures to Atmospheric Turbulence*, Item 76001, Engineering Sciences Data Unit, London.

ESDU (1976b) *Estimation of Fatigue Crack Growth Rates and Residual Strength of*

Components Using Linear Elastic Fracture Mechanics, Item 76019, Engineering Sciences Data Unit, London.

ESDU (1982) *Strong Winds in the Atmospheric Boundary Layer*, Part I: *Mean-Hourly Wind Speeds*, Item 82026, Engineering Sciences Data Unit, London.

ESDU (1985) *Characteristics of Atmospheric Turbulence Near the Ground*, Part II: *Single Point Data for Strong Winds* (*Neutral Atmosphere*), Item 85020, Engineering Sciences Data Unit, London.

ESDU (1986) *Characteristics of Atmospheric Turbulence Near the Ground*, Part III: *Variations in Space and Time for Strong Winds* (*Neutral Atmosphere*), Item 86010, Engineering Sciences Data Unit, London.

Esteva, L. (1967) Criteria for the construction of spectra for seismic design, *3rd Pan American Symp. on Structures*, Caracas, Venezuela.

Esteva, L. (1968) *Bases para la Formulacion de Decisiones de Diseno Sismico*, Rept. 182, Inst. Engng., Nat. Univ. Mexico.

Esteva, L. and Rosenblueth, E. (1964) Espectros de temblores a distancias moderados y grandes. *Bol. Soc. Mex. Ing. Sismica*, **2**, 1–18.

Ewing, W. M., Jardetzky, W. S. and Press, F. (1957) *Elastic Waves in Layered Media*, McGraw-Hill, New York.

Ewins, D. J. (1984) *Modal Testing: Theory and Practice*, Research Studies Press, John Wiley.

Eyre, R. and Tilly, G. P. (1977) Damping measurements on steel and composite bridges, *Symp. on Dynamic Behaviour of Bridges, Supplementary Report* 275, Transport and Road Research Laboratory, Crowthorne, UK.

Fessler, H., Mockford, P. B. and Webster, J. J. (1986) Parametric equations for the flexibility matrices of single brace tubular joints in offshore structures, *Proc. ICE*, **81**, 659–74.

Fisher, P. J. (1981) Summary of current design and fatigue correlation. Paper 12, *Proc. Conf. on Fatigue in Offshore Structural Steels*, Inst. of Civil Engineers, London.

Frahm, H. (1911) Device for damping vibrations of bodies, *US Patent* No. 989 958.

Goldman, D. E. and von Gierke, H. E. (1961) Effects of shock and vibration on man, in *Shock and Vibration Handbook*, Vol. 3 (eds C. M. Harris and C. E. Crede), McGraw-Hill, New York.

Goodman, L. E., Rosenblueth, E. and Newmark, N. M. (1955) Aseismic design of firmly founded elastic structures, *Trans. ASCE*, **120**, 782–802.

Gough, A. J. (1932) Corrosion fatigue, *J. Inst. Metals*, **49**, 17.

Graff, K. (1975) *Wave Motion in Elastic Solids*, Clarendon Press, Oxford.

Griffin, M. J. and Whitham, E. M. (1980) Discomfort produced by impulsive whole-body vibrations, *J. Acoust. Soc. America*, **68**, 1277–84.

Griffith, A. A. (1920) Phenomenon of rupture and flaws in solids, *Philos. Trans. Roy. Soc., Series A*, **221**, 163–98.

Guignard, J. C. and Guignard, E. (1970) *Human Response to Vibration: a Critical Survey of Published Work*, ISVR Memo. No. 373, Inst. of Sound and Vibration Research, Southampton.

Gumbel, E. J. (1958) *Statistics of Extremes*, Columbia Univ. Press.

Gurney, T. R. (1979) *Fatigue of Welded Structures*, Cambridge Univ. Press.

Guyan, R. J. (1965) Reduction of stiffness and mass matrices, *AIAA Journal*, **3**, 380.

Haas, E. and Labes, M. (1985) Dynamic behaviour of piping systems in a past strong earthquake (M6.6). Comparison with linear standard calculations. Paper K15/1, *8th Int. Conf. on Struct. Mech. in Reactor Technology*, SMIRT-8, Brussels, pp. 165–70.

Hahnkamm, E. (1933) Die Dampfung von Fundamentschwingungen bei veranderlicher Erregerfrequenz, *Ingenieur-Archiv*, **4**, 192–202.

Hansen, R. J., Reed, J. W. and Vanmarcke, E. H. (1973) Human response to wind-induced motion of buildings, *J. Str. Div., ASCE*, **99**, ST7, 1589–605.

Haroun, M. A. and Housner, G. W. (1982) Dynamic characteristics of liquid storage tanks, *J. Eng. Mech., ASCE*, EM5, 783–800.

Haroun, N. M. and Shepherd, R. (1986) Inelastic behaviour of X-bracing in plane frames, *J. Str. Div., ASCE*, **112**, ST4, 764–80.

Harris, R. I. and Deaves, D. M. (1980) The structure of strong winds. Paper 4, *Conf. on Wind Engineering in the Eighties*, Construction Industry Research and Information Association, London.

Heins, C. P. and Yoo, C. H. (1975) Dynamic response of a building floor system, *Building Science*, **10**, 143–53.

Hirai, A. and Ito, M. (1967) Response of suspension bridges to moving vehicles, *J. Faculty of Engng., Univ. of Tokyo*, (*B*), **29**, no. 2.

HMSO (1849) *Report of the Commissioners Appointed to Inquire into the Application of Iron in Railway Bridges*, Her Majesty's Stationery Office, London.

HMSO (1928) *Report of the Bridge Stress Committee*, Her Majesty's Stationery Office, London.

HMSO (1975) *Calculation of Road Traffic Noise*, Dept of the Environment, Welsh Office, Her Majesty's Stationery Office, London.

Hobbs, R. E. and Raoof, M. (1984) Hysteresis in bridge strand, *Proc. ICE*, Part 2. **77**, 445–64.

Holland, C. L. (1967) Performance effects of long-term random vertical vibration, *Human Factors*, **9**, 93–104.

Houbolt, J. C. (1950) A recurrence matrix solution for the dynamic response of elastic aircraft, *J. Aero. Sciences*, **17**, 540–50.

Hsieh, T. K. (1962) Foundation vibrations, *Proc. ICE*, **22**, 211–26.

Hurty, W. C. and Rubinstein, M. F. (1964) *Dynamics of Structures*, Prentice-Hall, Englewood Cliffs, New Jersey.

IAEE (1980) *Earthquake Resistant Regulations: A World List*, International Association of Earthquake Engineering, Tokyo.

Ibanez, P. and Ware, A. G. (1985) The experimental basis for parameters contributing to energy dissipation in piping systems. Paper K18/3, *8th Conf. on Struct. Mech. in Reactor Technology*, SMIRT-8, pp. 327–332, Brussels.

Inglis, C. E. (1934) *A Mathematical Treatise on Vibrations in Railway Bridges*, Cambridge Univ. Press.

Irons, B. M. (1966) Engineering application of numerical integration in stiffness method, *AIAA Journal*, **4**, 2035–7.

Irwin, A. W. (1978) Human response to dynamic motion of structures, *The Structural Engineer*, **56A**, 237–44.

Irwin, A. W. (1981a) *Live Load and Dynamic Response of the Extendable Front Bays of the Playhouse Theatre during The Who Concert*, Report for Lothian Region Architecture Dept.

Irwin, A. W. (1981b) Perception, comfort and performance criteria for human beings exposed to whole body pure yaw vibration and vibration containing yaw and translational components, *J. of Sound and Vibration*, **76**, 481–97.

Irwin, A. W. (1983) *Diversity of Human Response to Vibration Environments*, United Kingdom Group HRV, NIAE/NCAE, Silso, England.

Irwin, A. W. (1984) *Design of Shear Wall Buildings*, CIRIA Report 102, Construction Industry Research and Information Association, London.

Irwin, G. R. (1956) *Onset of Fast Propagation in High Strength Steels and Aluminium Alloys*, NRL Report 4763.

Irwin, G. R. (1958) Fracture, in *Encyclopaedia of Physics*, vol. 6, p. 551, Springer, Heidelberg.

Jeary, A. P. and Winney, P. E. (1972) Determination of structural damping of a large multi-flue chimney from the response to wind excitation, *Proc. ICE*, **53**, part 2, 569–77.

Jennings, A. (1967). A direct iteration method of obtaining latent roots and vectors of a symmetric matrix, *Proc. Cambridge Philos. Soc.*, **63**, 755–65.

Johnson, R., Bretherton, I., Tomkins, B., Scott, P. M. and Silvester, D. R. V. (1978) The effect of sea water corrosion on fatigue crack propagation in structural steel. Paper 15, *European Offshore Steels Research Seminar*, Welding Institute, UK.

Jones, R. T., Pretlove, A. J. and Eyre, R. (1981) Two case studies of the use of tuned vibration absorbers on footbridges, *The Structural Engineer*, **59B**, 27–32.

Kishan, H. and Traill-Nash, R. W. (1973) Calculation of response and loading of highway bridges from modal coordinates, *Pub. Int. Assoc. Bridge and Struct. Engng.*, **33**-II, 113–30.

Kolmogorov, A. N. (1941) Energy dissipation in locally isotropic turbulence, *Dokl. Akad. Nauk SSR*, **32**, 19–21.

Komori, A., Ichihashi, I., Masuda, Y., Uchiyama, U., Saruyama, I., Fujii, S., Shibata, H. and Okamura, H. (1985) Seismic qualification tests of active components for nuclear power plants (active components test program). Paper K14/1, *8th Int. Conf. on Struct. Mech. in Reactor Technology*, SMIRT-8, pp. 111–16, Brussels.

Konon, W. and Schuring, J. R. (1985) Vibration criteria for historic buildings, *J. Constr. Eng. Manag., ASCE*, **111**, 208–15.

Kwok, K. C. S. (1984) Damping increase in building with tuned mass damper, *J. Eng. Mech., ASCE*, **110**, EM11, 1645–8.

Lam, L. C. H. and Lam, R. P. (1979) An experimental study on the dynamic behaviour of a multi-storey steel-framed building, *Proc. ICE*, part 2, **67**, 707–20.

Lamb, H. (1904) On the propagation of tremors over the surface of an elastic solid, *Philos. Trans. Roy. Soc., Series A*, **203**, 1–42.

Lanczos, C. (1950) An iterative method for the solution of the eigenvalue problem of linear differential operators, *J. Res. Nat. Bureau of Standards*, **45**, 255–82.

Lawson, T. V. (1980) *Wind Effects on Buildings*, Applied Science, London.

Lee, D. H. (1973) A review of impact and fatigue for prestressed concrete railway bridges, *Proc. ICE*, **55**, part 2, 87–107.

Lenzen, K. H. (1966) Vibration of steel joist concrete slab floors, *AISC. Eng. Jour.*, **3**, 133–6.

Lenzen, K. H. and Murray, T. M. (1969) *Vibration of Steel Beam Concrete Slab Floor Systems*, Report No. 29, Univ. of Kansas, Lawrence, Kansas.

Leonard, D. R. (1966) *Human Tolerance for Bridge Vibrations*, TRRL Report LR34, Transport and Road Research Laboratory, Crowthorne, UK.

Lew, H. S., Leyendecker, E. V. and Dikkers, R. D. (1971) *Engineering Aspects of the* 1971 *San Fernando Earthquake*, Nat. Bur. Standards, USA, Bldg. Sci. Ser. no. 40.

Love, A. E. H. (1926) *Some Problems of Geodynamics*, Cambridge Univ. Press, Cambridge.

Lysmer, J. and Kuhlemeyer, R. L. (1969) Finite dynamic model for infinite media, *Proc. ASCE*, **95**, EM4, 859–77.

Mack, J. D., Stevens, J. C. and Stevens, S. S. (1960) Growth of sensation on seven continua as measured by force of handgrip, *J. Expt. Psych.*, **59**, 60.

Maddox, S. J. (1974) *Fatigue of Welded Joints Loaded in Bending*, TRRL Supplementary Report 84 UC, Transport and Road Research Laboratory, Crowthorne, UK.

Maheri, M. R. (1987) *Hydrodynamic Investigations of Cylindrical Structures and Other Fluid–Structure Systems*. PhD Dissertation, Univ. of Bristol.

Major, A. (1962) *Vibration Analysis and Design of Foundations for Machines and Turbines*, Collets, London.

Martin, D. J. (1978) *Low Frequency Traffic Noise and Building Vibration*, TRRL Supplementary Report 429, Transport and Road Research Laboratory, Crowthorne, UK.

Matsuishi, M. and Endot, T. (1968) *Fatigue of Materials Subjected to Varying Stress*, Japan Society of Mechanical Engineers, Fukuoka, Japan.

Matsumoto, Y., Nishioka, T., Shiojiri, H. and Matsuzake, K. (1978) Dynamic design of footbridges, *Proc. Int. Assoc. Bridge and Struct. Engng.*, P-17/78, Aug., pp. 1–15.

Matthews, S. W. (1973) This changing earth, *Nat. Geog. Mag.*, **143**, Jan., 1–37.

McDonald, A. and Thomson, J. F. (1980) The fatigue strength of large-scale welded tubular T joints, Paper 34, *European Offshore Steels Research Seminar*, Welding Institute, London.

Medearis, K. (1978) Rational damage criteria for low-rise structures subjected to blasting vibrations, *Proc. ICE*, part 2, **65**, 611–21.

Meister, F. J. (1935) *Die Empfindlichkeit des Menschen gegen Erschutterungen*, Forschung auf dem Gebiete des Ingenieurwesens, Berlin, vol. 6, p. 116.

Merovich, A. T., Nicoletti, J. P. and Hartle, E. (1982) Eccentric bracing in tall buildings, *J. Struct. Div. ASCE*, **108**, ST9, 2066–80.

Milford, R. V. (1982) Appraisal of methods for predicting the cross-wind response of chimneys, *Proc. ICE*, part 2, **73**, 313–28.

Miner, M. A. (1945) Cumulative damage in fatigue, *Trans. Amer. Soc. Mech. Engrs.*, **67**, A159–64.

Morril, B. J. (1971) *Evidence of Record Vertical Accelerations at Kagel Canyon during the Earthquake*, Geol. Survey Prof. Paper 733, US Govt., Printing Office, Washington, DC.

Moss, D. S. (1982) *Bending Fatigue of High-Yield Reinforcing Bars in Concrete*, TRRL Special Report SR 748, Transport and Road Research Laboratory, Crowthorne, UK.

NASTRAN (1970) *Theoretical Manual* (ed. R. H. McNeal), The McNeal–Schwendler Corp., Los Angeles.

Nath, B. and Potamitis, S. G. (1982) Coupled dynamic behaviour of realistic arch dams including hydrodynamic and foundation interaction, *Proc. ICE*, part 2, **73**, 587–607.

NBC (1985) *National Building Code of Canada*, National Research Council of Canada, Ottawa, NRCC No. 17303.

NBC *Supplement to the National Building Code of Canada* (1985) National Research Council of Canada, Ottawa, NRCC No. 23178.

Newmark, N. M. (1959) A method of computation for structural dynamics, *Trans. ASCE*. **127**, 1406–35.

Newmark, N. M. and Hall, W. J. (1973) *Procedures and Criteria for Earthquake Resistant Design. Building Practices for Disaster Mitigation*, Building Science Series 46, Nat. Bur. of Standards, Feb., pp. 209–237.

Nicholls, H. R., Johnson, C. F. and Duvall, W. I. (1971) *Blasting Vibrations and their Effects on Structures*, US Bureau of Mines, Bulletin 656.

Nieto-Ramirez, J. A. and Veletsos, A. S. (1966) *Response of Three–Span Continuous Highway Bridges to Moving Vehicles*, Univ. of Illinois, Eng. Expt. Station, Bulletin 489.

Nordby, G. M. (1958) Fatigue of concrete – a review of research, *ACI Journal, Proceedings*, **55**, 191–219.

Nour–Omid, B., Parlett, B. N. and Taylor, R. L. (1983) Lanczos versus subspace iteration for solution of eigenvalue problems, *Int. J. Num. Methods. Engng.*, **19**, 859–71.

Novak, M. and El Hifnawy, L. (1983) Vibration of hammer foundations, *Int. J. Soil Dyn. and Earthqu. Engng.*, **2**, 43–53.

Oehler, L. T. (1957) Vibration susceptibilities of various highway bridge types, *Proc. ASCE*, ST4, **83**, 1318.

Okada, A., Yamashita, T. and Fukuda, K. (1971) City vibration and its threshold limits, *Environmental Research*, **4**, 471–7.

Onysko, D. M. and Bellosillo, S. B. (1978) *Performance Criteria for Residential Floors. Report to Central Mortgage and Housing Corp., Ottawa, Ontario*, Eastern Forest Products Lab.

Ormondroyd, J. and Den Hartog, J. P. (1928) The theory of the dynamic vibration absorber, *Trans. ASME*, **50**, APM-SO-7, 9–22.

Palmgren, A. G. (1924) Die Lebensdauer von Kugellargen (The durability of ball bearings), *VDI-Zeitschrift des Vereines Deutscher Ingenieure*, **68**, 339–41.

Paris, P. C. (1964) The fracture mechanics approach to fatigue, in *Fatigue – an Interdisciplinary Approach*, Syracuse Univ. Press, New York.

Park, R. and Paulay, T. (1975) *Reinforced Concrete Structures*, John Wiley, New York.

Park, R. and Paulay, T. (1980) Concrete structures. Chap. 5 in *Design of Earthquake Resistant Structures* (ed. E. Rosenblueth), Pentech Press, London.

Pernica, G. and Allen, D. E. (1982) Floor vibration measurements in a shopping centre, *Can. J. Civ. Engng. (CDN)*, **9**, 149–55.

Polonsek, A. (1970) Human response to vibration of wood joist floors, *Wood Science*, **3**, 111–19.

Postlethwaite, F. (1944) Human susceptibility to vibration, *Engineering, London*, **157**, 61.

Priestley, M. J. N. (1980) Masonry. Chap. 6 in *Design of Earthquake Resistant Structures* (ed. E. Rosenblueth), Pentech Press, London.

Pritchard, B. N. (1984) Steel chimney oscillations: a comparative study of their reported performance versus predictions using existing design techniques, *Eng. Struct.*, **6**, 315–23.

Przemieniecki, J. S. (1968) *Theory of Matrix Structural Analysis*, McGraw-Hill, New York.

Raithby, K. D. (1979) Flexural fatigue behaviour of plain concrete, *Fatigue of Engng. Mat. Struct.*, **2**, 269–78.

Rayleigh, Lord (1885) On waves propagated along the plane surface of an elastic solid, *Proc. London Math. Soc.*, **17**, 4–11.

Reid, H. F. (1910) The mechanics of the earthquake, in *The Californian Earthquake of April* 18, 1906. *Report of the State Earthquake Investigation Commission*, vol. 2, Carnegie Institution of Washington, DC.

Reiher, H. and Meister, F. J. (1931) *Die Empfindlichkeit des Menschen gegen Erschutterungen*, Forschung auf dem Gebiete des Ingenieurwesens, Berlin, vol. 2, no. 11.

Reissner, E. (1936) Stationare, axialsymmetrische durch eine schuttelnde Masse erregte Schwingungen eines homogenen elastischen Halbraumes, *Ingenieur-Archiv*, **7**, 381–96.

Richter, C. F. (1958) *Elementary Seismology*, W. H. Freeman and Co., San Francisco.

Ritchie, R. O. and Knott, J. F. (1973) Mechanisms of fatigue crack growth in low alloy steel, *Acta Met.*, **21**, 639.

Rooke, D. P. and Cartwright, D. J. (1975) *Method of Compounding Stress Intensity Factors for Complex Configurations*, Royal Aircraft Establishment Tech. Rep. 75063, June.

Rooke, D. P. and Cartwright, D. J. (1976) *Compendium of Stress Intensity Factors*, Her Majesty's Stationery Office, London.

Roshko, A. (1961) Experiments on the flow past a circular cylinder at very high Reynolds numbers, *J. Fluid Mech.*, **10**, 345–56.

Salter, R. J. (1974) *Highway Traffic Analysis and Design*, Macmillan, London.

Sando, F. D. and Batty, V. (1974) Road traffic and the environment, *Central Statistical Office Social Trends*, **5**, 64–69, Her Majesty's Stationery Office, London.

SAP7 (1980) *A General Purpose Finite Element Program–User Manual*, Dept. of Civil Engng., Univ. of Southern California, Los Angeles.

Schilling, C. G. (1982) Impact factors for fatigue design, *Proc. ASCE*, **108**, ST9, 2034–44.

Schleicher, F. (1926) Zur Theorie des Baugrundes, *Bauingenieur*, **7**, 931–52.

Scruton, C. and Flint, A. R. (1964) Wind-excited oscillations of structures, *Proc. ICE*. **27**. 673–702.

Seed, B. H., Lysmer, J. and Hwang, R. (1975). Soil–structure interaction analyses for seismic response, *J. Geotech. Engng. Div., ASCE*, **101**, 439–57.

Seed, H. B., Ugas, H. and Lysmer, J. (1976) Site dependent spectra for earthquake resistant design, *Bull. Seism. Soc. Amer.*, **66**, 221–43.

Sehmi, N. S. (1986) The Lanczos algorithm applied to Kron's method, *Int. J. Num. Methods. Engng.*, **23**, 1857–72.

Seide, P. (1975) *Small Elastic Deformations of Thin Shells*, Noordhoff Int. Publishing, Leyden, The Netherlands.

Selby, A. and Severn, R. T. (1972) An experimental assessment of the added mass of some plates vibrating in water, *Int. J. Earthqu. Engng. and Struct. Dyn*, **1**, 189–200.

Selna, L. G. and Tso, W. K. (1980) *Twisting failure of the New Society Hotel. Reinforced Concrete Structures Subjected to Wind and Earthquake Forces*, ACI Special Pub. SP-63-19, 459–96.

Severn, R. T., Jeary, A. P. and Ellis, B. R. (1980) Forced vibration tests and theoretical studies on dams, *Proc. ICE*, part 2, **69**, 605–34.

Sheikh, S. A. and Uzumeri, S. M. (1982) Analytical model for concrete confinement in tied columns, *J. Struct. Div., ASCE*, **108**, ST12, 2703–22.

Siess, C. P. and Vincent, G. S. (1958) Deflection limitations of bridges, *Proc. ASCE*, **84**, ST3, paper 1633.

Simiu, E. (1974) Wind spectra and dynamic along-wind response, *J. Struct. Div., ASCE*, **100**, ST9, 1897–1910.

Simiu, E. and Scanlan, R. H. (1977). *Wind Effects on Structures: an Introduction to Wind Engineering*, Wiley, New York.

Siskind, D. E., Stachura, V. J., Stagg, M. S. and Kopp, J. W. (1980) *Structure Response and Damage Produced by Airblast from Surface Mining*, US Bureau of Mines, Report of Investigations 8485.

Siskind, D. E., Stagg, M. S., Kopp, J. W. and Dowding, C. H. (1980) *Structure Response and Damage Produced by Ground Vibration from Surface Mine Blasting*, US Bureau of Mines, Report of Investigations RI 8507.

Sisodiya, R. G. and Cheung, Y. K. (1971). A higher order in-plane parallelogram element and its application to skewed curved box-girder bridges, *Conf. on Developments in Bridge Design and Construction*, Cardiff.

Skorecki, J. (1966) The design and construction of a new apparatus for measuring the vertical forces exerted in walking: A gait machine, *J. Strain Analysis*, **1**, no. 5.

Smith, J. W. (1969) *The Vibration of Highway Bridges and the Effects on Human Comfort*, PhD Thesis, Univ. of Bristol, UK.

Smith, J. W. (1973a) The prediction of ground vibration produced by traffic on a rough road surface, *Symp. on Environment and Transport Technology*, Loughborough, UK.

Smith, J. W. (1973b) Finite strip analysis of the dynamic response of beam and slab highway bridges, *Int. J. of Earthqu. Engng. and Struct. Dyn.*, **1**, 357–70.

Sneddon, I. N. (1946) Distribution of stress in the neighbourhood of a crack in an elastic solid, *Proc. Roy. Soc., Series A*, **187**, 229–60.

Snowdon, L. C. (1970) *The Static and Fatigue Performance of Concrete Beams with High Strength Deformed Bars*, CIRIA Report 24, Construction Industry Research and Information Association, London.

Splittgerber, H. (1978) Effects of vibrations on buildings and on occupants of buildings, *Conf. on Instrumentation for Ground Vibration and Earthquakes*, pp. 147–52, Inst. of Civil Engineers, UK.

Srinivasulu, P. and Vaidyanathan, C. V. (1976) *Handbook of Machine Foundations*, Tata McGraw-Hill, New Delhi.

SSRAP (1985) *Use of Past Earthquake Experience Data to Show Seismic Ruggedness of Certain Classes of Equipment in Nuclear Power Plants*, Report by the Senior Seismic Review Advisory Panel, Washington, DC.

Steffens, R. J. (1974) *Structural Vibration and Damage*, Her Majesty's Stationery Office, London.

Stodola, A. (1927) *Steam and Gas Turbines*, McGraw-Hill, New York.

Stokes, G. G. (1883) Discussion of a differential equation relating to the breaking of railway bridges, *Mathematical and Physical Papers*, Cambridge Univ. Press, Cambridge, Vol. 2, p. 178.

Streeter, V. L. and Wylie, E. B. (1979) *Fluid Mechanics*, 7th ed., McGraw-Hill, New York.

Subudhi, M., Bezler, P., Wang, Y. K. and Alforque, R. (1984) *Alternate Procedures for the Seismic Analysis of Multiply Supported Piping Systems*, NUREG/CR-3811, Nuclear Regulatory Commission, USA.

Tappin, R. G. R. and Clark, P. J. (1985) Jindo and Dolsan bridges: design, *Proc. ICE*, part 1, **78**, 1281–1300.

Tilly, G. P. (1977) Damping of highway bridges: a review, *Symp. on Dyn. Behaviour of Bridges*, TRRL Supp. Report 275, 1–9, Transport and Road Research Laboratory, Crowthorne, UK.

Tilly, G. P. (1978) Fatigue problems in highway bridges, *Bridge Engineering, Transportation Research Record*, no. 664, 93–101.

Tilly, G. P. (1979) Fatigue of steel reinforcement bars in concrete: a review, *J. Fatigue of Engng. Mats. and Structs.*, **2**, 251–68.

Tilly, G. P. and Nunn, D. E. (1980) Variable amplitude fatigue in relation to highway bridges, *Proc. Inst. Mech. Engng.*, **194**, 259–67.

Tilly, G. P., Cullington, D. W. and Eyre, R. (1984) Dynamic behaviour of footbridges, *Int. Assoc. for Bridge and Struct. Engng., IABSE Periodica* S-26/84, 13–24.

Timoshenko, S. P. (1928) Vibration of bridges, *Trans. ASME*, **50**.

Timoshenko, S. P. (1955) *Vibration Problems in Engineering*, Van Nostrand, Princeton, New Jersey.

Timoshenko, S. and Goodier, J. N. (1951) *Theory of Elasticity*, McGraw-Hill, New York.

Timoshenko, S. P. and Young, D. H. (1965) *Theory of Structures*, 2nd ed., McGraw-Hill, New York.

Timoshenko, S. P. and Woinowsky-Krieger, S. (1959) *Theory of Plates and Shells*, McGraw-Hill, New York.

Toledo-Leyva, J. and Veletsos, A. S. (1958) *Effects of Roadway Unevenness on Dynamic Response of Simple Span Highway Bridges*, 8th Progress Report of the Highway Bridge Impact Investigation, Univ. of Illinois.

Tracey, D. M. (1974). Finite elements for three-dimensional elastic crack analysis. *Nuclear Engineering and Design*, **26**, 282–90.

Trifunac, M. D. and Brady, A. G. (1975) On the correlation of seismic intensity scales with the peaks of recorded strong ground motion, *Bull. Seism. Soc. Amer.*, **65**, 139–62.

Trott, J. J. and Whiffin, A. C. (1965) Measurements of the axle-loads of moving vehicles on trunk roads, *Roads and Road Constr.*, **43**, 511.

Turner, M., Clough, R., Martin, H. and Topp, L. (1956) Stiffness and deflection analysis of complex structures, *J. Aero. Sci.*, **23**, 805–23.

UBC (1982) *Uniform Building Code*, International Conference of Building Officials, Whittier, California.

Van der Hoven, I. (1957) Power spectrum of horizontal wind speed in the frequency range from 0.0007 to 900 cycles per hour, *J. Met.*, **14**, 160–64.

Van Eldik Thieme, H. C. A. (1961) *Passenger Riding Comfort Criteria and Methods of Analysing Ride and Vibration Data*, Society for Automotive Engineers paper 295A for SAE Int. congress and Exposition of Auto. Eng.

Vashi, K. M., Schelling, D. R. and Heins, C. P. (1970) *Impact Factors for Curved Highway Bridges*, Highway Res. Rec. No. 302, Highway Research Board, Washington, DC.

Veletsos, A. S. and Huang, T. (1970) Analysis of dynamic response of highway bridges, *Proc. ASCE*, EM5, **96**, 593–620.

Vickery, B. J. (1965) *On the Flow behind a Coarse Grid and its Use as a Model of Atmospheric Turbulence in Studies Related to Wind Loads on Buildings*, Nat. Phys. Lab. Aero. Report 1143.

Vickery, B. J. (1978) A model for the prediction of the response of chimneys to vortex shedding, *Proc. 3rd Int. Chimney Des. Symp.*, Munich, October.

Vickery, B. J. and Clark, A. (1972) Lift or across-wind response of tapered stacks, *J. Struct. Div., ASCE*, **98**, ST1, 1–20.

von Kármán, T. (1934) Turbulence and skin friction, *J. Aeronaut. Sci.*, **1**, 1.

Walker, W. H. (1968) *Model Studies of the Dynamic Response of a Multi-Girder Highway Bridge*, Univ. of Illinois, Eng. Expt. Station, Bulletin 495.

Waller, R. A. (1966) Building on springs, *Nature*, **211**, 794–6.

Waller, R. A. (1969) *Building on Springs*, Pergamon, Oxford.

Walsh, E. G. (1960) Sensations aroused by rhythmically linear motion–phase relationships, *J. Physiol., Proc. Physiol. Soc.*, **155**, 53.

Warburton, G. B. (1976) *The Dynamical Behaviour of Structures*, 2nd ed., Pergamon, Oxford.

Warburton, G. B. (1978) Soil–structure interaction for tower structures, *Int. J. Earthqu. Engng. Struct. Dyn.*, **6**, 535–56.

Warburton, G. B., Richardson, J. D. and Webster, J. J. (1971) Forced vibrations of two masses on an elastic half space, *Trans. ASME, J. Appl. Mech.*, 148–56.

Wargon, A. (1983) Design and construction of Sydney Tower, *The Structural Engineer*, **61A**, 273–81.

Wastling, M. A. and Smith, J. W. (1987) An instrument for detecting arbitrary peaks and troughs of a fluctuating stress signal, *Strain, Jour. of British Soc. of Strain Measurement*, August.

Watts, G. R. (1984) *Vibration Nuisance from Road Traffic – Results of a 50 Site Survey*, TRRL Report LR 1119, Transport and Road Research Laboratory, Crowthorne, UK.

Weingarten, V. I., Ramanathan, R. K. and Chen, C. N. (1983) Lanczos eigenvalue algorithm for large structures on a minicomputer, *Comp. and Structs.*, **16**, 253–7.

Wen, R. K. and Toridis, T. (1962) Dynamic behaviour of cantilever bridges, *Proc. ASCE*, **88**, EM4, 27–43.

Westergaard, H. M. (1933) Water pressures on dams during earthquakes, *Trans. ASCE*, **98**, 418–33.

Whale, L. (1983) *Vibration of Timber Floors – a Literature Review*, TRADA Research Report 2/83, Timber Research and Development Assoc., UK.

Wheeler, J. E. (1982) Prediction and control of pedestrian induced vibration in footbridges, *J. Struct. Div., ASCE*, **108**, ST9, 2045–65.

Whiffin, A. C. and Leonard, D. R. (1971) *A Survey of Traffic-Induced Vibrations*, TRRL Report LR 418, Transport and Road Research Laboratory, Crowthorne, UK.

Williams, C. (1983) Some dynamic characteristics of a tall building, *Proc. ICE*, part 2, **75**, 751–9.

Wilson, E. L., Der Kiureghian, A. and Bayo, E. P. (1981) A replacement for the SRSS method in seismic analysis, *Int. J. Earthqu. Engng. and Struct. Dyn.*, **9**, 187–94.

Wilson, E. L., Farhoomand, I. and Bathe, K.-J. (1973) Nonlinear dynamic analysis of complex structures, *Int. J. Earthqu. Engng. and Struct. Dyn.*, **1**, 241–52.

Wiss, J. F. and Parmelee, R. A. (1974) Human perception of transient vibration, *J. Struct. Div., ASCE*, **100**, ST4, 773–87.

Wohler, A. (1871) Tests to determine the forces acting on railway carriage axles and the capacity of resistance of the axles, *Engineering*, **11**, 199.

Wooton, L. R., Warner, M. H., Sainsbury, R. N. and Cooper, D. H. (1972) *Oscillations of Piles in Marine Structures: A Report of the Full Scale Experiments at Immingham*, CIRIA Tech. Note 40, Construction Industry Research and Information Association, London.

Wordsworth, A. C. and Smedley, G. P. (1978) Stress concentrations at unstiffened tubular joints, Paper 31, *European Offshore Steels Research Seminar*, Welding Institute, UK.

Wyatt, T. A. (1980) Evaluation of gust response in practice, Paper 7, *Conf. on Wind Engineering in the Eighties*, CIRIA, London.

Wyatt, T. A. and Best, G. (1984) Case study of the dynamic response of a medium-height building to wind-gust loading, *Eng. Struct.*, **6**, 256–61.

Yamada, K. and Hirt, M. A. (1982) Fatigue crack propagation from fillet weld toes, *J. Struct. Div., ASCE*, **108**, ST7, 1526–40.

Yanev, P. I. and Smith, P. D. (1985) *The Effects of the Great Chile Earthquake of* 1985. *Report to Seismic Qualification Utilities Group* by EQE Inc., San Francisco, California.

Zienkiewicz, O. C. (1977) *The Finite Element Method*, 3rd ed. McGraw-Hill UK.

Zienkiewicz, O. C. and Cheung, Y. K. (1964) The finite element method for analysis of elastic isotropic and orthotropic slabs, *Proc. ICE*, **28**, 471–88.

Zienkiewicz, O. C. and Nath, B. (1963) Earthquake hydrodynamic pressures on arch dams – an electrical analogue solution, *Proc. ICE*, **25**, 165–75.

Zienkiewicz, O. C., Parekh, C. J. and King, I. P. (1968) Arch dams analysed by a linear finite element shell solution program, *Proc. Symp. Arch Dams, Inst. of Civil Engrs.*, London.

Index